Lecture Notes in Electrical Engineering

Volume 1202

The book series *Lecture Notes in Electrical Engineering* (LNEE) publishes the latest developments in Electrical Engineering—quickly, informally and in high quality. While original research reported in proceedings and monographs has traditionally formed the core of LNEE, we also encourage authors to submit books devoted to supporting student education and professional training in the various fields and applications areas of electrical engineering. The series cover classical and emerging topics concerning:

- Communication Engineering, Information Theory and Networks
- Electronics Engineering and Microelectronics
- Signal, Image and Speech Processing
- Wireless and Mobile Communication
- Circuits and Systems
- Energy Systems, Power Electronics and Electrical Machines
- Electro-optical Engineering
- Instrumentation Engineering
- Avionics Engineering
- Control Systems
- Internet-of-Things and Cybersecurity
- Biomedical Devices, MEMS and NEMS

For general information about this book series, comments or suggestions, please contact leontina.dicecco@springer.com.

To submit a proposal or request further information, please contact the Publishing Editor in your country:

China

Jasmine Dou, Editor (jasmine.dou@springer.com)

India, Japan, Rest of Asia

Swati Meherishi, Editorial Director (Swati.Meherishi@springer.com)

Southeast Asia, Australia, New Zealand

Ramesh Nath Premnath, Editor (ramesh.premnath@springernature.com)

USA, Canada

Michael Luby, Senior Editor (michael.luby@springer.com)

All other Countries

Leontina Di Cecco, Senior Editor (leontina.dicecco@springer.com)

**** This series is indexed by EI Compendex and Scopus databases. ****

Daniel-Ioan Stroe · Nasimuddin ·
Shahedul Haque Laskar · Shivendra Kumar Pandey
Editors

Emerging Electronics and Automation

Select Proceedings of the 3rd International Conference, E2A 2023, Volume 2

Springer

Editors
Daniel-Ioan Stroe
Faculty of Engineering and Science
Aalborg University
Aalborg, Denmark

Nasimuddin
SRO Department
Institute for Infocomm Research, A*STAR
Singapore, Singapore

Shahedul Haque Laskar
Department of Electrical
and Instrumentation Engineering
National Institute of Technology Silchar
Silchar, Assam, India

Shivendra Kumar Pandey
Department of Electrical
and Instrumentation Engineering
National Institute of Technology Silchar
Silchar, Assam, India

ISSN 1876-1100 ISSN 1876-1119 (electronic)
Lecture Notes in Electrical Engineering
ISBN 978-981-97-3089-6 ISBN 978-981-97-3090-2 (eBook)
https://doi.org/10.1007/978-981-97-3090-2

This Springer imprint is published by the registered company Springer Nature Singapore Pte Ltd.
The registered company address is: 152 Beach Road, #21-01/04 Gateway East, Singapore 189721, Singapore

Contents

About the Editors

Daniel-Ioan Stroe received a Dipl.-Ing. degree in automatics from "Transilvania" University of Brasov, Romania, in 2008, and an M.Sc. degree in wind power systems from Aalborg University, Aalborg, Denmark, in 2010. He has been with the Department of Energy at Aalborg University (AAU-Energy) since 2010, from where he obtained his Ph.D. degree in lifetime modeling of Lithium-ion batteries in 2014. He is currently an Associate Professor with AAU Energy and the leader of the Batteries research group. He was a Visiting Researcher with RWTH Aachen, Germany, in 2013, and since 2022, he has been a visiting professor at Czech Technical University in Prague, Czechia. He has co-authored over 200 scientific peer-reviewed publications, most of them on topics related to Lithium-ion battery performance and lifetime modeling and battery state estimation. Furthermore, he is serving as a special issue editor and topic editor for various journals. Daniel's current research interests are in energy storage systems for grid and e-mobility, Lithium-based battery testing and modeling, and lifetime estimation and diagnostics of Lithium-ion batteries.

Nasimuddin received his B.Sc. degree in 1994 from Jamia Millia Islamia, India, and his M.Tech. (Microwave Electronics) and Ph.D. degrees in 1998 and 2004, respectively, from the University of Delhi, India. Dr. Nasimuddin has worked as a Senior Research Fellow (1999–2003) in DST sponsored project and CSIR grant as a senior Research Fellow in Engineering Science at the Department of Electronic Science, University of Delhi, India. He has worked as an Australian Postdoctoral Research Fellow (2004–2006) in awarded a Discovery project grant from the Australian Research Council at Macquarie University, Australia. Currently, he is working as a principal scientist at the Institute for Infocomm Research (I2R), A*STAR, Singapore. He has published 235 journal and conference technical papers on microstrip-based microwave antennas and components. Three US/SG patents have been granted and four filed on antenna technologies. He has edited two books and contributed a chapter to the book "Microstrip Antennas" published in 2011. He has ranked in the top 2% of World Scientists in 2023 for career-long impact and single-year impact.

Dr. Shahedul Haque Laskar was born in Nitainagar (Hailakandi), Assam (India) in 1962. He attended Irfan Ali Memorial H. S. School, Nitainagar, and Public H. S. School, Hailakandi. He did Pre-university from Cotton College, Guwahati (1977–1979) and Engineering courses from graduate up to research from Z. H. College of Engineering and Technology, AMU, Aligarh. He joined teaching as a lecturer in the Electrical Engineering Department, REC (presently NIT), Silchar in 1991 and now working as a Professor in the Electronics and Instrumentation Engineering Department, NIT Silchar. In addition to life member of ISTE (India), he is in the elite group of elected IEEE Senior Members and a member of the IEEE Instrumentation and Measurement Society. He is also a life member of the Instrument Society of India. He reviewed a good number of papers in some international Journals and conferences. He has been working in the domain of instrumentation including measurements and sensing.

Shivendra Kumar Pandey graduated in Electronics and Communication Engineering from RGPV Bhopal, India. He received a Ph.D. in Electrical Engineering from IIT Indore, India in Jan 2018. He received an International Travel Grant from SERB-DST, Government of India. Since June 2018, he has been working as an Assistant Professor of Electronics and Instrumentation Engineering at the National Institute of Technology Silchar, Assam, India. He has published research articles in Applied Physics Letters, *Journal of Applied Physics*, Review of Scientific Instruments, Physica Status Solidi-Rapid Research Letters, etc. His current research interests include Emerging memory technologies (PCM, RRAM), Nanoelectronics, Electronic Devices, and Sensors. He obtained an Early Career Research Award from the Science and Education Research Board (SERB), Department of Science and Technology (DST), Government of India. He is the Principal Investigator of the Nanoelectronics and Emerging Memory (NEEM) Research Laboratory at NIT Silchar, Assam, India.

Advancing Biomechanics: Utilizing IMU Sensor Technology to Quantify Ground Reaction Forces and Temporal Parameters

Karan Kanchan, Harsh Holalad, Ami Mehta, and Prathamesh Mantri

Abstract This paper presents the remarkable capabilities of the Equilibrium (EQ) system, developed by the authors, towards accurately capturing kinetic and temporal data for various activities. The EQ system demonstrated its superiority over existing devices as documented in the literature through a comprehensive study. Notably, it overcomes the limitations of traditional gait analysis labs by offering a cost-effective, time-efficient, portable, and scalable solution. The findings highlight the significant advancements achieved by the EQ system in the field of gait analysis, providing a practical and comprehensive solution for diverse applications. The key contributions of this paper will show the capabilities of the Equilibrium (EQ) system by addressing all the major drawbacks of the functionalities as compared to a standard gait/biomechanics analysis lab, computing Vertical Ground Reaction Force (VGRF) for multiple activities with better accuracy than the other wearable solutions presented in the literature and computes temporal parameters such as stance-swing times in a gait test and full-body air time in a jump test.

Keywords Gait analysis · Wearable technology · Ground reaction forces · Temporal parameters · Kinetics · Biomechanics

1 Introduction

Biomechanics is an interdisciplinary field that combines the principles from biology, physics, and engineering to understand the human body functions in motion, stability, and the forces acting upon them. The thrust area of biomechanics that deals with human walking and motions exhibited is the "Gait analysis". Studying a person's gait allows us to recognize their distinctive movements, establish typical gait patterns, identify pain-causing conditions, and assess treatments to address irregularities. The

K. Kanchan · H. Holalad (✉) · A. Mehta · P. Mantri
Vayu Technology, Mumbai, India
e-mail: harshholalad@vayu.tech

D.-I. Stroe et al. (eds.), *Emerging Electronics and Automation*, Lecture Notes in Electrical Engineering 1202, https://doi.org/10.1007/978-981-97-3090-2_1

Gait analysis is helpful for multiple applications, including identification of anomalies, guiding therapeutic decision-making, or for athletic performance enhancement. The various methods of Gait analysis [1] include visual observations by an expert [2], video analysis [3], electromyography signals [4], and force plates [5]. The literature presents methods of measuring the gait using a lab, controlled environment, or wearable sensor method.

1.1 Gait Analysis Labs

Most conventional gait analysis is conducted in a gait analysis lab with force plates and high frame rate motion capture camera systems. The labs provide crucial objective information, such as movement, force, and spatial and temporal gait parameters. However, using a gait analysis lab has few constraints despite its high precision, dependability, and repeatability. Some limitations [6] include the demand for a controlled environment and lack of portability, the need for a skilled operator, the difference in output between a controlled environment and an open field or overground, and expensive.

1.2 Inertial Measurement Unit (IMU)-Based Wearable Solutions

Inertial Measurement Unit (IMU)-based technologies are used to overcome the drawbacks of traditional gait analysis labs [7]. IMUs measure acceleration, angular velocity, magnetic fields, motion, orientation, and heading when coupled with sensor fusion software. These devices are low-cost, highly accessible, and can be worn all day without interrupting the subject's motion.

It is evident that both the existing systems for the Gait analysis have limitations. There is a need for end-to-end product solutions driven by a combination of wearable sensors and smart software that provide accurate, precise, objective, and repeatable data for conducting gait analysis.

Further sections of the paper describe the details of one such end-to-end product solution (EQ system) developed by authors that addresses the challenges of the existing systems. The developed solution is detailed in Sect. 2, the experimental setup in Sect. 3, and the results in Sect. 4.

2 Equilibrium System

The Equilibrium (EQ) system developed by the authors is an end-to-end product solution driven by wearable sensors and AI software to provide accurate, precise, objective, and repeatable data for conducting gait analysis without force plates or high-frame-rate motion capture camera systems.

The developed EQ system can be tested in any natural environment. It's a portable and low-cost solution compared to traditional gait analysis labs. The system can compute forces, 3D movement data, segment and joint angles, and spatial and temporal parameters. The product solution also provides a brilliant reporting framework that can be customized entirely to deliver diagnostic information and individualized and tailored exercise recommendations via an immediate feedback mechanism. This whole process takes a few minutes, which makes the product solution a scalable offering that enables continuous and periodic monitoring of users.

2.1 Vertical Ground Reaction Force (VGRF)

Another significant advantage of the EQ system is the use of Vertical Ground Reaction Force (VGRF). VGRF is the force the body applies to the supporting surface through the foot that is equal in magnitude and directed in the opposite direction from the ground reaction force (GRF). The VGRF is one of the most significant components out of the three GRF components and is an essential factor for the analysis of human behavior and the study of biomechanics. Analyzing this parameter can help us determine injury risk and performance optimization during activities such as running, jumping, squatting, walking, etc.

Many wearable solutions have been developed to compute VGRF over the last decade. Some use a single sensor, some multiple sensors, and others a complete body tracking system. The authors in [6] compare different authors computing GRF using different methods for different activities. The scope of this paper covers three activities: runs, vertical jumps, and squats.

Running

Using statistical models, the authors in [8] estimated the VGRF during walking and running. They mention using a single bi-axial accelerometer at a sampling frequency of 40 Hz and a measuring acceleration range of ± 67 g. The most significant observed inaccuracy was 17.5% for running. The average absolute difference between the anticipated VGRF and the one directly measured by the force plate was 9% for running activity. In subsequent work [9], they recorded accelerations from a single three-axis accelerometer over the right iliac crest. This new device could record data at a sampling frequency of 100 Hz, and the maximum acceleration range was ± 6 g. In this work, they calculated the VGRF during braking and propulsion. They reported an average absolute error between actual and predicted of 8.3% and 18% for VGRF during propulsion and VGRF during braking, respectively.

The approach by authors in [10] employed a single tri-axial accelerometer mounted on the upper torso. The force plate output was compared with the VGRF calculated using the author's approach as the subjects completed various running exercises. The absolute error for a single measurement over several trials ranged between 12 and 24%.

The authors in [10] studied the validity and reliability of ViPerform (by Dorsavi) to measure the peak VGRF. A tri-axial accelerometer placed on the tibia with an output range of ± 13 g was used to gather samples at 100 Hz. They conclude that ViPerform does not measure VGRF compared to the golden standard force plates. And, over multiple trials, they reported an error of 16.04%. The authors in [11] also studied the validity of ViPerform to measure the peak VGRF. They used two three-axis accelerometers fixed on the medial tibia of each leg to measure the tibial acceleration during running. The two commercial accelerometers had a range of ± 24 g and a sampling frequency of 100 Hz. They were fixed on both tibias, along the tibial axis in the midpoint between the lower edge of the medial malleolus and the medial joint line of the knee. The average RMSE error (logarithmic approximation) of VGRF computed from acceleration, compared to the direct measurement from the force plate, was 8.28% across different running speeds.

The authors of [12] predicted the VGRF using two IMU sensors mounted on the shank (one on each leg) and compared VGRF estimates against the force plate. They claim their prediction was unreliable for every participant and reported an average absolute error of 33.3%.

Vertical Jump

The authors of [13] estimate VGRF using a single accelerometer sensor mounted close to the body center of mass. They have reported comparing the VGRF of the accelerometer and the force plate for two types of jumps: Countermovement jump (CMJ) and Drop jump (DJ). The loading and landing phases' average absolute VGRF errors for CMJ were reported as 4.8% and 35.8%, respectively. Also, they reported a 53.6% error on landing in DJ.

In [14], the authors have conducted an accelerometer vs. force plate comparison for different vertical jumps in two population groups of teen and adolescent kids: healthy and osteogenesis imperfecta (OI type 1). They used a single accelerometer positioned on the right waist. As compared to the force plate, the maximum absolute error across multiple types of jumps and both population groups was 31%. The single two-legged vertical jump errors recorded were 14% and 8%, respectively, for healthy and OI type 1 populations. The errors reported most likely occurred during the loading phase. It is not explicitly mentioned in the paper.

Squat

The GRF during a squat motion was measured by [15] by comparing and examining the IMU performance against the force plate. The squat motion was modeled on the sagittal plane as a three-segment linkage and was recorded using three IMUs placed on the lower back, thigh, and shank. The VGRF estimation result showed a high correlation and low root mean square error (RMSE) (R = 0.93, RMSE < 0.02 of body weight).

Also, squat tasks were investigated by authors in [16]. Their goal was to estimate internal joint moments and GRF. The IMUs were custom-made wireless sensors with a sampling frequency of 100 Hz. The average RMSE error in VGRF was 15N, and max VGRF was 40N. There was no mention of absolute errors.

3 Experimental Setup

This section will describe the experimental setup and specifications of the developed EQ system, the force plate, and the video motion capture system. The following section further describes the process of data acquisition and analysis.

3.1 Data Acquisition

The sensor system used in this study is the EQ system by Vayu Technology. It comprises 13 sensors, with each sensor having a dimension of 47 mm x 30 mm x 13 mm and a weight of 10 gm, as shown in Fig. 1a. The sensor has a ± 16 g accelerometer range, ± 2000 deg/s gyroscope range, and ± 480mT magnetometer range. And the sensor samples at 100 Hz.

The sensors are placed on 13 segments, as shown in Fig. 1b, using custom-designed velcro straps. The EQ system maps the following segments: forearms, upper arms, head, torso, lumbar (L5), thighs, shins, and feet (Fig. 1b).

The proprietary mobile application by Vayu Technology facilitates a wireless data collection process. Figure 2 depicts the interface between the equilibrium sensor and the software for connecting, recording, and/or visualization. The data is collected via the 13 EQ sensors in a synchronized manner and sent to the mobile application via Bluetooth. The mobile application encrypts and sends this data to the cloud servers via a secure connection, and the cloud server processes the data using proprietary algorithms to provide kinetic and kinematic data. The solution also provides movement, performance, and injury risk prediction insights.

3.2 Force Plate and Video Capture System

The research-grade lab with Kistler force plates, portable Kistler force plates, and Panasonic & Optitrak high-speed camera systems was used to acquire the data to compare the EQ system's kinetic and kinematic data. This paper discusses only the comparison of kinetic data.

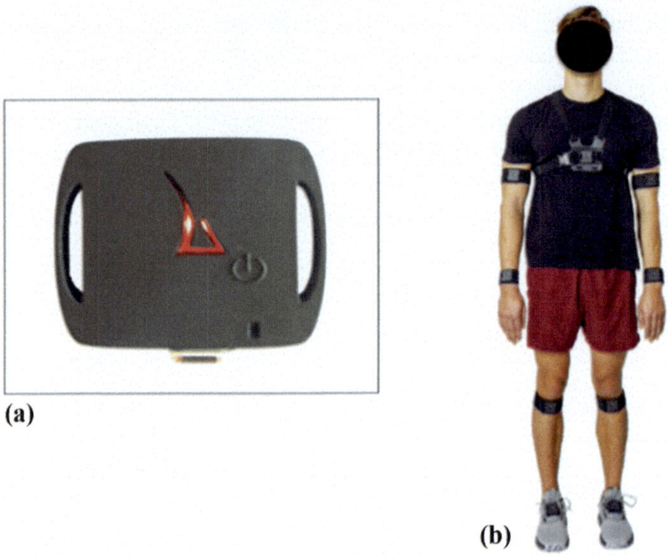

Fig. 1 **a** Equilibrium sensor **b** Equilibrium system on a user

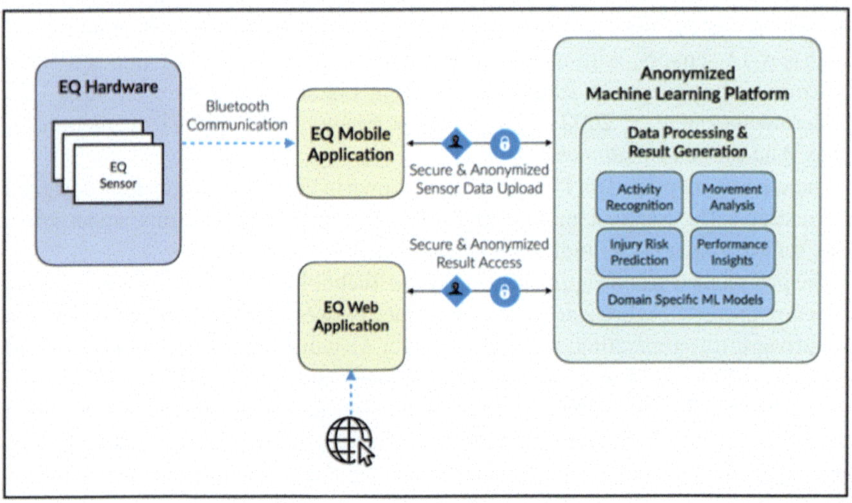

Fig. 2 Equilibrium system on a user

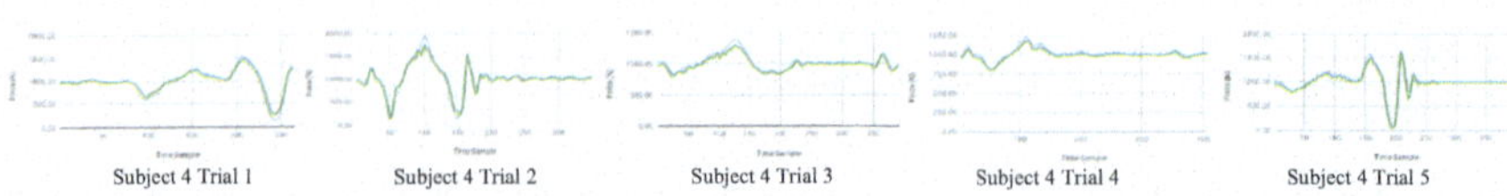

Fig. 3 Squat force plate data (dotted blue) overlapped on EQ force plate data (solid green)

3.3 Participants

A total of six participants performed in two studies. The first session was performed in a lab with two males and two females. The second session was conducted on a running track with portable force plates. The average age of the six participants was 26.17 ± 2.4.

3.4 Procedures

The purpose of the first session was to validate the kinetic and temporal data (across three different activities) captured using the EQ system with respect to research-grade force plates. The second session aimed to validate the kinetic and temporal data captured by the EQ system and compare it with the force plate measurements during a block start in a sprint test. After obtaining informed consent, the subjects' profiles were created using their weight and height information. The EQ sensors were placed on the subjects, and they were taught the EQ calibration and testing protocols. Subjects were asked to practice the protocol and were prepped for testing. The subjects were given information about the force plate location. They were asked to try to hit the force plate during the activity. The trial was ignored if partial foot contact was on the force plate. In the first session, each subject performed multiple vertical jump trials and multiple run trials while wearing the EQ system on the force plate, and each trial was recorded by the high-speed motion capture system. Finally, the data acquisition was concluded when each user had two jump and two run trials with EQ, force plate, and motion capture data. The fourth user performed multiple squat trials, also included in the study. In the second session, each subject performed three block start sprints. The force plate was positioned to capture the first foot contact as soon as the subject got off the block. The first-step data captured by the force plate was compared against the first-step data captured by the EQ system.

The four subjects from the first session were given the Subject IDs 1, 2, 3, and 4, and the two subjects from the second session were given the Subject IDs 5 and 6. The EQ sensors sampled at 100 Hz, while the force plate and the video capture system sampled at 1200 Hz (both the lab and the portable track force plate) and 1500 Hz, respectively. So, the force plate and the video data were down-sampled and synchronized appropriately with the EQ system data for analysis. The force plate data below 50N was considered noise, per the authors in [14]. So, the contact and take-off points were considered at the 50N mark during the force data comparison. The absolute error between the peak forces computed from the EQ system and the force plate and signal correlations between the EQ system and the force plate is used to quantify the error observed. The temporal parameters, stance time in runs and air time in jumps, from the EQ system and the force plate are also compared to quantify the accuracy of the EQ system.

4 Results and Discussion

The kinetic and temporal parameters obtained from the force plate will be compared with the EQ system, and three metrics are discussed in the results section, and they are:

- The correlation coefficient between the force calculated by EQ and the force plate data for all three activities—squat, jump, and run.
- The absolute peak error between the force calculated by EQ and the force plate data for all three activities—squat, jump, and run.
- The air time for jumps and stance time for runs.

Table 1 shows the correlation coefficients calculated between the force plate's force signal and the EQ system's force signal for all three activities from the first and second sessions. The table has the coefficients tabulated for each trial, and we can see that the squat activity has the best average of 0.980, followed by jump and run activities. Three trials (out of 27) went lower than a 0.9 correlation coefficient—one for the jump activity with 0.865 and the other two for the run trial with a block start of 0.856 and 0.885. The results in Table 1 show that the force signal from the EQ system and force signal from the force plate are strongly correlated, and the EQ system captures all the significant patterns in the signal for all activities except for the off-the-block sprint trials, where it does not capture the initial peak of the foot contact in the first step off the block as seen in Fig. 6. The images in Figs. 3, 4, and 5 all show the EQ system's force signal overlapping the force plate's force signal for squat, jump, and run trials, respectively. The graphs contain trials for all the subjects who were part of the study, and they clearly demonstrate how well the EQ system force signal overlaps with the force plate force signal.

4.1 Absolute Peak Error (APE)

The tabular columns in Tables 2, 3, and 4 show the absolute peak error in percentage between the force calculated by the EQ system and the force plate for squat, jump, and run activities, respectively. The absolute peak error is calculated by taking the absolute difference between the peak of the force plate data and the peak of the EQ force data, dividing it by the reference, and multiplying by 100 to get a percentage number. The force plate data is considered the reference data.

Squats

Table 2 shows the VGRF data of the force plate and the EQ system in Newtons (N) for all 5 trials conducted by Subject 4, along with the absolute difference in Newtons and the absolute percentage error. The average error observed was $6.84 \pm 2.69\%$, and the maximum observed error was 11.10%, as shown in Table 2.

Table 1 Correlation coefficient between the EQ system force signal and the force plate force signal

Activity	Subject ID	Trial ID	Correlation coefficient	Average	Standard Deviation	Min	Max
Squat	4	1	0.995	0.980	0.009	0.972	0.995
		2	0.973				
		3	0.981				
		4	0.972				
		5	0.977				
CMJ	1	1	0.949	0.946	0.011	0.865	0.961
		2	0.929				
	2	1	0.953				
		2	0.953				
	3	1	0.865				
		2	0.961				
	4	1	0.935				
		2	0.945				
Run	1	1	0.990	0.945	0.04	0.856	0.992
		2	0.978				
	2	1	0.974				
		2	0.932				
	3	1	0.989				
		2	0.992				
	4	1	0.915				
		2	0.971				
	5	1	0.856				
		2	0.949				
		3	0.926				
	6	1	0.945				
		2	0.885				

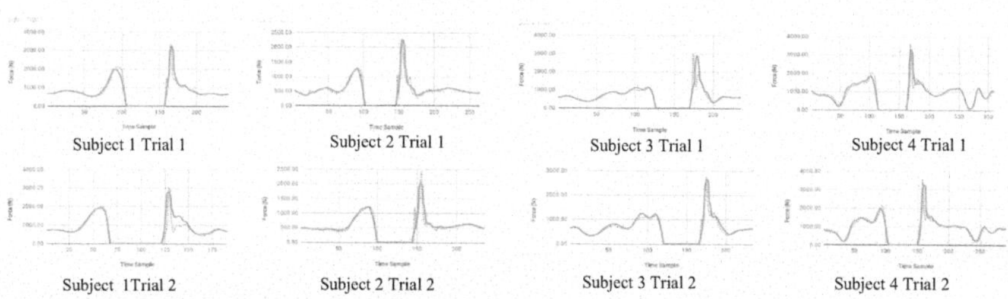

Fig. 4 Jump force plate data (dotted blue) overlapped on EQ force plate data (solid green)

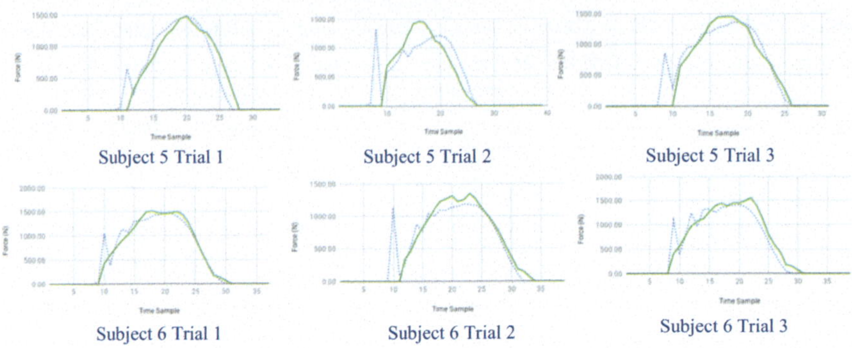

Fig. 5 First session Run force plate data (dotted blue) overlapped on EQ force plate data (solid green)

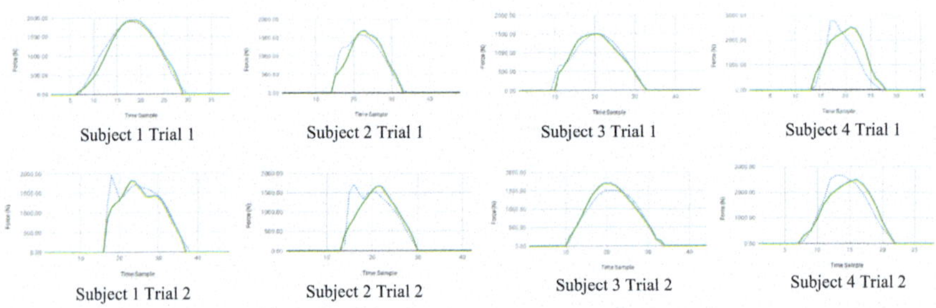

Fig. 6 Second session Run force plate data (dotted blue) overlapped on EQ force plate data (solid green)

Table 2 Absolute VGRF error in % for squat activity for subject 4

Trial ID	Force plate (N)	EQ (N)	Absolute difference	Absolute error (%)	Average error (%)	Standard deviation (%)	Min (%)	Max (%)
1	1541	1479	62	4.02	6.84	2.69	4.02	11.1
2	1895	1685	210	11.10				
3	1385	1282	102	7.42				
4	1239	1172	66	5.37				
5	1593	1496	97	6.11				

Jumps

Table 3 shows the VGRF data of the force plate and the EQ system in Newtons for all four subjects (two trials per subject) during the loading and landing phases of the jump activity. It also shows the absolute difference in Newtons and the absolute percentage error. The average error reported by the EQ system during the loading phase is 4.27 ±

Table 3 Absolute VGRF error in % for jump activity

Activity	Subject ID	Trial ID	Force plate (N)	EQ (N)	Absolute difference	Absolute error (%)	Average error (%)	Standard deviation (%)	Min (%)	Max (%)
CMJ loading	1	1	2125	1995	130	6.14	4.27	2.69	1.20	7.70
		2	1984	1902	81	4.12				
	2	1	1302	1285	17	1.33				
		2	1230	1215	14	1.20				
	3	1	1160	1176	16	1.38				
		2	1190	1252	61	5.17				
	4	1	2005	1851	154	7.70				
		2	2135	1984	151	7.10				
CMJ landing	1	1	3354	3229	125	3.73	6.48	3.54	1.47	12.47
		2	3010	2776	234	7.78				
	2	1	2297	2264	33	1.47				
		2	2310	2022	288	12.47				
	3	1	3067	2914	153	4.99				
		2	2778	2665	113	4.07				
	4	1	3611	3287	324	8.97				
		2	3575	3276	298	8.35				

Table 4 Absolute VGRF error in % for run contact phase activity

Subject ID	Trial ID	Force plate (N)	EQ (N)	Absolute difference	Absolute error (%)	Average error (%)	Standard deviation (%)	Min (%)	Max (%)
1	1	1899	1941	42	2.23	5.96	3.85	0.07	13.06
	2	1934	1818	116	6.01				
2	1	1581	1689	107	6.83				
	2	1716	1667	48	2.85				
3	1	1504	1503	1	0.09				
	2	1505	1701	196	13.06				
4	1	2783	2528	254	9.16				
	2	2665	2490	175	6.57				
5	1	1326	1461	135	10.18				
	2	1475	1470	1	0.07				
	3	1362	1454	92	6.75				
6	1	1466	1514	48	3.27				
	2	1260	1349	89	7.06				
	3	1421	1555	134	9.43				

2.69%, and the average error during the landing phase is 6.48 ± 3.54%. The loading phase error reported by the EQ system is slightly lower than what is reported in the literature by [9], i.e., 4.8%, and the landing phase error reported by the EQ system is also significantly lower than what is reported in the literature by [13, 14], and [15], each reporting an error of 35.8%, 14%, and 19%, respectively. Furthermore, [14] reports a maximum error of 31%, significantly higher than the maximum error observed by the EQ system of 12.47%, as shown in Table 3.

Runs

Table 4 shows the VGRF data of the force plate and the EQ system in Newtons for all six subjects, i.e., four subjects from the first session (two trials per subject) and two subjects from the second session (three trials per subject) during the contact phase of the run activity. The average error reported by the EQ system is 5.96 ± 3.85%, which is significantly lower than that mentioned in the literature. Authors [7, 10, 11], and [12] report an average error of 9%, 16.04%, 8.28%, and 33.3%, respectively. Also, the maximum error for a single measurement recorded by the EQ system is 13.06%, which is significantly lower than the maximum error reported in the literature [8, 9], and [7] of 17.5%, 18%, and 24%, respectively.

The dynamic nature of the run activity shows a higher standard deviation in the reported average error percentage when compared to jumps and squats. The standard deviation of the average error for runs is the highest 3.85%, followed by a jump at 2.69% for loading and 3.54% for landing, and finally, for squats at 2.69%. Similarly, the maximum error of 13.06% reported for a single trial is the highest for a run, followed by 12.47% for a jump and 11.10% for a squat.

4.2 *Temporal Parameters*

This section compares the temporal results between the EQ system force data and force plate force data for jump and run trials. Table 5 summarizes air time differences in the jump trials, and a maximum error of 0.02 s is observed. The start and end of air time are chosen at the 50N mark from the force plate data as per [17–21]. The same 50N threshold is chosen for runs as well. Table 6 summarizes ground time differences in the run trials, which are part of the lab and track studies. The first session shows a maximum difference of 0.01 s error where the middle of the run was considered the step of interest.

The second session, where the first step off the block is considered, shows a maximum difference of 0.03 s error as the sensor system cannot recognize the initial peak due to the initial contact. The average absolute error (in seconds) for all the 14 trials is 0.011 s.

Table 5 Absolute error in seconds for air time in jump activity

Activity	Subject ID	Trial ID	Force Plate (s)	EQ (s)	Absolute difference
Run ground phase	1	1	0.23	0.22	0.01
		2	0.21	0.20	0.01
	2	1	0.19	0.18	0.01
		2	0.16	0.16	0.00
	3	1	0.23	0.22	0.01
		2	0.23	0.23	0.00
	4	1	0.13	0.14	0.01
		2	0.14	0.14	0.00
	5	1	0.19	0.16	0.03
		2	0.17	0.15	0.02
		3	0.17	0.15	0.02
	6	1	0.21	0.2	0.01
		2	0.23	0.21	0.02

Table 6 Absolute error in seconds for ground time in run activity

Activity	Subject ID	Trial ID	Force plate (s)	EQ (s)	Absolute difference
Run ground phase	1	1	0.23	0.22	0.01
		2	0.21	0.20	0.01
	2	1	0.19	0.18	0.01
		2	0.16	0.16	0.00
	3	1	0.23	0.22	0.01
		2	0.23	0.23	0.00
	4	1	0.13	0.14	0.01
		2	0.14	0.14	0.00
	5	1	0.19	0.16	0.03
		2	0.17	0.15	0.02
		3	0.17	0.15	0.02
	6	1	0.21	0.2	0.01
		2	0.23	0.21	0.02
		3	0.21	0.22	0.01

5 Conclusion

In conclusion, the study’s findings unequivocally demonstrate that the EQ system, developed by Vayu Technology, surpasses other reported devices in the literature to accurately capture kinetic and temporal data across a range of activities. The EQ system offers a remarkable improvement in accuracy when compared to existing alternatives. Moreover, it effectively resolves the critical limitations of traditional gait analysis labs, providing a cost-effective, time-efficient, portable, and scalable solution. By addressing these drawbacks, the EQ system represents a significant advancement in the field of gait analysis, offering a comprehensive and practical solution for various applications.

References

1. Baker R (2007) The history of gait analysis before the advent of modern computers. Gait Posture 26(3):331–342
2. Khandelwal S, Wickström N (2014) Identification of gait events using expert knowledge and continuous wavelet transform analysis. In: 7th international conference on bio-inspired systems and signal processing (BIOSIGNALS 2014), Angers, France, March 3–6, 2014. SciTePress
3. Michelini A, Eshraghi A, Andrysek J (2020) Two-dimensional video gait analysis: a systematic review of reliability, validity, and best practice considerations. Prosthet Orthot Int 44(4):245–262
4. Harris GF, Wertsch JJ (1994) Procedures for gait analysis. Arch Phys Med Rehabil 75(2):216–225
5. Quinn MM et al (2007) Evaluation of agreement between numerical rating scales, visual analogue scoring scales, and force plate gait analysis in dogs. Veterinary Surg 36(4):360–367
6. Ancillao A et al (2018) Indirect measurement of ground reaction forces and moments by means of wearable inertial sensors: a systematic review. Sensors 18(8):2564
7. Wundersitz DWT et al (2013) Validity of an upper-body-mounted accelerometer to measure peak vertical and resultant force during running and change-of-direction tasks. Sports Biomech 12(4):403–412
8. Neugebauer JM, Hawkins DA, Beckett L (2012) Estimating youth locomotion ground reaction forces using an accelerometer-based activity monitor. PLoS ONE 7(10):e48182
9. Neugebauer JM, Collins KH, Hawkins DA (2014) Ground reaction force estimates from ActiGraph GT3X+ hip accelerations. PLoS ONE 9(6):e99023
10. Raper DP et al (2018) Use of a tibial accelerometer to measure ground reaction force in running: a reliability and validity comparison with force plates. J Sci Med Sport 21(1):84–88
11. Charry E et al (2013) Study on estimation of peak ground reaction forces using tibial accelerations in running. In: 2013 IEEE eighth international conference on intelligent sensors, sensor networks and information processing. IEEE
12. Thiel DV et al (2018) Predicting ground reaction forces in sprint running using a shank mounted inertial measurement unit. Proceedings 2(6):MDPI
13. Howard R, Conway R, Harrison AJ (2014) Estimation of force during vertical jumps using body fixed accelerometers 102–107
14. Pouliot-Laforte A et al (2014) Validity of an accelerometer as a vertical ground reaction force measuring device in healthy children and adolescents and in children and adolescents with osteogenesis imperfecta type I
15. Min S, Kim J (2015) Inertial sensor based inverse dynamics analysis of human motions. In: 2015 IEEE international conference on advanced intelligent mechatronics (AIM). IEEE
16. Kodama J, Watanabe T (2016) Examination of inertial sensor-based estimation methods of lower limb joint moments and ground reaction force: results for squat and sit-to-stand movements in the sagittal plane. Sensors 16(8):1209
17. Cavanagh PR, Lafortune MA (1980) Ground reaction forces in distance running. J Biomech 13(5):397–406
18. Malisoux L, Gette P, Delattre N, Urhausen A, Theisen D (2022) Spatiotemporal and ground-reaction force characteristics as risk factors for running-related injury: a secondary analysis of a randomized trial including 800+ recreational runners. Am J Sports Med 50(2):537–544

19. Zhang Q, Jin T, Cai J, Xu L, He T, Wang T, Tian Y, Li L, Peng Y, Lee C (2022) Wearable triboelectric sensors enabled gait analysis and waist motion capture for IoT-based smart healthcare applications. Adv Sci 9(4):2103694
20. Channa A, Popescu N, Faisal M (2022) Parkinson's disease gait evaluation leveraging wearable insoles and deep learning approach. In: 2022 8th international conference on control, decision and information technologies (CoDIT) 2022, vol 1. IEEE, pp 543–549
21. Voisard C, De l'Escalopier N, Ricard D, Oudre L, Automatic gait events detection with inertial measurement units: healthy subjects and moderate to severe impaired patients.

A Comb-Type Capacitive Sensor for Moisture and 2-FAL Detection in Transformer Oil

Abdul Suboor, Shahwaiz, Mohammad Hasir, Obaidur Rahman, Mohd Mohsin Sohail, and Shakeb A. Khan

Abstract The insulation system of a transformer comprises solid Kraft paper insulation and liquid transformer oil. However, due to the continuous influence of thermal and electrical stresses, the insulation deteriorates. Two crucial degradation parameters that may be utilized to assess the age of the insulation system or the state of the transformer's health are 2-FAL (2-furfuraldehyde) and moisture. These parameters can predict the remaining life of the insulation of the transformer. In order to protect the transformer, continuous monitoring of the insulation system of the transformer is needed. This paper proposes a comb-type capacitive sensor for detecting moisture and 2-FAL concentration in oil. A simulation-based study is performed in Ansys Maxwell Software to analyze the feasibility and characteristics of the sensor in different sensing films. A silicon oxide and titanium dioxide film is used as a sensing film to increase the sensitivity of the sensor. Silicon Oxide (SiO_2) and Titanium dioxide (TiO_2) film show a good response for detecting moisture and 2-FAL in oil. The proposed sensor is a simple, low-cost, and effective method of online monitoring of transformer insulation health.

Keywords Transformers · Ansys Maxwell software · Moisture · 2-FAL · Capacitive sensors · Sensitivity · Transformer insulation

---

A. Suboor · Shahwaiz · M. Hasir · O. Rahman
Department of Electrical Engineering, Aligarh Muslim University, Uttar Pradesh, Aligarh 202001, India

M. M. Sohail
Department of Electrical Engineering, Maulana Mukhtar Ahmad Nadvi Technical Campus, Malegaon Nashik, Maharashtra 423203, India

S. A. Khan (✉)
Department of Electrical Engineering, Jamia Millia Islamia, New Delhi 110025, India
e-mail: Skhan@jmi.ac.in

D.-I. Stroe et al. (eds.), *Emerging Electronics and Automation*, Lecture Notes in Electrical Engineering 1202, https://doi.org/10.1007/978-981-97-3090-2_2

1 Introduction

The operation of power systems is closely related to the performance of transformers and their insulation systems. As the failure of the insulation system can cause a breakdown of the transformer and a disruption of the power system; hence, it is necessary to carry out preventive maintenance of the transformer. Furthermore, the time-consuming procedure of estimating degradation by-products using traditional approaches provides little assistance. However, online assessment of insulation breakdown by-products via sensors is a simple, low-cost answer to the difficult, lengthy technique. Because sensor-based estimate offers several advantages over other frequently used offline estimation strategies.

As a result, various sensors have been investigated for online condition monitoring of power transformer insulation. Moreover, the evolution of chemical sensors involves the development of various capacitive sensors for the on-site assessment of transformer insulation degradation by-products [1].

The insulation system of the transformer, comprising cellulose-based solid kraft paper and mineral oil, chemically decomposes under electrical and thermal stress, producing different gases, furan compounds, and moisture as degradation by-products. Furthermore, these pollutants are dispersed in the transformer oil/mineral oil during dynamic operation [2]. As a result, Dissolved Gas Analysis (DGA) has become one of the most important diagnostic tools in power transformer status monitoring. However, the greater time intervals between results and the possibility of sample tampering have prompted research into sensor-based condition monitoring of transformer oil [3].

The concentration of specific dissolved insulation degradation by-products in mineral oil is an essential indicator of the condition of the transformer insulation system. As one of the several furanic compounds created in the oil during insulation degradation, the most stable by-product formed by the decomposition of cellulosic paper insulation is 2-furfuraldehyde (2-FAL) [4]. As a result, its concentration can be directly related to the solid insulation's aging, which can provide information on the power transformers' remaining useful life. Water is produced as a byproduct of the degradation of paper insulation. Furthermore, moisture ingresses from the outside via the breather, reducing the dielectric strength of the insulation and hastening its breakdown [5].

There are two sensor-based estimating approaches for transformer oil degradation compounds: optical and electrical. The optical methods for measuring furfurals mentioned in the literature include liquid chromatography [6], spectroscopic analysis [7], and colorimetry [8]. These optical sensors are expensive to maintain, require sophisticated equipment, and require expert workers. Other electrical sensor detection strategies for 2-FAL include [9] and the Polyvinyl Chloride (PVC) electrode-based potentiometric detection methodology [10]. These electrical approaches, however, do not provide a quantitative measurement of 2-FAL.

Capacitance-based sensors for transformer oil testing have recently been developed. [4] proposes a thin-film Molecularly Imprinted Polymer (MIP)-based capacitive sensor for 2-FAL detection. Later, in [11], the sensor structure was changed to a comb-type shape to improve the sensor's sensitivity. These capacitive sensors have only been designed to detect 2-FAL. Contactless cross-capacitive sensors have recently been developed for real-time monitoring of transformer oil by measuring moisture and 2-FAL content [12–14]. Cross-capacitive sensors have been claimed to be useful in assessing 2-FAL and humidity in transformer oil, however, sensitivity and selectivity are the key difficulties.

In this paper, the comb-type structured sensor for the estimation of insulation degradation by-products that is 2-FAL and moisture in the transformer oil is proposed. Silicon dioxide and titanium dioxide are the two different sensing films used to increase the sensitivity of the proposed sensor for moisture and 2-FAL detection, respectively. Different finger comb type sensors are developed for optimal measurement of moisture and 2-FAL in mineral oil. A simulation-based study is carried out on Ansys Maxwell software to examine the feasibility an2d performance of the proposed sensors.

The concept of the comb-type sensor structure and its equivalent electrical model is given in Sect. 2. Section 3 provides details of the simulation of the proposed sensor. Results obtained in the simulation studies are discussed in Sects. 4 and 5 concludes the paper.

2 Theory and Electrical Equivalent Model of the Proposed Sensor

The proposed sensor is based on the fringe field idea. When compared to a single electrode sensor, the comb structure increases the total active area of the sensor. In reaction to humidity and 2-FAL, the larger fringe causes a greater shift in sensor capacitance. As a result, the sensor obtained high sensitivity.

The indicated sensor's electrical equivalent circuit can be obtained by examining the equivalent circuits of similar systems stated in the literature [15]. The copper electrodes in the sensor sandwich an FR4 fiber glass epoxy. The sensor's electrical equivalent circuit is made up of various circuit elements such as C_{epo}, R_{epo}, Co, Ro, C_M, and Cs. Figure 1a depicts the physical facts that have led to the consideration of these circuit elements. The developed sensor produces two field paths when it is poured into the oil samples. The field originates from one copper electrode and ends at the other. A straight field passes through the epoxy, and a fringing field passes through the oils surrounding the electrodes. Through the transformer oil surrounding the sensor, the fringing field at the sensor's borders creates another curved electric field channel that connects one copper electrode to the other electrode.

The development of a fringing field around each finger results, in the formation of transformer oil capacitance (Co) and resistance (Ro), due to the coating layer on

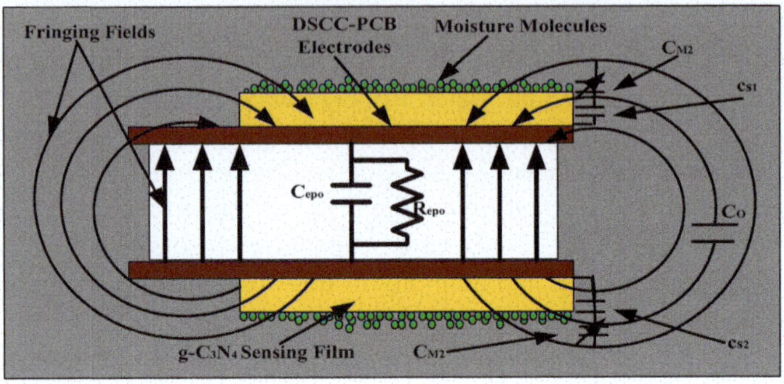

Fig. 1 Physical phenomena due to which different circuit elements come into the electrical equivalent circuit

each side Cs1 and Cs2 are the two coating capacitances. Two variable capacitances C_{M1} and C_{M2} are also formed by the adsorption layer of the by-product molecule of transformer oil on either side of the sensor. The simplified circuit after ignoring resistance Repo and Roil because of exceedingly high (G order), as shown in Fig. 2a. C_E is the capacitance equivalent of C_P (parasitic capacitance) and C_{epo}. The circuit is simplified further in terms of total capacitance C_R, which is the equivalent capacitance of C_E and C_N, as shown in Fig. 2b.

The Eqs. (1–5) show the relationship between several electrical equivalent circuit parameters. It is vital to remember that the first number in each double-subscripted parameter refers to the finger number (1, 2, 3, 4, and 5), while the second number represents the electrode on either side of the finger 1 or finger 2 of the developed sensor.

$$C_E = C_p + C_{epo} \tag{1}$$

$$C_{S_{qn}} = \frac{C_{S_{q1}} * C_{S_{q2}}}{C_{S_{q1}} + C_{S_{q2}}} \tag{2}$$

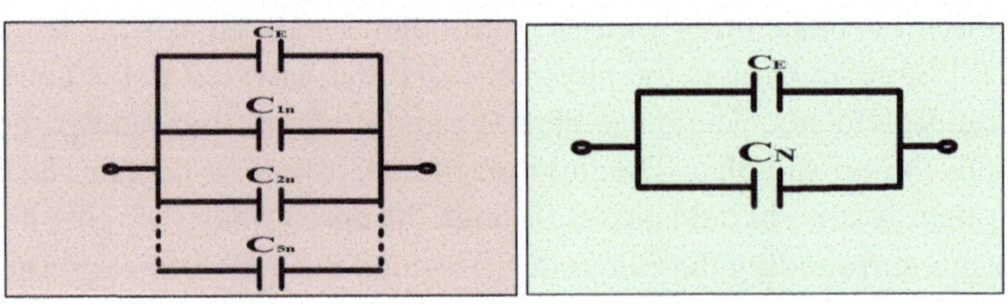

(a) (b)

Fig. 2 **a** Electrical equivalent circuit of the sensor neglecting all the resistive elements due to high values. **b** Electrical equivalent circuit of the proposed sensor

$$C_{M_{qn}} = \frac{C_{M_{q1}} * C_{M_{q2}}}{C_{M_{q1}} + C_{M_{q2}}} \tag{3}$$

$$\begin{aligned} C_{qn} &= \frac{C_{sq_n} * C_o * C_{M_{qn}}}{C_{S_{qn}} * C_o + C_o * C_{M_{qn}} + C_{M_{qn}} * C_{sq_n}} \\ &= \frac{C_{S_{qn}} * C_o}{C_{S_{qn}} + C_o + \frac{C_{Sqn} * C_o}{C_{M_{qn}}}} \end{aligned} \tag{4}$$

$$\begin{aligned} C_R &= C_E + C_{1n} + C_{2n} + C_{3n} + C_{4n} + C_{5n} \\ &= C_E + C_N = C_E + 5C_{1n} \end{aligned} \tag{5}$$

where

C_{sq1} represents the capacitance due to the qth finger near electrode 1 and C_{sq2} represents the capacitance due to the qth finger near electrode 2;

C_{Mn1} is the capacitance due to the sensing layer of qth finger near electrode 1 and C_{Mn2} is the capacitance due to the sensing layer of the qth finger near electrode 2;

Cs_{qn} is the resultant capacitance of Cs_{q1} and Cs_{q2};

C_{Mqn} is the resultant of C_{Mq1} and C_{Mq2};

C_{qn} is the resultant of Cs_{qn}, C_o, and C_{Mqn}; and.

C_N is the resultant capacitance of C_{1n}, C_{2n}, C_{3n}, C_{4n}, and C_{5n}.

As the concentration of impurities (2-FAL and moisture) is increased in the transformer oil, the capacitance Cs_l increases, which in turn increases the overall capacitance C_R. Hence, by the measurement of the capacitance of the sensor, the concentration of 2-FAL and moisture is estimated.

3 Simulation of the Capacitive Sensor

The geometry feature of the Ansys Maxwell software was used to create the necessary comb-type structure. Among electrostatic, magnetostatic, etc., an electrostatic interface is chosen for this work and the study is kept in stationary mode. The steps followed for simulation are given in this section.

3.1 Creation of Sensor Structure

The sensor body was created using the Ansys Maxwell Software. Firstly, four different dimensions of the substrate block are cut and the dimensions of these blocks are shown in Table 1.

Table 1 Dimensions of the proposed comb-type sensor structure

Dimensions (mm)	2 fingers	3 fingers	4 fingers	5 fingers
Length	50	50	50	50
Breath	9.5	15	20.5	26
Thickness	1.3	1.3	1.3	1.3
Gap	1.5	1.5	1.5	1.5
Each finger	4	4	4	4

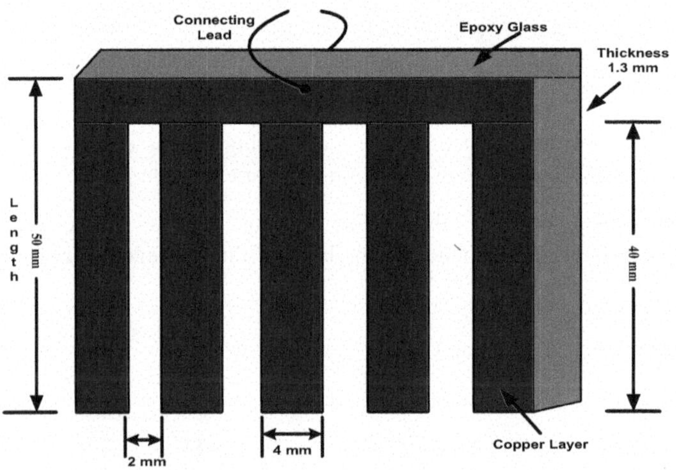

Fig. 3 Schematic design of comb sensor

3.2 Material Assignment

The necessary materials (Copper, FR4 Epoxy, Transformer Oil, Silicon oxide, Titanium oxide, and Air) have been chosen from the Add Material section and then assigned one at a time to the blocks of the newly constructed construction. As the value of 2-FAL and moisture in mineral oil changes, so does the dielectric constant of the oil. As the mixture level of oil contaminants varies from 0 to 40 ppm, an appropriate value of the dielectric constant is assigned to simulate an oil sample. Figure 3 depicts the structure of the newly constructed sensor following the material assignment.

3.3 Excitation Assignment

The two electrodes of the capacitive sensor are given a small excitation to develop the electric field around the sensor which would be used to compute the capacitance

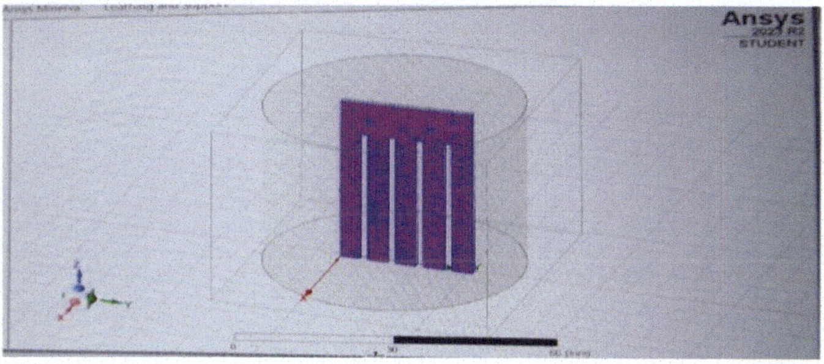

Fig. 4 Mesh operation into the proposed sensor

of the sensor. Electrode 1 is given + 1 V potential, while -1 V potential is given to electrode 2. These electrodes are given excitation through the connecting leads at the top of the sensor.

3.4 Mesh Operation

Once the structure of the sensor is created and excitation is provided, then the sensor is enclosed in a cylindrical structure to provide meshing. Meshing is the operation of the software which generates finite elements of the whole structure. It is done by selecting the mesh section, from which the build mesh option is selected. The whole structure is discretized into small elements as shown in Fig. 4.

3.5 Electric Field Distribution

The comb-type sensor is simulated for moisture and 2-FAL detection at different ppm levels in mineral oil. Titanium dioxide (TiO_2) and Silicon oxide (SiO_2) sensing film are used for increasing the sensitivities in the detection of moisture and 2-FAL in mineral oil is used in the transformer for cooling purposes.

The Electric Field Distribution (EFD) at 0 and 40 ppm level moisture and 2-FAL in mineral oil for both titanium oxide and silicon oxide are shown.

The electric field distribution of the sensor at 40 ppm for moisture and 2-FAL concentration with titanium dioxide as a sensing film is shown in Fig. 5, while Fig. 6 shows the electric field distribution of the sensor at the same ppm level but a different sensing film is silicon oxide for both moisture and 2-FAL. The electric field distribution is clearly different for both the sensing film variants of the sensor. And, the electric field distribution changes with the concentration of 2-FAL and moisture.

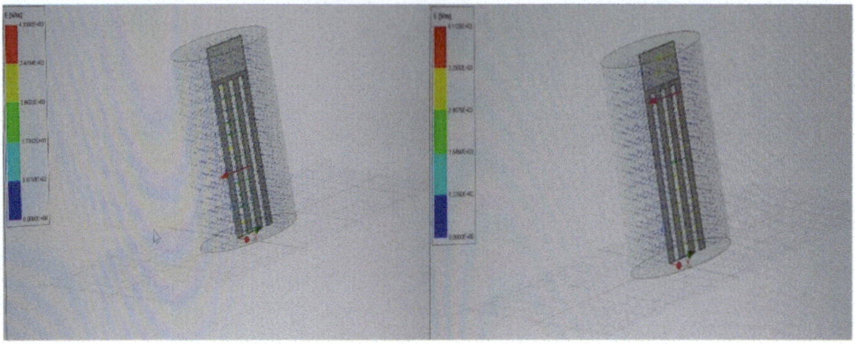

(a) (b)

Fig. 5 Electric Field Distribution with TiO_2 sensing film at **(a)** 40 ppm moisture concentration **(b)** 40 ppm 2-FAL concentration

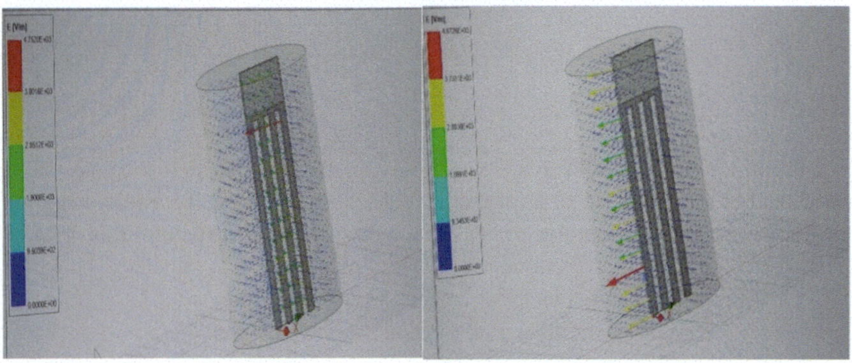

(a) (b)

Fig. 6 Electric Field Distribution with SiO_2 sensing film at **a** 40 ppm moisture concentration **b** 40 ppm 2-FAL concentration

4 Results and Discussion

The proposed sensor has been simulated in Ansys Maxwell Software, which has been described in detail in the previous section. Firstly, by the use of simulation, find out the optimal sensor design for the 2-FAL and moisture detection in mineral oil. Silicon oxide and titanium dioxide are used as a sensing film in this particular work to increase the sensitivity of the comb-type sensor for 2-FAL and moisture detection in mineral oil. The concentration of 2-FAL and moisture in oil is measured through a sensor in terms of change in capacitance. The detection of impurities by the proposed sensor is compared, among the three variants, in terms of normalized capacitance and sensitivity of the sensor to change in the amount of by-products of insulation deterioration that are present. The detection of these by-products is described below.

Table 2 Optimal comb-type sensor structure for moisture detection in mineral oil

No of fingers	Sensitivities
2	0.00832
3	0.01478
4	0.02175
5	0.02921

Table 3 Optimal comb-type sensor structure for 2-FAL detection in mineral oil

No of fingers	Sensitivities
2	0.01229
3	0.02078
4	0.03035
5	0.03991

4.1 Selection of Fingers in Comb Structure

A fringing field in a sensor increased by increasing the number of fingers in comparison to a single electrode structure. Hence, it also increases the sensor sensitivity by increasing the sensor capacitance values. The optimal sensor structures for moisture and 2-FAL detection in mineral oil are shown in Table 1 and Table 2, respectively. For finding the optimal sensor structure initially sensor is simulated without sensing film.

As shown in Tables 2 and 3, the five-finger structure has the highest sensitivities compared to two-, three-, and four-finger structures in mineral oil samples for both moisture and 2-FAL detection. In the simulation, the sensitivities of the five-finger comb sensor in mineral oil samples are 0.03991 pf/ppm for moisture and 0.02921 pf/ppm for 2-FAL, which contains moisture and 2-FAL from 0 to 40 ppm.

4.2 Sensor Response in Mineral Oil for Moisture Detection

For the detection of moisture in transformer oil, a comb-type sensor is created, and titanium dioxide and silicon oxide are used as a sensing film to improve the sensitivity. Both the films have highest moisture absorption capacity and show good results. In this study, three different thickness of sensing film is used to find the relation of thickness with sensitivities.

Figure 7a shows the normalized capacitance responses of the sensor for different concentrations of moisture, while Fig. 7b shows the normalized capacitance responses of the sensor for different 2-FAL concentrations for different thicknesses of titanium dioxide sensing film. Sensor with titanium dioxide sensing film of thickness 0.01 mm has the highest changes for both moisture and 2-FAL concentrations, respectively, as shown in Fig. 7a and b.

Response of sensor for different thickness of silicon oxide sensing film in transformer oil samples that contains moisture and 2-FAL contents as shown in Fig. 8a and b. Sensor with sensing film silicon oxide of thickness 0.01 mm has been shown the highest sensitivities for both moisture and 2-FAL concentration as shown in the above Fig. 8a and b, respectively.

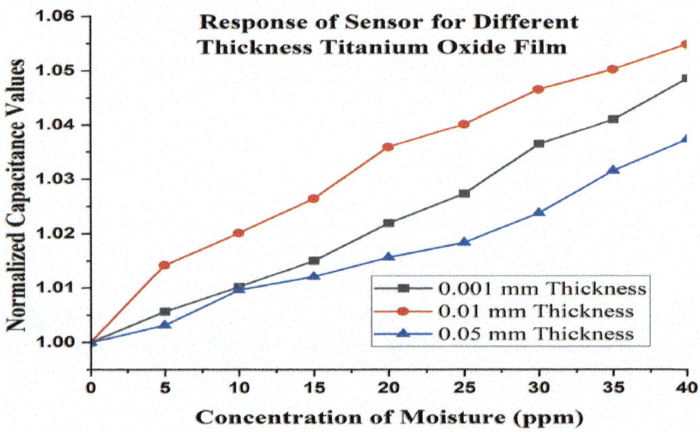

(a)

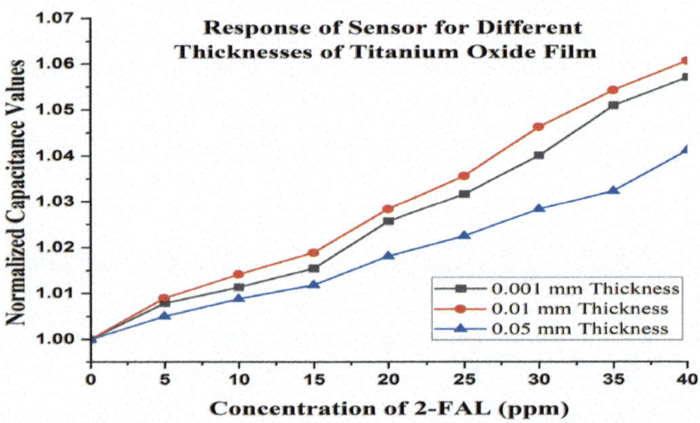

(b)

Fig. 7 **a** Response of sensor for different concentrations of moisture **b** Response of sensor for different concentrations of 2-FAL

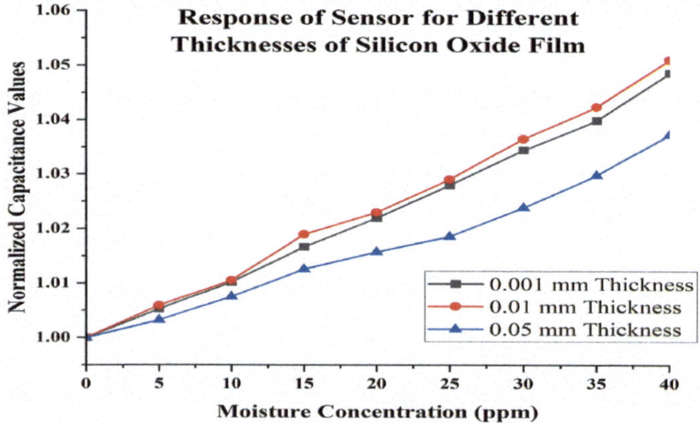

(a)

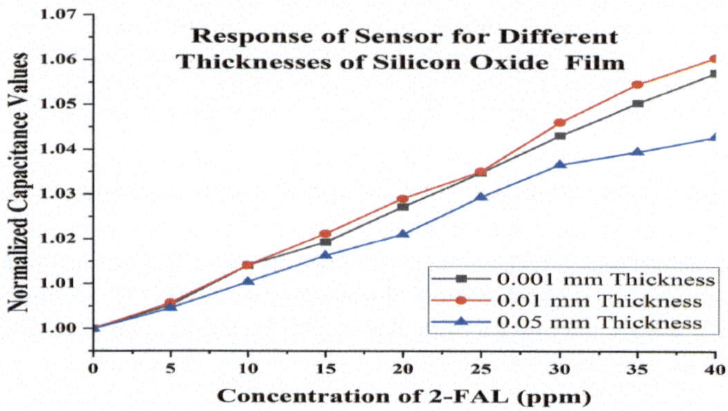

(b)

Fig. 8 **a** Response of sensor for different concentrations of moisture **b** Response of sensor for different concentrations of 2-FAL

4.3 Sensitivities

Sensor sensitivity is the term used to determine how well a sensor can detect the changes in moisture levels in the environment it's monitoring. A highly sensitive sensor can detect even slight variations in impurities content, while a less sensitive sensor might only respond to significant changes.

The sensitivities of the proposed sensor in moisture and 2-FAL concentration at different thicknesses of titanium dioxide and silicon oxide sensing film as shown in Tables 4 and 5, respectively.

Table 4 shows the sensitivities of the titanium dioxide sensing film-based sensor with different thickness values in moisture and 2-FAL concentration. Sensor with

Table 4 Sensitivities of sensors with different thicknesses of TiO_2 sensing material used for moisture and 2-FAL detection

Thickness of film (mm)	Sensitivities	
	Moisture concentration	2-FAL concentration
0.001	0.0428496	0.05081253
0.01	0.0439738	0.05296743
0.05	0.0301275	0.03523291

Table 5 Sensitivities of sensors with different thicknesses of SiO_2 sensing material used for moisture and 2-FAL detection

Thickness of film (mm)	Sensitivities	
	Moisture concentration	2-FAL concentration
0.001	0.0428067	0.05057643
0.01	0.0436235	0.05202775
0.05	0.0303461	0.03306174

film thickness 0.01 mm has shown the highest sensitivities in moisture and 2-FAL impurities are 0.0439738 pf/ppm and 0.05296743 pf/ppm, respectively.

A designed sensor with different widths of silicon oxide sensing film is used to detect moisture and 2-FAL content is shown in Table 5. The sensitivity of the proposed sensor is 0.0436235 pf/ppm for moisture and 0.05202775 pf/ppm for 2-FAL in mineral oil is the highest of thickness 0.01 mm among other silicon dioxide sensing film thicknesses.

5 Conclusion

Here, an effort is being made to find the optimal sensor structure for the identification of impurities in mineral oil. This work also improves the sensitivity and selectivity of a comb-type sensor by measuring moisture and 2-FAL in mineral oil using thin-film sensor technologies. A sensor's modeled finite elements allow for an efficient investigation of the selectivity and sensitivity of the sensor. Selectivity and sensitivity are effectively investigated using a sensor's finite elements modeled, with titanium oxide and silicon oxide sensing material coating. The objective is to quantify the moisture and 2-FAL contents of mineral oil to periodically assess the level of contaminants in aged oil and power transformers submerged in oil. Silicon dioxide and titanium oxide-coated sensors are used for impurities estimation (moisture and 2-FAL) to achieve higher selectivity and sensitivity. The simulation findings indicate that the sensor's capacitance value is significantly affected by variations in moisture and 2-FAL concentrations within the 0–40 ppm range. The five-finger structure is the optimal sensor structure for both moisture and 2-FAL in mineral oil. The five-finger sensor structure shows good sensitivities for both moisture and 2-FAL in transformer oil. The sensitivity of a comb-type sensor with 0.01mm thickness of titanium dioxide film for moisture in mineral oil detection is 0.0439738 pf/ppm and 0.05296743 pf/ppm in 2-FAL identification. 0.0436235 pf/ppm and 0.05202775 pf/ppm are the sensitivities in oil, using silicon oxide film for moisture and 2-FAL detection, respectively.

References

1. Wani SA, Nezami MM, Khan SA, Sohail S (2020) A capacitive sensor for detecting insulation degradation by sensing 2-FAL in transformer oil. IEEE Trans Dielectr Electr Insul 27(6):2179–2187. https://doi.org/10.1109/TDEI.2020.008889
2. Sun C, Ohodnicki PR, Stewart EM, Member S (2017) Chemical sensing strategies for real-time monitoring of transformer oil: a review. IEEE Sens J 17(18):5786–5806
3. Wani SA, Rana AS, Sohail S, Rahman O, Parveen S, Khan SA (2021) Advances in DGA based condition monitoring of transformers: a review. Renew Sustain Energy Rev 149:111347. https://doi.org/10.1016/j.rser.2021.111347
4. Nezami MM, Wani SA, Khan SA, Khera N, Sohail S (2018) An MIP-based novel capacitive sensor to detect 2-FAL concentration in transformer oil. IEEE Sens J 18(19):7924–7931. https://doi.org/10.1109/JSEN.2018.2864793
5. Mandlik M, Ramu TS (2014) Moisture aided degradation of oil impregnated paper insulation in power transformers. IEEE Trans Dielectr Electr Insul 21(1):186–193
6. Pahlavanpour B, Duffy G (1993) Development of a rapid spectrophotometry method for analysis of furfuraldehyde in transformer oil as an indication of paper ageing. In: Annual report - conference on electrical insulation and dielectric phenomena, vol 6 (1993) pp 493–498. https://doi.org/10.1109/ceidp.1993.378924
7. Rico-Yuste A, González-Vallejo V, Benito-Peña E, De Las Casas Engel T, Orellana G, Moreno-Bondi MC (2016) Furfural determination with disposable polymer films and smartphone-based colorimetry for beer freshness assessment. Anal Chem 88(7):3959–3966. https://doi.org/10.1021/acs.analchem.6b00167.
8. Laskar S, Bordoloi S (2013) Monitoring of moisture in transformer oil using optical fiber as sensor. J. Photonics 2013:1–7. https://doi.org/10.1155/2013/528478
9. Bosworth T, Setford S, Heywood R, Saini S (2001) Pulsed amperometric detection of furan compounds in transformer oil. Anal Chim Acta 450:253–261
10. Shamsipur M, Akbar A, Beigi M, Teymouri M, Rasoolipour S, Asfari Z (2009) Highly Sensitive and Selective Poly (vinyl chloride) - membrane potentiometric sensors based on a Calix [4] arene derivative for 2-Furaldehyde. Anal Chem 81(16):6789–6796
11. Nezami MM, Wani SA, Khan SA, Khera N, Sohail S (2020) A comb-type capacitive 2-FAL sensor for transformer oil with improved sensitivity. IEEE Trans Instrum Meas 69(7):4524–4532. https://doi.org/10.1109/TIM.2019.2942519
12. Rahman O, Islam T, Member S, Khera N, Khan SA (2021) A novel application of the cross-capacitive sensor in real-time condition monitoring of transformer oil. IEEE Trans Instrum Meas 70:1–12
13. Rahman O, Islam T, Ahmad A, Parveen S, Kheraand N, Khan SA (2021) Cross capacitance sensor for insulation oil testing. IEEE Sens J 1. https://doi.org/10.1109/JSEN.2021.3096463

14. Hussain E, Rahman O, Suboor A, Masood A, Islam T, Khan SA (2023) Condition monitoring of oil immersed transformer using thin dielectric film coated cross capacitance sensor. In: 2023 international conference on recent advances in electrical, electronics and digital healthcare technologies (REEDCON), New Delhi, India, pp 443–448. https://doi.org/10.1109/REEDCON57544.2023.10150676
15. Panasyuk TL, Mirsky VM, Piletsky SA, Wolfbels OS (1999) Electropolymerized molecularly imprinted polymers as receptor layers in capacitive chemical sensors. Anal Chem 71(20):4609–4613

Anomaly Detection System for Smart Lighting

Mohd Ahsan Siddiqui, C. Rama Krishna, and Mala Kalra

Abstract A smart lighting system leverages the power of IoT (Internet of Things) for optimal management of lighting facilities and substantial energy savings. To ensure the uninterrupted functionality of these systems, it is essential to prioritize detecting anomalies in time. The manual identification of individual anomalous smart lights is challenging when thousands of lights are in the entire system. For this purpose, we have proposed the integration of the LDR sensing system into each smart light to predict the illumination failure in it. LDR module measures the intensity of light emitted by the bulb and facilitates the detection of idle lights and prompt maintenance interventions. This paper presents two approaches for smart light anomaly detection leveraging IoT. One is the point anomaly detection system, which indicates individual lights that are not functioning; the other is the collective anomaly detection system, which sends alerts whenever there is an illumination-related issue with one or more lights in the system but does not specify individual anomalous lights. The performance of the proposed methods is validated by detecting the illumination failure of smart lights in a smart lighting system simulated using the Porteous Professional 8.11 simulator.

Keywords Anomaly Detection System (ADS) · Internet of Things (IoT) · Smart light · Illumination detection · Visible light failure detection · Illumination failure detection

1 Introduction

Smart lighting is an innovative technology that enables users to control their lighting automatically using IoT [1]. These systems comprise LED bulbs embedded with software, also known as firmware, that communicate with a mobile app, a smart home assistant, or other smart accessories. Automating or remotely controlling lighting

M. A. Siddiqui (✉) · C. Rama Krishna · M. Kalra
Department of CSE, National Institute of Technical Teachers Training and Research, Chandigarh 160019, India
e-mail: ahsan@manuu.edu.in

D.-I. Stroe et al. (eds.), *Emerging Electronics and Automation*, Lecture Notes in Electrical Engineering 1202, https://doi.org/10.1007/978-981-97-3090-2_3

eliminates the need for traditional wall switches. Some of its benefits are lights may be controlled remotely through an app or voice; by dynamically adjusting lighting levels based on real-time data, it ensures that energy is used efficiently, reducing electricity consumption and promoting sustainability; it offers a change in color and levels of illuminance as per user preferences, It provides a facility for synchronization with different devices for entertainment to create pleasant environments.

According to a Grand View Research [2] report, the global smart lighting market was valued at USD 9.32 billion in 2020 and is expected to grow at a Compound Annual Growth Rate (CAGR) of 19.9% from 2021 to 2028. It also states that key factors driving this market growth are increasing demand for smart homes and smart cities, technological advancements in wireless communication and LED technology, and rising awareness about energy conservation.

Smart lighting is becoming increasingly popular and widely adopted in smart homes, cities, offices, and industries due to its advantages and convenience. Ensuring the failure-free operation of such systems is crucial for maintaining continuity, which further instills user confidence, enabling the broader adoption of this transformative technology [3]. The manual identification of individual anomalous smart lights is challenging when thousands of lights are in the entire system. For this purpose, we have proposed integrating an LDR sensing system into each smart light to predict its illumination failure [4]. LDR module measures the intensity of light emitted by the bulb and facilitates the detection of idle lights and prompt maintenance interventions [5]. This paper presents two approaches for smart light anomaly detection leveraging IoT. One is the point anomaly detection system, which indicates individual lights that are not functioning [6]; the other is the collective anomaly detection system, which sends alerts whenever there is an illumination-related issue with one or more lights in the system but does not specify individual anomalous lights [7]. The performance of the proposed methods is validated by detecting the illumination failure of smart lights in a smart lighting system simulated using the Porteous Professional 8.11 simulator [8].

This paper's contributions are (a) proposed a framework designed to detect illumination-related anomalies in smart lights. It provides an effective and efficient solution for identifying deviations from the expected functionality of lights; (b) demonstrated the practical feasibility of the proposed framework by implementing a simulated system that integrates the developed anomaly detection algorithm into smart lighting infrastructure.

The paper is divided into five Sections. Section 1 provides an introduction, and Sect. 2 presents the literature survey on anomaly detection techniques for smart lighting systems. Section 3 discusses the proposed methodology and Sect. 4 offers the experiment setup and results. Finally, Sect. 5 provides concluding remarks.

2 Literature Review

The literature survey is focused on findings of the existing anomaly detection techniques for smart lighting systems.

The study in [3] suggests an interior lighting management system that responds to daylight and human presence to save energy. Ceiling luminaires and workstation desks are equipped with occupancy and light sensors to get information regarding user presence and illumination level, which is sent to a lighting controller. The performance of the system is validated by developing an office lighting testbed. The findings demonstrate that the system successfully adjusts the dimming levels of the luminaires to provide the required levels of illumination at the workplace, hence offering an effective and economical method of managing the lighting.

In [9], authors proposed a technique for identifying cyberattacks on IoT devices located in the home. The approach models user behavior as a series of user events depending upon time and temperature measurements and learns event sequences to spot irregular activities. The suggested technique can identify more than 90% of anomalies, with a false positive rate that seldom exceeds 10% across most devices. The proposed solution is tested on a network of home IoT devices in the lab. Validating the usefulness of the suggested strategy in real-world circumstances would need further study on a larger-scale network with more users and devices.

The approach in [10] uses machine learning models based on deviations in power consumption of IoT devices to identify anomalies in smart homes. Their experimental setup comprises a smart home with Smart Cameras. Distributed Denial of Services (DDoS) and brute-force attacks are performed to find deviations in power usage. The performance and execution time of four popular classification methods, namely Support Vector Machine (SVM), Linear Regression (LR), Random Forest (RF), and K-Nearest Neighbor (KNN), have been investigated on the collected dataset. Ten-fold cross-validation is used to ensure the consistency of the methods. SVM has shown the best performance out of all the models, with an accuracy of 94.04%. The dataset was used to train and test the machine learning models in a controlled setting. Thus, the models' performance may differ with more complicated and varied data in real-world situations.

In [11], authors have compared various machine learning models, namely Artificial Neural Network (ANN), DT, SVM, LR, and RF, to predict anomalies in an IoT environment. Thirteen features of network traffic have been used for predicting anomalies. Different attacks, such as Denial of Services (DoS) and spoofing, have been launched to deviate the network traffic to create anomalies. DT, RF, and ANN have obtained the highest accuracy of 99.4%. The proposed model can only detect network traffic-based anomalies, and there is no provision to detect novel attacks on network traffic as there is no ML updating (training for new data) arrangement for the same.

In [12], using Power Line Communication technology, the authors suggest a method for monitoring smart lighting networks for abnormalities. The system employs time series analysis and recurrent neural networks to forecast the values of the examined time series and identify odd behavior that may indicate an attack or abuse attempt. The average accuracy of the proposed system is 94.33%. The average False Positive Rate (FPR) is 4.34. The suggested approach may have limited applicability to non-smart lighting communications network infrastructures due to its testing on a single such network. Both the scalability of the proposed system and its performance under varying circumstances are addressed in the study.

In [13], authors have proposed an anomaly detection system for the IoT environment. The anomaly detection is based on features of network traffic. Further fog and cloud computing have been incorporated for efficient and optimal operation of the Intelligent System for Anomaly Detection (ISAD). The algorithm is 96% accurate in detecting network anomalies. This model is not able to see the sensor failure or malfunction. Further accuracy may be improved.

In [14], authors have proposed a machine learning-based hybrid intrusion detection system to protect home devices such as intelligent bulbs/lights, smart appliances, temperature controllers, etc. The proposed framework utilizes five ML models (XGBoost, Random Forest, Decision Tree, and K-Nearest Neighbor) to test the performance in terms of accuracy. The hybrid model has adopted a mixture of ML models and misused techniques (signature-based). The average accuracy of Random Forest, XGBoost, Decision Tree, and K-Nearest Neighbor are 97%, 97.16%, 95.75%, and 96.67%, respectively. The deployed method cannot detect the anomalies in the sensors or devices for wrong readings.

In [15], a family guard has been proposed to detect network traffic-based anomalies. For easy monitoring and security, authors have suggested an extra layer in the existing framework of the smart home environment. One class classifier has been used to test the accuracy of the proposed model. The average accuracy of one class classifier is 73.56%. This ML model has been compared with the other two models (i.e., local outlier factor and isolation forest) with an average accuracy of 88.55% and 74.62%, respectively. Among these three models, local outlier factor ML performed well—extra suggestion of layer results in more complex architecture. Sensor failure anomaly and sensor reading value anomaly detection are not possible.

In [16], an Event Verification System (EVS) has been proposed for events occurring regarding notification or state change of the actuators or sensors value. It is a variant of the anomaly detection system. Fingerprints of events have been designed. Further, event rules have been formulated based on these fingerprints. Two distinct state-of-the-art EVS are deployed in two innovative home environments containing 12 actuators and 16 sensors to evaluate the efficacy of their approach. According to the results provided in the paper, 71% of their physical signatures are susceptible to evasion. The authors suggest that after EVS generates fingerprints in a smart home, the defense can detect evadable fingerprints and protect them against evasion attacks. The proposed model is not capable of detecting the anomaly that is based on network traffic. Sensor failure detection is not possible.

Some papers have focused on the intelligent control of the lighting system for maintaining a pleasant environment and energy savings. In contrast, others have detected anomalies based on network traffic or power consumption. Detecting abnormalities in bright lights related to illumination failure when the entire lighting system is working correctly (i.e., input supply, output supply, Wi-Fi) is highly required in some applications, such as systems comprising a vast number of lights, lights located in accident-prone areas and high-security zones so that defective lights can be timely replaced.

3 Proposed Methodology

An intelligent lighting system for a smart home, commercial building, and smart city has been considered in the suggested technique. Figure 1 depicts the smart light anomaly detection system for a smart city. Each light comprises two sensors: a voltage sensor and an LDR sensor module. The voltage sensor provides the input mains status (i.e., 0, 1) while LDR offers 0 or 1 value as output in case of absence or presence of illumination. The input mains voltage sensor's value is kept at logic 1 for illumination anomaly detection. Voltage and LDR sensor values of each light L_i will be fed to the connected AND gate whose output ω_i will be providing detection of individual defective bulbs (point anomaly) as shown in Fig. 1. On the other hand, the β function will take input as voltage and LED sensor values of all the lights in the intelligent lighting system. If the value of $\beta = 0$, it will alert the system that there is an illumination failure and denotes a collective anomaly.

The values of the LDR and voltage sensors of each smart light have been sent to the receiver/controller. The smart lights can be linked to the controller unit using Wi-Fi/Ethernet or LORAWAN. The controller panel comprises an LED corresponding to each smart light for displaying its status. Whenever there is an illumination failure despite input voltage, the respective LED will start glowing, indicating a point anomaly. An alternative cost-effective model is also proposed where sensors' values are collectively fed into another AND gate on the controller/receiver, which

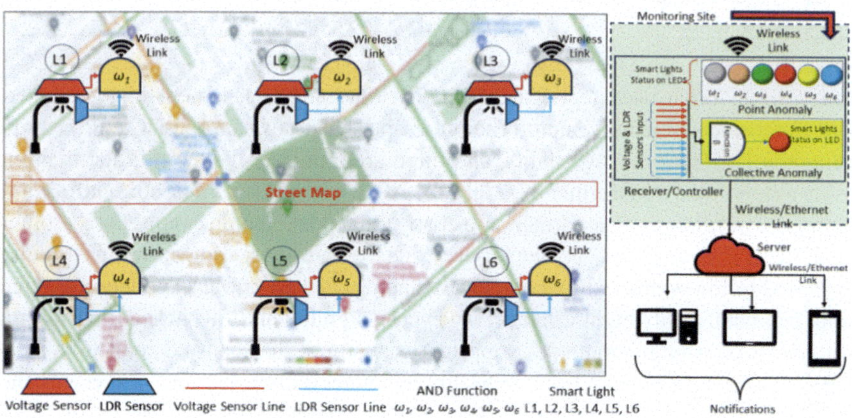

Fig. 1 Proposed smart light anomaly detection system

is represented as β and detects collective anomaly by changing the status of the LED to high. There is also a provision for transmitting these sensors' values to the server for worldwide monitoring through the app

Algorithm 1 presents anomaly detection in the illumination of smart lights. Input given is n lights of the smart lighting system (L_1, L_2, … L_n). The voltage and LDR sensor outputs corresponding to each light are read and kept in the variables v_{Li} and $LDR_{Li,}$ respectively. Further logical AND is performed on these values, and output is stored in ω_i. If the value of ω_i is zero, it indicates that Light L_i is not functioning.

The subsequent sections discuss various components of the proposed systems in detail, i.e., an AC mains input sensing module for positive logic generation, an LDR sensing module, anomalous smart bulb detection (point anomaly), and collective anomaly detection circuits.

- **AC mains sensing module and positive logic (0,1) generation for a smart bulb/ light**

Low-voltage DC (generated from the 220v or 110v Alternating Current (AC) mains) is essential for positive logic (0,1) and has many practical applications. The smart light can only be used if the mains power is on. Electronic circuits, LED lights, toys, and many other home devices require low-voltage Direct Current (DC), such as 5v, 6v, 9v, and 12v.

Algorithm I Illumination Failure Detection

Input: L_1, L_2, L_3, ... L_n
Output: Illumination failure

For i=1 to n
    Read input voltage of Light $v_{Li} \longleftarrow L_{i.}$
    Read the output of the LDR sensing module $LDR_{Li} \longleftarrow L_{i.}$
    $\omega_i \longleftarrow v_{Li}$ AND LDR_{Li}
    If (ω_i ==0)
        Print ("Light Li is not functioning")
    End if
End for

Batteries are often used to power them, but constant replacement of batteries is both expensive and inconvenient. The other option is to use an AC–DC converter to create the DC from the AC mains, and there are numerous of them in the market. Using a step-down transformer to reduce the AC's voltage is the most straightforward method; nevertheless, transformers are costly, bulky, and inconvenient. On the other hand, in a transformer-less power supply, high-voltage alternating current is directly converted to low-voltage direct current. Transformer-less power is widely used due to its compact, lightweight circuit. However, a significant drawback of this supply is no isolation from the main AC supply. As a result, it may cause damage to a low-voltage DC circuit or even life due to direct contact with it. Figure 2 represents the basic transformer-less input main voltage sensing module.

A voltage-dropping capacitor, also known as an X-rated capacitor, is the primary component in a transformer-less power supply circuit. This X-rated capacitor is

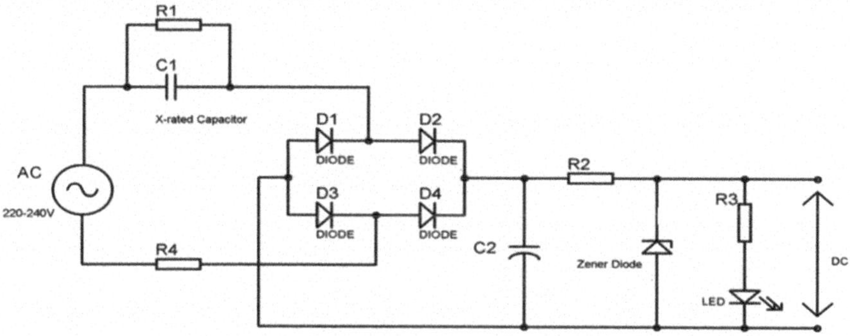

Fig. 2 Voltage sensing module design

connected in series with the line to reduce the voltage on an AC phase line. Capacitor power supplies are a common alternative to conventional power transformers. Choosing the suitable capacitor for decreasing voltage depends on the desired current drain and the capacitor's reactance. Equation 1 is used to calculate the capacitor's reactance, where X is the capacitor's reactance, f is the frequency of AC mains supply (i.e., 50 Hz for India), and C is the Capacitance of an X-rated capacitor. Finally, the current (I) in the circuit (Fig. 2) can be calculated with the help of Eq. 2.

$$CapX = \frac{1}{2\pi fC} \tag{1}$$

$$I = \frac{V}{X} \tag{2}$$

where I is the current in the circuit and V is the amplitude of AC supply voltage.

- **LDR Sensing Module**

The LDR sensing module comprises an input voltage (V_{in}), an LDR sensor (LDR), and a load resistor (RL). The LDR sensor has been connected in series with the input voltage (V_{in}) and load resistor (RL). The LDR sensing module needs to be attached to the bulb so that its light falls on it. To ensure optimal and reliable working of the LDR circuit, the LDR is covered with a black layer or box such that only the bulb's light falls on it. When the LDR sensor detects the light, output voltage (V_{out}) is generated. In the absence of light on the surface, the output voltage (V_{out}) is 0, which helps identify a defective bulb while having an input supply. Figure 3 shows a typical LDR circuit design.

- **Point Anomaly Detection Circuit**

Figure 4 shows the primary circuit consisting of AND gates through which AC mains voltage and LDR sensor readings have been passed. The discrete values of 0 and 1 have been adopted for noise-free operation. The circuit is compact, cost-effective,

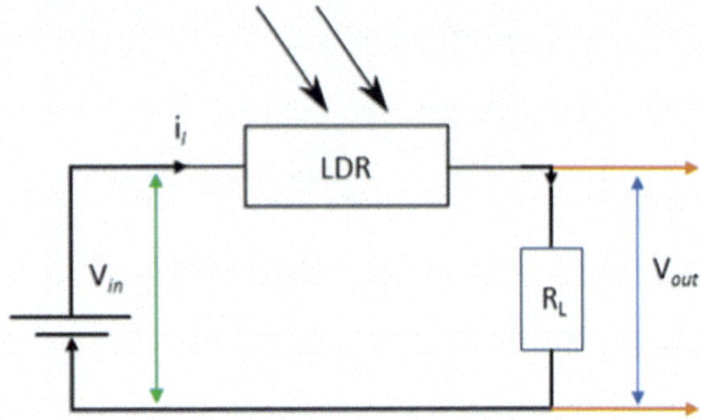

Fig. 3 LDR circuit design for illumination detection

and can be easily integrated into a smart bulb/light during manufacturing. If there is illumination failure with a particular bulb or tube light, the output of that associated AND becomes 0, and the defective bulb can be easily identified.

- **Collective Anomaly Detection Circuit**

Figure 5 provides an alternative failure detection circuit design that is more cost-effective than the one given above. It is also useful when the number of lights is huge or remotely installed (i.e., street lights, etc.). In the proposed circuit, AC mains voltage and LDR sensor readings of the n lights are collected and fed into the AND gate. If the output of the AND gate is 0, it points to collective anomaly detection, which will be reflected at the monitoring site by glowing the LED.

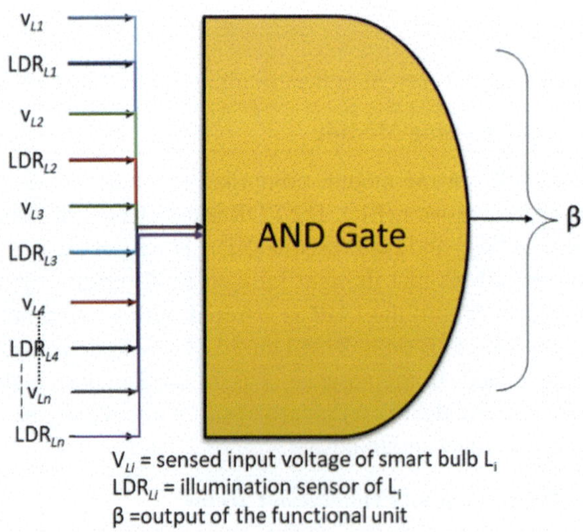

Fig. 4 Collective illumination failure detection design

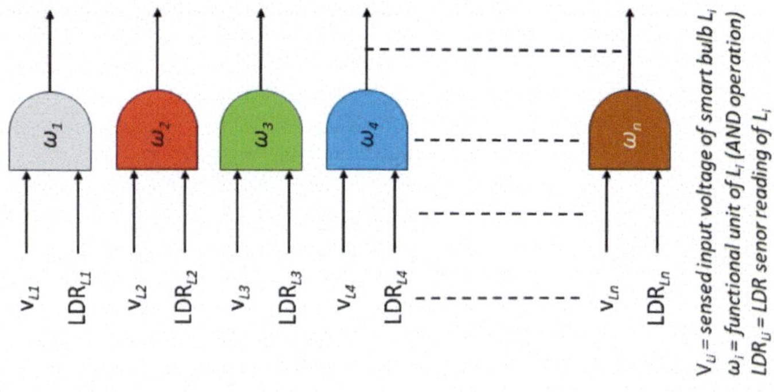

Fig. 5 Anomalous bulb detection design

4 Experimental Setup and Results

The functional block diagram of the proposed methodology is shown in Fig. 6. Voltage and LDR sensors are connected to each bulb. The voltage sensor is based on the transformer-less power supply where the AC main supply is sensed. Then, the amplitude is decreased with the help of an X-rated capacitor and converted into DC voltage with the help of a bridge rectifier. Further, using an electrolyte capacitor, the DC voltage is conditioned for any AC ripples. In the final stage, a Zener diode regulates the output voltage. Finally, an approximately 5-V DC has been received, representing logic 1. The final output voltage is also given to the LDR sensor circuit to energize it. The LDR circuit consists of an input supply in series with LDR and load resistance. When the input supply is given, the output voltage and current across the load resistor vary according to the light intensity. In the absence of light on LDR, 0-V DC is received across the load resistor, which shows the logic 0. This logic 0 shows the illumination failure condition.

Figure 7 depicts the simulation setup on Proteus Professional 8.11 software. The simulation has been performed by considering two light bulbs. A pair of voltage and LDR sensor readings (v_{L1}, LDR $_{L1}$) from the first bulb is given to AND gate U1. Similarly, the second bulb's voltage and LDR sensor readings are provided to AND gate U2. The output of the U1 and U2 is given as input to the AND gate U3. The circuit consists of a 0.47uF voltage-dropping capacitor linked in series with the AC phase line. Different values of this type of capacitor have been shown in Table 1. Since capacitors of this kind are not polarized, they may be plugged in from either end. An additional precaution against electric shock is provided by a 470k ohm resistor connected in parallel with the capacitor. This kind of resistance is known as Bleeder resistance. A bridge rectifier (a set of four diodes) is utilized to get rid of the negative half of the AC. The term for this procedure is rectification. To smooth out the resulting wave, a 1000 μF/50 v capacitor is utilized for filtering. Finally, a 5.0-V-1-W Zener diode is employed as a voltage regulator. Since the output voltage of this circuit is around 12 V, the Zener diode flows back the excess current to bring the voltage down to approximately 5.0 V. A Zener diode with a different value, such

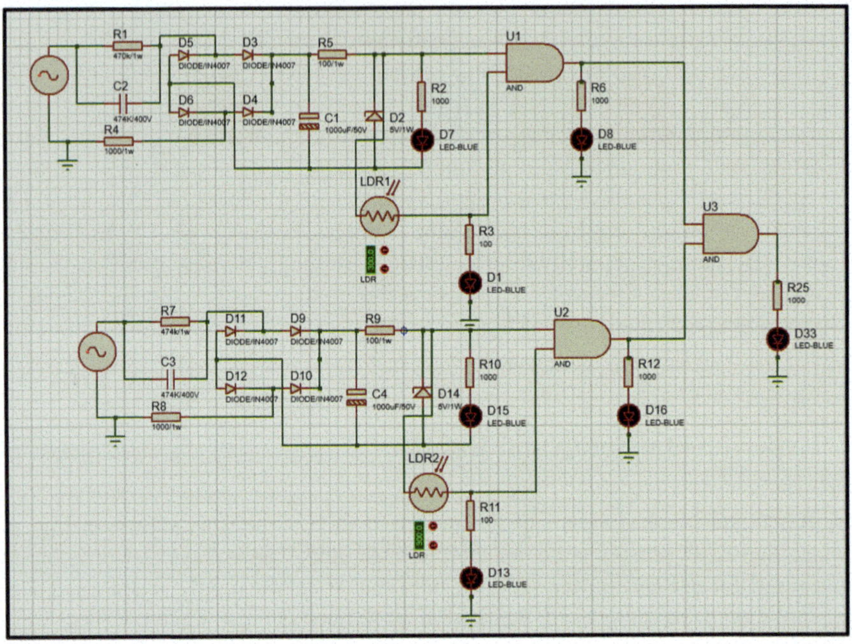

Fig. 6 Experiment setup on proteus simulator. **a** Detection of point anomaly in individual light

as 6.1 V, 8 V, etc., may be used instead to get the necessary voltage. For monitoring and diagnostics, an LED has been wired in. R3 (100 oh Fig. 2) is put into play as a current-limiting component. Resistor R4 (Fig. 2) should have 1 W or higher (5w) wattage to avoid burn. The typical resistor is thicker than this one. Different combinations of the values of the input sensors have been given to the circuit. These values are shown in Table 3. The LEDs (D8, D16, D33) are connected to the output of AND gates (U1, U2, U3). LEDs will act as an alarm for anomaly detection. An LED will glow whenever the production of the connected AND gate is true; that is, a defective bulb is detected.

Table 1 shows the X-rated capacitor code, capacity in microfarad, voltage, and current. In the simulation, capacitor code (474K) has been utilized, providing the circuit's 12-V AC and 25 mA current. It is sufficient to fire the AND gates (U1, U2), as shown in Fig. 7. Table 2 lists all the components used in the simulation. They may vary as per the system design under consideration.

Table 3 shows the simulation results of the LDR circuit. When the light intensity increases, the resistance decreases, and electrons begin to move from one layer to another. Hence, the voltage across the load resistance and current begin to increase. The value of the positive logic, approximately 2.2-V DC, is selected, while the rest of the values below it are rejected. When LDR detects no light, there is no current flow or voltage development across the resistor (R_L). Thus, the failure of illumination

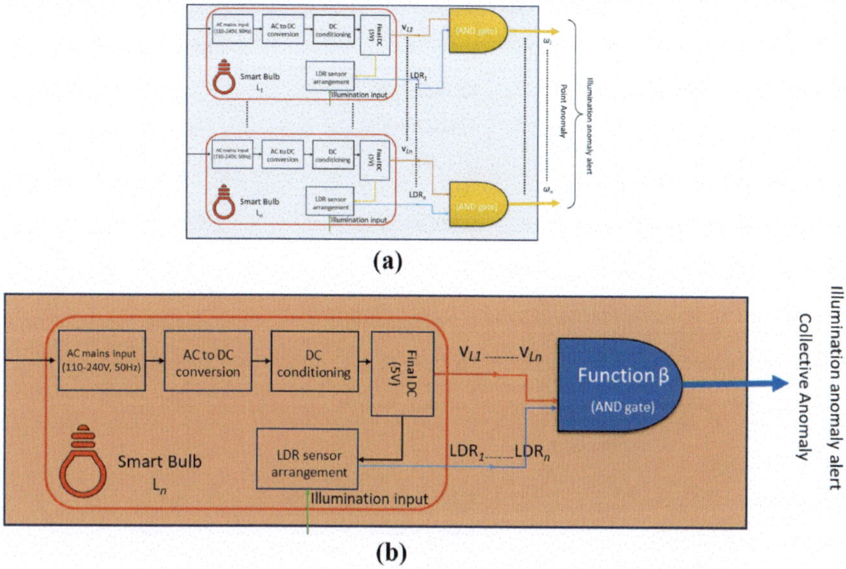

Fig. 7 **a**, **b** Functional block diagram of **a** Point and **b** Collective anomaly detection systems

Table 1 X-rated capacitor code, value, voltage, and current

Capacitor code (k)	Capacitor value (uF)	Voltage (v)	Current (mA)
104	0.1	Four	8
334	0.33	10	22
474	0.47	12	25
684	0.68	18	100
105	1	24	40
225	2.2	24	100

(light) has been detected. This will help to overcome the situation where there are hundreds of lights are installed. It is very difficulty each lights check manually.

The input supply is available, the circuit is working, and the Wi-Fi status is functional, but the bulb is not glowing. The illumination landscape becomes very complex when the smart lights are installed on the street, and the number is in the thousands.

Table 4 shows the truth table of the simulated circuit. It consists of four input sensor values (v_{L1}, LDR_{L1}, v_{L2}, LDR_{L2}) of two bulbs/tube lights. The output (ω_1, ω_2) has been estimated by giving the various combinations of 0s and 1s. The functions ω_1 and ω_2 show the illumination failure of the specific bulb, while β represents the illumination failure of bulb/bulbs in the smart lighting system.

Table 4 depicts the timing diagram of the Truth Table. It represents the various activities of input logic (0,1) based on the timing of activation and deactivation of

Table 2 Description of the components

Part Name	Qty	Value	Make	Remarks
Resistors R1, R7	2	470KΩ/1 W	Generic	Built-in with simulator
X Rated-Capacitors C2, C3	2	474KΩ/400V	Generic	
Diodes D3, D4, D5, D6, D9, D10, D11, D12	8	IN4007	Generic	
Resistors R4, R8	2	1000Ω/1 W	Generic	
Capacitors Electrolyte C1, C4	2	1000 μF/50V	Generic	
Resistors R5, R9	2	100Ω/1W	Generic	
Zener Diodes D2, D4	2	5.0 V/49mA	Generic	
Resistors R2, R10	2	1000Ω/1 W	Generic	
Diodes D1, D7, D8, D13, D15, D16, D33 (LEDs Blue color)	7	Forward Voltage 2.2 v DC	Generic	
Resistors R3, R11	2	100Ω/1 W	Generic	
AND Gates U1, U2, U3	3	2 input and one output	Generic	
Resistors R6, R12, R25	3	100Ω/1 W	Generic	
Light Dependent Resistors LDR1, LDR2	2	10KΩ	Generic	

Table 3 Lux versus voltage versu current of LDR circuit

Light intensity (in Lux)	Current in the LDR circuit (in mA)	Voltage output of LDR circuit (in mV)
10	0.17	2070
20	0.29	2150
30	0.41	2180
40	0.52	2210
50	0.63	2230
75	0.87	2260
100	1.11	2290
125	1.33	2320
100	1.54	2340
175	1.74	2370
200	1.93	2390
225	2.12	2410
250	2.31	2430
275	2.49	2450
300	2.66	2460

Table 4 Truth table of the proposed methodology

Bulb L_1		Bulb L_2		ω_1 (Point anomaly in L_1)	ω_2 (Point anomaly in L_2)	β (Collective anomaly)
v_{L1}	LDR_{L1}	v_{l2}	LDR_{L2}	v_{L1} AND LDR_{L1}	v_{L2} AND LDR_{L2}	ω_1 AND ω_2
0	0	0	0	0	0	0
0	0	0	1	0	0	0
0	0	1	0	0	0	0
0	0	1	1	0	1	0
0	1	0	0	0	0	0
0	1	0	1	0	0	0
0	1	1	0	0	0	0
0	1	1	1	0	1	0
1	0	0	0	0	0	0
1	0	0	1	0	0	0
1	0	1	0	0	0	0
1	0	1	1	0	1	0
1	1	0	0	1	0	0
1	1	0	1	1	0	0
1	1	1	0	1	0	0
1	1	1	1	1	1	1

0 = illumination Failure, 1 = illumination Available

inputs (voltages and LDRs). It verifies that the output of different functions is as per the expectations of the simulation.

5 Conclusion

This paper has suggested two design approaches for detecting illumination-related anomalies in a smart lighting system. The first approach targets detecting the point anomaly, whereas the second is oriented toward collective anomaly detection. The second approach alerts the user of some faulty bulbs, unlike the first one. Specifically points out the individual bulbs that are malfunctioning. The proposed solutions deal with the scenario where, despite having all inputs available to the smart bulb, i.e., input voltage, current, Wi-Fi status (i.e., connected), etc., it is not glowing due to a faulty LED. The system is simulated using Proteus Professional 8.11 for performance validation and has been tested successfully on different inputs (Fig. 8).

The potential use cases for the designed system are remote monitoring of many street lights and lights in commercial buildings, hospitals, etc., leading to timely

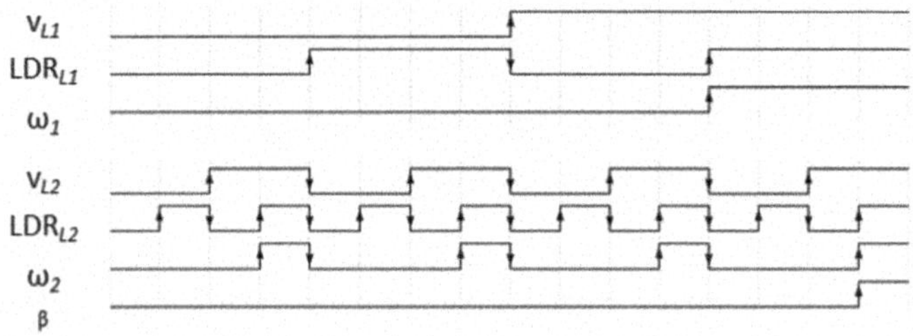

Fig. 8 Timing diagram of the truth table in Table 4

replacement of defective bulbs. This setup is handy for accident-prone areas and high-security zones for continuous patrolling.

The proposed solution can also be used with conventional bulbs/lights by mounting the developed anomaly detection module on the driver circuit. For sending Illumination failure detection alert.

5.1 Limitation and Future Work

The limitation of the work is that it is a simulation-based proposed methodology. Actual implementation will present the actual behavior of the anomaly detection system. Further precise detection of light rays is essential; otherwise, the system can produce false positives.

Machine learning techniques can be incorporated into the proposed methods for accurate, reliable, and optimal system control.

References

1. Majib Y, Barhamgi M, Heravi BM, Kariyawasam S, Perera C (2023) Detecting anomalies within smart buildings using do-it-yourself internet of things. J Ambient Intell Humans Comput 14(5):4727–4743. https://doi.org/10.1007/s12652-022-04376-w
2. Sierra Wireless (2023) Benefits of smart lighting for a modern city. https://www.ankurlighting.com/blogs/general/benefits-of-smart-lighting (accessed Jun. 23, 2023)
3. Caicedo D, Li S, Pandharipande A (2017) Smart lighting control with workspace and ceiling sensors. Light Res Technol 49(4):446–460. https://doi.org/10.1177/1477153516629531
4. Du H, Wang Y, Xu X, Liu M on the Web, and undefined 2023, "Niffler: Real-time device-level anomalies detection in smart home," dl.acm.org, Accessed: Nov. 21, 2023. https://doi.org/10.1145/3586073
5. Chetana M, Prasad D, Bhargavi N, Navya G, Automatic street light using LDR. journal-dogorangsang.in. 10.36893.DRSR.2023.V13I0.0262–0268
6. Senthilkumar DS, Prathap N, Senthilkumar T, Parthasaradi V (2023) Energy saving solar powered LED street lights with automatic intensity control. IRE J 6(10):1–4. Accessed: Nov. 21, 2023. https://www.topflightvalet.com/formatedpaper/1704218.pdf

7. Bin Iskandar MA, Saaidin S, Kutty SB, Kassim M (2023) Brightness controlled solar powered intelligent street light. In: 2023 19th IEEE international colloquium on signal processing and its applications, CSPA 2023 - Conference Proceedings, 2023, pp 156–161. https://doi.org/10.1109/CSPA57446.2023.10087420.
8. S. M. Shifa, T. M. Mridul, S. S. Rafat, M. H. Monjur, M. R. Islam, and M. K. Hassan, "Design and Implementation of Smart Wheelchair with Advanced Control Interfaces," in International Conference on Robotics, Electrical and Signal Processing Techniques, 2023, pp 155–159. https://doi.org/10.1109/ICREST57604.2023.10070067
9. Yamauchi M, Ohsita Y, Murata M, Ueda K, Kato Y (2020) Anomaly detection in smart home operation from user behaviors and home conditions. IEEE Trans Consum Electron 66(2):183–192. https://doi.org/10.1109/TCE.2020.2981636
10. Dilraj M, Nimmy K, Sankaran S (2019) Towards behavioral profiling based anomaly detection for smart homes. In: TENCON 2019 - 2019 IEEE region 10 conference (TENCON), IEEE, Oct. 2019, pp 1258–1263. https://doi.org/10.1109/TENCON.2019.8929235
11. Hasan M, Islam MM, Zarif MII, Hashem MMA (2019) Attack and anomaly detection in IoT sensors in IoT sites using machine learning approaches. Internet Things (Netherlands) 7:100059. https://doi.org/10.1016/j.iot.2019.100059
12. Andrysiak T, Saganowski Ł (2020) Anomaly detection for smart lighting infrastructure with the use of time series analysis. JUCS - J. Univers. Comput. Sci. 26(4):508–527. https://doi.org/10.3897/jucs.2020.027
13. Moreira DAB, Marques HP, Costa WL, Celestino J, Gomes RL, Nogueira M (2021) Anomaly detection in smart environments using AI over fog and cloud computing. In: 2021 IEEE 18th annual consumer communications and networking conference (CCNC). IEEE, pp 1–2. https://doi.org/10.1109/CCNC49032.2021.9369449.
14. Alghayadh F, Debnath D (2021) A hybrid intrusion detection system for smart home security based on machine learning and user behavior. Adv Internet Things 11(01):10–25. https://doi.org/10.4236/ait.2021.111002
15. de Melo PHAD, Miani RS, Rosa PF (2022) Family guard: a security architecture for anomaly detection in home networks. Sensors (Basel) 22(8):2895. https://doi.org/10.3390/s22082895
16. Ozmen MO, Song R, Farrukh H, Celik ZB (2023) Evasion attacks and defenses on smart home physical event verification. In: Proceedings 2023 network and distributed system security symposium, Internet Society, Reston. https://doi.org/10.14722/ndss.2023.23070

Design and Implementation of an Optimal Energy-Efficient Dual-Axis Solar Tracking System

Ankur Gupta and P. S. Pravin

Abstract The pressing need for sustainable energy solutions has triggered the rapid development of solar technologies. Among these advancements, dual-axis solar tracking systems have emerged as a promising approach to augment solar energy capture. The dual-axis solar tracking system operates by dynamically orienting solar panels along both the azimuth and elevation axes, allowing them to precisely follow the sun's position throughout the day. This continual alignment significantly increases the absorption of solar irradiance, thereby optimizing energy output. This paper delves into the design and implementation of automated dual-axis solar tracking system showcasing the performance enhancement compared to a traditional stationary solar panel. The system employs sensors and servo motors to detect the sun's position, utilizing this information to dynamically adjust the orientation of solar panels. By evaluating the latest research findings, this paper underscores the potential of dual-axis solar tracking systems to revolutionize solar energy generation, making a compelling case for their widespread adoption in the renewable energy sector.

Keywords Solar energy · Dual-axis tracking system · Renewable energy · Optimization

1 Introduction

In the quest for sustainable and renewable energy sources, solar power stands out as a promising and increasingly vital solution. Among the numerous advancements in solar technology, the development of efficient solar tracking systems plays a pivotal role in maximizing energy output. Traditional fixed-tilt solar systems are limited in their ability to harness the full potential of solar irradiance due to their static positioning. In contrast, dual-axis solar tracking systems dynamically orient solar panels to follow the sun's path, significantly enhancing energy capture throughout the

A. Gupta · P. S. Pravin (✉)
Department of Electronics and Instrumentation Engineering, National Institute of Technology Silchar, Assam 788010, India
e-mail: pravin@ei.nits.ac.in

D.-I. Stroe et al. (eds.), *Emerging Electronics and Automation*, Lecture Notes in Electrical Engineering 1202, https://doi.org/10.1007/978-981-97-3090-2_4

day [1]. The utilization of dual-axis solar tracking systems has garnered considerable interest due to their potential to substantially increase energy yield compared to fixed-tilt systems. As the technology evolves and becomes more economically viable, it becomes essential to investigate the design intricacies, control algorithms, and implementation challenges to harness its full potential [2].

Solar tracking systems should be designed to continuously orient solar panels to follow the sun's path throughout the day, maximizing exposure to solar irradiance. This dynamic alignment significantly enhances the energy output of PV systems compared to fixed-tilt installations. Dual-axis tracking systems, in particular, offer the advantage of tracking the sun's position in both the azimuth (East–West) and elevation (North–South) axes, ensuring precise alignment with the sun at all times [3]. The azimuth axis aligns the panels along the east–west direction to track the sun's daily east-to-west movement, while the elevation axis tilts the panels up or down to follow the sun's varying height above the horizon throughout the day and across seasons. As the demand for solar energy integration continues to grow, the need for such systems has become increasingly important.

Traditional fixed-solar panel setups are limited in their ability to capture maximum sunlight as the sun's position changes. The automated solar tracking system addresses this limitation by integrating sensors (here LDR sensors are used) and servo motors that enable continuous adjustment of the solar panels to track the sun's movement. This dynamic tracking ensures that solar panels are always oriented perpendicularly to the incoming sunlight, maximizing the exposure of photovoltaic cells and, consequently, improving energy harvesting efficiency [4]. By actively tracking the sun's position, the solar panels can receive sunlight more directly, optimizing energy production. Maximizing the energy output leads to improved power generation compared to fixed solar panels. The system is adaptable to varying weather conditions and seasons, ensuring maximum efficiency year-round.

The implementation of dual-axis solar tracking systems involves intricate control mechanisms, sensor technologies, and algorithms to accurately and efficiently direct the panels' movement. These systems offer a higher efficiency rate in energy generation compared to fixed-tilt or single-axis tracking systems, particularly in locations where sunlight angles vary significantly throughout the day and across seasons. Their application finds relevance in various sectors, including commercial and industrial solar installations, concentrated solar power plants, and residential solar setups [5]. However, while offering increased energy yields, the cost and complexity of dual-axis systems have traditionally limited their widespread adoption [6]. Continued research and advancements in materials, control systems, and tracking mechanisms aim to make dual-axis solar tracking systems more cost-effective and practical, contributing to the ongoing evolution of solar energy technology and its broader integration into the renewable energy landscape.

In this work, an automated optimal energy-efficient solar tracking system is designed and implemented to optimize the efficiency of solar panels by ensuring their alignment with the sun throughout the day. Performance assessment is carried out with the physical setup by turning OFF and ON the dual-axis solar tracking system for various directions of sunlight.

The remainder of this paper is classified as follows: The system description along with the various components used is presented in Sect. 2, while the detailed optimal algorithm and logic are discussed in Sect. 3. The results and discussion for the energy variations for various directions of sunlight are discussed in Sect. 4. Finally, the concluding remarks of the work are detailed in Sect. 5.

2 System Description

A dual-axis solar tracking system using light-dependent resistors (LDRs) and servo motors is designed to increase the efficiency of solar panels by orienting them towards the sun for maximum exposure. The main components of the system are described below.

2.1 Components

Light-dependent resistors (LDRs): LDRs are light-sensitive devices whose resistance changes with the intensity of light falling on them. Figure 1 depicts the photograph of the setup of LDRs installed for sensing the light intensity. They are used to detect the direction and intensity of sunlight. At the core of the automated solar tracking system are sensors that detect the sun's position in the sky. These sensors, often based on light intensity measurements, continuously monitor the sun's azimuth and elevation angles. The resistor works on the principle of photo conductivity which states that when the light falls on the surface of an LDR, the conductivity of the material reduces as the photons in the incident light have energy greater than the band gap of the semiconductor material used in the LDR.

Servo motor/Actuation mechanism: Servo motors are devices that can precisely rotate to specific angles. In this system, they are used to adjust the orientation of solar panels in both the azimuth and elevation axes. The control unit communicates with an actuation mechanism responsible for physically adjusting the position of the solar panels. This mechanism can include servos that tilt and rotate the panels with precision according to the calculated angles. Servo motors might experience wear and tear due to continuous movement, requiring occasional maintenance.

Control Unit: A control unit generally includes an electronic device such as an Arduino, Raspberry Pi, or a dedicated controller, which processes the data from the LDRs and controls the servo motors based on the sunlight's position. This sophisticated control unit processes real-time data from the sensors and calculates the optimal orientation of the solar panels. This unit acts as the brain of the system, making dynamic adjustments to ensure the panels are constantly aligned with the sun. In this work, Arduino Uno has been used as the control unit in which the ATmega328P microcontroller works as the primary control unit.

Fig. 1 Setup of Light-Dependent Resistors

Solar Panels: The solar panels are the energy-harvesting devices that are adjusted by the servo motors to be optimally aligned with the sun's position for maximum energy capture. They are generally made up of photovoltaic cells, typically constructed from semiconductor materials, commonly crystalline silicon. These cells are responsible for converting sunlight into electricity through the photovoltaic effect. A 1-watt solar panel has a power output of 1 W under standard test conditions (STC). This value is a standardized measure that allows for consistent comparison between different solar panels. In this paper for the purpose of demonstration, we have chosen a 6V 1W Solar Panel. The voltage and current produced by the solar panel depend on its design and the characteristics of the individual cells. In a 1-watt panel, the voltage and current will combine to yield the 1-watt power output. The voltage is usually in the range of a few volts, and the current is in the order of a fraction of an ampere. The physical size of a 1-watt solar panel can vary, but it is generally small and compact. Dimensions may be on the order of a few square inches or centimeters, making it suitable for portable applications. A 1-watt solar panel is commonly used for small-scale applications, such as charging small electronic devices (e.g., smartphones, power banks), powering LED lights, etc.

Power Supply: For the continuous operation of the automated tracking system, the automated solar tracking system requires a reliable power supply. This can be generally achieved through a combination of solar-powered batteries and grid connection, ensuring sustainability and reliability even in varying weather conditions. The system will consume some power for operation, which should be accounted for in the overall energy balance.

3 Algorithm

The dual-axis solar tracking system with LDR and servo motors is implemented following an algorithm that continuously adjusts the position of solar panels to maximize solar energy collection. A screenshot of the circuit connection with the sensors, the actuation devices, and the controller is given in Fig. 2.

The schematic of the physical design of the dual-axis solar tracking system is depicted in Fig. 3.

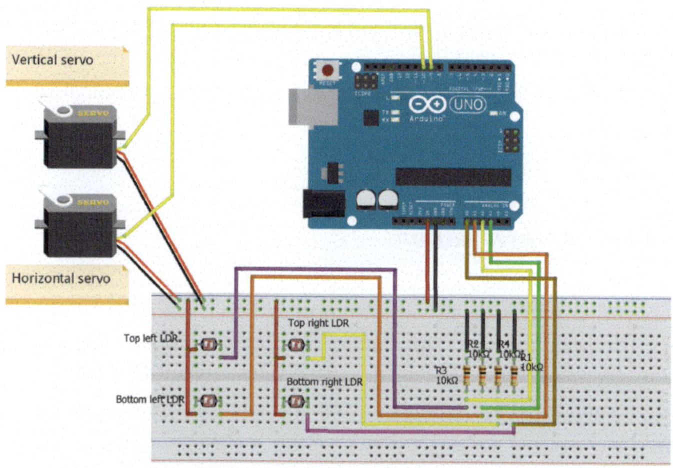

Fig. 2 Circuit diagram of the setup

Fig. 3 Physical setup of the dual-axis solar tracking system

3.1 Initialization

In the initialization step, we set up the microcontroller and connect the LDR sensors and servo motors. The threshold values for light intensity are defined in order to determine the optimal position of the solar panels.

3.2 Reading LDR Sensor Values

Multiple LDRs are positioned on the solar panel system to detect the intensity of light in various directions. In this step, the analog values from the LDR sensors are continuously monitored and these values will indicate the amount of light received by each sensor.

3.3 Calculation of Deviation

The microcontroller processes the information gathered from the LDRs to calculate the precise position of the sun. This calculation involves both azimuth (horizontal) and elevation (vertical) angles. The deviation from the optimal light intensity is calculated by comparing the readings from the LDR sensors with the predefined threshold values.

$$Deviation = Desired\ Light\ Intensity - Current\ Light\ Intensity \quad (1)$$

3.4 Adjusting the Servo Motors

The direction and amount of adjustment needed for both the azimuth (horizontal) and elevation (vertical) axes are determined based on the aforementioned deviation. The microcontroller then sends signals to the servo motors, commanding them to adjust the solar panels in both the azimuth and elevation axes. The azimuth axis corresponds to the east–west movement, while the elevation axis refers to the north–south tilt. Adjusting each servo motor involves several calculations as follows:

3.5 Calculating the Average Light Intensity

$$Left\ average = \frac{(LDR\ 1 + LDR\ 2)}{2} \tag{2}$$

$$Right\ average = \frac{(LDR\ 3 + LDR\ 4)}{2} \tag{3}$$

$$Top\ average = \frac{(LDR\ 1 + LDR\ 3)}{2} \tag{4}$$

$$Bottom\ average = \frac{(LDR\ 2 + LDR\ 4)}{2} \tag{5}$$

3.6 Comparing the Average Readings to the Average Intensity

If Left average > Right average, turn horizontal servo to the right
If Left average < Right average, turn horizontal servo to the left
If Top average > Bottom average, turn vertical servo upward
If Top average < Bottom average, turn vertical servo downward.

3.7 Adjusting Panel Orientation

This event continuously monitors the LDR sensor readings and makes corresponding adjustments to the servo motors to ensure the solar panels are aligned optimally with the sun's position.

3.8 Optimization and Calibration

This step involves fine-tuning the threshold values and servo motor movements for smoother and more accurate tracking. A feedback mechanism ensures that the system corrects any errors in tracking thereby achieving optimal tracking of the sun's energy to generate power.

3.9 Safety Measures

Various safety features are to be implemented in order to protect the system from damage during extreme weather conditions (strong winds, heavy rain, etc.).

4 Results and Discussion

The physical setup is implemented starting with designing a sturdy frame to support the solar panels and the tracking mechanism. The frame is designed in such a way as to be able to support the weight of the solar panels and tracking mechanisms, yet in a way that minimizes wind resistance. The minute-wise profiles for the power generated for different orientations of the solar panel while it is stationary and while it is mounted on the solar tracker system when the initial position of solar panel was facing towards the east direction are shown in Fig. 4.

In a similar fashion, the profiles for the power generated for different orientations of the solar panel while it is stationary and while it is mounted on the solar tracker system when the initial position of solar panel was facing towards the south direction are shown in Fig. 5.

Again, the profiles for the power generated for different orientations of the solar panel while it is stationary and while it is mounted on the solar tracker system when

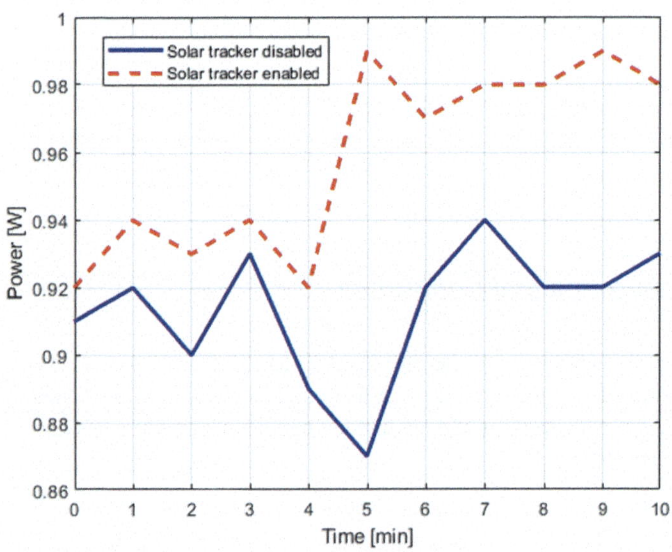

Fig. 4 Profiles of power generated with and without auto-tracking system (East direction)

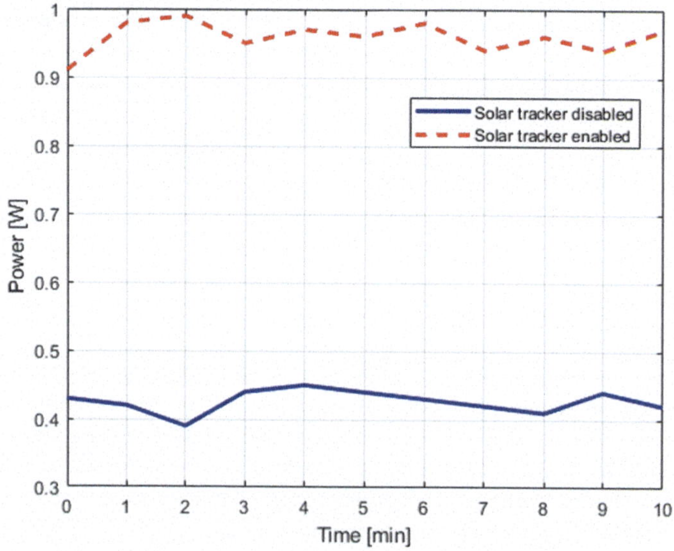

Fig. 5 Profiles of power generated with and without auto-tracking system (South direction)

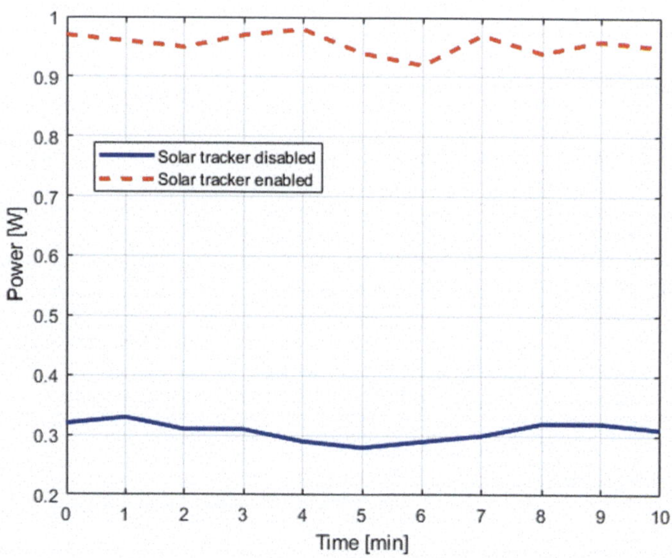

Fig. 6 Profiles of power generated with and without auto-tracking system (West direction)

the initial position of solar panel was facing towards the west direction are shown in Fig. 6.

It was observed that implementing a dual-axis solar tracking system offers numerous advantages in maximizing the efficiency and energy output of solar panels as compared to a fixed version. The power generated by the designed 6V 1W solar panel was observed to be more as compared to the fixed panel power. This is mainly due to the dual-axis solar tracking system that continuously orients the solar panels to face the sun directly, optimizing energy capture throughout the day. By dynamically following the sun's movement, the energy efficiency of the panel can be improved substantially as observed in the results.

5 Conclusion

In conclusion, the implementation of dual-axis solar tracking systems offers a substantial increase in energy production, making them a compelling choice for those seeking higher efficiency and output from their solar installations. By adjusting both the azimuth (east–west) and elevation (up–down) angles, they capture more sunlight, significantly enhancing energy production compared to fixed panels. These systems are well-suited for locations where maximizing solar energy production is a priority, offering a more sustainable and efficient alternative to traditional fixed solar panel setups.

References

1. Muthukumar P, Manikandan S, Muniraj R, Jarin T, Sebi A (2023) Energy efficient dual axis solar tracking system using IOT. Measur: Sens 28:100825
2. Huang Y, Li M, Xu Y (2020) Design and implementation of solar tracking system with dual-axis active control. IEEE Trans Sustain Energy 11(3):1567–1575
3. Baouche FZ, Abderezzak B, Ladmi A, Arbaoui K, Suciu G, Mihaltan TC, Raboaca MS, Hudişteanu SV, Ţurcanu FE (2022) Design and simulation of a solar tracking system for PV. Appl Sci 12:9682
4. Awasthi A, Shukla AK, Murali Manohar SR, Dondariya C, Shukla KN, Porwal D, Richhariya G (2020) Review on sun tracking technology in solar PV system. Energy Rep 6:392–405
5. Tchao ET, Asakipaam SA, Fiagbe YA, Yeboah-Akowuah B, Ramde E, Agbemenu AS, Kommey B (2022) An implementation of an optimized dual-axis solar tracking algorithm for concentrating solar power plants deployment. Sci African 16
6. Jose PJ, Akbari P, Dhokiya J, Ajmera Z, Pandya V, Jani YN, Ravalia A, Keshvani MJ (2022) Solar tracking: the best alternative to obtain more solar power output. Mater Today: Proc 67:921–926

A Simple Capacitive Sensor for Stress Measurement in Concrete

Uroosa Ghazi, Kazi Jabed Akram, and Tarikul Islam

Abstract It is very crucial to measure the strength of any infrastructure to prevent casualties due to the collapse of the structure. Concrete is one of the main elements in construction. Structural engineers want to use concrete that reaches high strength in a short time. However, measurement and monitoring of the strength of concrete in the initial phase is a tedious job and costly affair. Due to stress, there is a formation of microcrack that leads to weakness of the structure. In this paper, a simple inexpensive capacitive sensor is designed and fabricated for the measurement of stress for the pressure range of 0–1 MPa. The variation in stress during the hydration of cement concrete causes a change in the thickness of the dielectric layer of the sensor leading to a change in capacitance value. With an increase in compressive stress, the capacitance value increases, while, in tensile stress, the capacitance value decreases. The maximum sensitivity of the sensor used for concrete stress measurement is 28.8 pF/MPa.

Keywords Hydration of concrete · Capacitive sensor · Stress measurement · Response parameters

1 Introduction

Concrete strength is a critical mechanical attribute that should be checked for quality assurance. Cement is a constituent of concrete and a binder used in construction. The concrete consists of lime and silica [1]. The property of concrete that it is the best binder material makes it irreplaceable. Various oxides are formed in cement after sintering, for example, calcium oxide, silicon oxide, aluminum oxide, iron oxide, and magnesium oxide. On heating, calcination oxides get converted to silicates,

U. Ghazi · T. Islam (✉)
Jamia Millia Islamia, New Delhi, India
e-mail: tislam@jmi.ac.in

K. J. Akram
Indian Institute of Technology, Madras, India

D.-I. Stroe et al. (eds.), *Emerging Electronics and Automation*, Lecture Notes in Electrical Engineering 1202, https://doi.org/10.1007/978-981-97-3090-2_5

aluminates, and other compounds [2]. Alite, balite, aluminate, and ferrite phases are the four main forms of minerals found in the anhydrous condition (without water). When water is introduced, the cement particles are surrounded by water, and the surface of the mineral grains in clinker reacts with water quickly, generating a sequence of new compounds and releasing heat, this reaction is known as hydration.

Conventionally, the strength of in situ concrete was monitored using techniques such as the rebound hammer or Schmidt hammer (ASTM C805), penetration resistance test (ASTM C803), ultrasonic pulse velocity (ASTM C597), pullout test (ASTM C900), drilled core (ASTM C42), cast-in-place cylinders (ASTM C873), etc., for decades [3]. But they do come with certain drawbacks and limitations like destructive testing, sampling variability, time consumption, inaccuracy, cost, etc. Several methods for measuring cement strength during hydration have already been proposed. Strength measurement can be done using a variety of methods and devices, including fringing field impedance sensors [1, 4], PZT-based sensors [5], Concrete Vibration sensors (CVS), resin jacketed piezo-sensor (RJP) [2], capacitive sensor [6], cement/aluminum embedded PZT [7], etc. Structural Health Monitoring (SHM) systems use a variety of sensors, such as vibration and curing sensors, which have a major influence on safety and structural safety, to gather data on the performance and condition of structures [8]. On the contrary, capacitive sensors have several intrinsic benefits over their counterparts, including high sensitivity, outstanding stability, fast response, high resolution, strong frequency response, low loading effect due to high input impedance, cost-effectiveness, efficiency, ease of operation, and low power consumption [8]. A succession of papers on the capacitive principle used in sensors has sparked a lot of interest in non-destructive (NDT) on reinforced concrete in this decade. The electrical capacitive tomography (ECT) approach is perhaps the most well-known example [9]. Technologies such as barcodes, RFID, BIM, IoT ecosystems, and virtual reality modeling language are part of the construction industry's digital revolution [10]. In order to understand the microstructure and composition of fly ash mixed with Portland cement, its hydration is studied at different temperatures using SEM, XRD, and pore solution analysis [11]. Recently, civil engineering has been advancing towards the building of intelligent structures, with sensors installed on the surface of, or incorporated in, such structures to allow real-time monitoring of active vibration control [12], temperature [13], and lighting [14]. However, self-sensing concrete has emerged as a promising alternative for SHM [15].

In this paper, the development of stresses during the hydration process has been successfully examined with the help of a flexible parallel plate capacitive sensor. The purpose of this work is to fabricate and test a capacitive sensor for the strength measurement of cement during the initial phase of hydration. The capacitive sensor responds to the change in the gap between the parallel plates due to stress developed during concrete formation. Two types of stresses namely tensile stress and compressive stress are developed due to the reaction of cement powder with water. Since the concrete structure during hydration contains an excess amount of water, the response of the sensor must be free from water molecules. Therefore to avoid the cross-sensitivity due to water in cement structure, the sensing film must be flexible having high Young's modulus elasticity but at the same time, it should be hydrophobic.

Therefore Poly Di-methyl Siloxane (PDMS) film is selected for the fabrication of the sensor. Different capacitive sensors with different geometrical parameters are fabricated and initial experiments are conducted by applying known weight to each sensor. The best sensor was used to monitor the stress of hydration by measuring the change in capacitance value.

2 Methodology

2.1 Basic Principle of Capacitive Pressure Sensor

The structure of the parallel plate capacitive sensor is shown in Fig. 1. It consists of a flexible dielectric layer separated by metal electrodes. Flexible dielectric layer is made of PDMS film of thickness “d”. Top electrode is a flexible thin film and is fixed on the flexible insulating layer. The capacitance value for the dielectric medium of permittivity ε_r can be given by

$$C = \varepsilon_0 \varepsilon_r \frac{A}{d}, d << A \tag{1}$$

where A is the cross-sectional area of the electrode ($l \times w$) and d is the distance between the two electrodes.

Due to stress developed during hydration, the gap between the electrodes changes causing change in the capacitance value. Since the gap between the electrodes is very small, the deflection of the top electrode is uniform and the change in the area of cross section is negligible. For smaller displacement of the flexible electrode, however, the area is kept constant, and the capacitance change owing to gap variation is favored. With the application of stress (compressive stress), the capacitance value can be given by

$$C\prime = \varepsilon_0 \varepsilon_r \frac{A}{d - \Delta d} \tag{2}$$

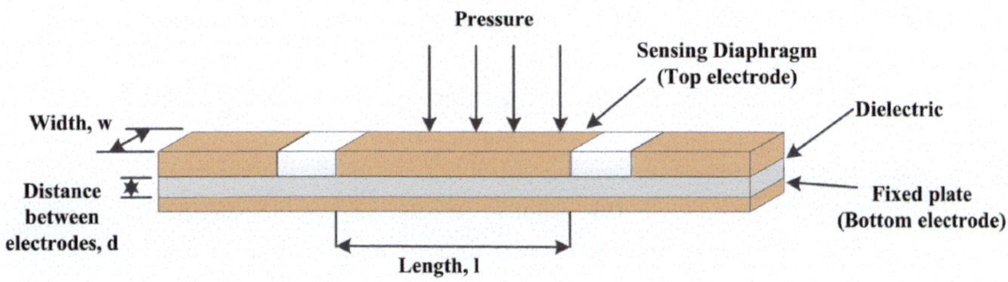

Fig. 1 A schematic view of a basic capacitive pressure transducer

where Δd is the variation of separation of the electrodes.

Expanding the expression (2) using the Taylor series of expansion, the capacitance value can be written as

$$C\prime = \varepsilon_0 \varepsilon_r \frac{A}{d\left(1 - \frac{\Delta d}{d}\right)} = \varepsilon_0 \varepsilon_r \frac{A}{d}\left[1 + \frac{\Delta d}{d} - \left(\frac{\Delta d}{d}\right)^2 + \left(\frac{\Delta d}{d}\right)^3 - \ldots\ldots\ldots\ldots\right] \quad (3)$$

Neglecting higher order terms of $\Delta d/d$,

$$C\prime = \varepsilon_0 \varepsilon_r \frac{A}{d}\left[1 + \frac{\Delta d}{d}\right] = C_0\left[1 + \frac{\Delta d}{d}\right] \quad (4)$$

So, the change in capacitance value due to small deflection can be given by

$$\Delta C = C\prime - C_0 = C_0 \frac{\Delta d}{d}$$

Finally,

$$\frac{\Delta C}{C_0} = \frac{\Delta d}{d} \quad (5)$$

The expression shows that the change in C is directly proportional to the change in "Δd" for tiny changes in the gap [16].

According to the Hooke's law, stress (S) that is force per unit area (pressure) is directly proportional to the strain. It can be given as

$$S = Y \frac{\Delta d}{d} \quad (6)$$

where Y is Young's modulus of elasticity of PDMS film and $\frac{\Delta d}{d}$ is the strain. Therefore, stress developed during hydration can be given by

$$S = Y \frac{\Delta C}{C_0} \quad (7)$$

Young's modulus of elasticity of the PDMS film is Y = 2.145 MPa, Eq. (7) can be written as

$$\frac{\Delta C}{C_0} = 0.4662 \times S \quad (8)$$

It indicates that for a small change in stress of the film, the normalized change in capacitance value linearly increases with increase in stress developed in the concrete. For concrete during hydration the stress varies from 0 to 0.45 MPa.

The structure was simulated using MATLAB R2019a software for the parallel plate capacitor with A $= 225$ mm^2, d $= 2.695$ mm, and $\varepsilon_r = 2.69$. The base capacitance value without stress is 1.98 pF. Figure 2a shows a plot between deflection in the dielectric thickness and change in applied pressure. Figure 2b shows the variation of capacitance value with increase in pressure.

The sensitivity which is the change in capacitance value due to a change in pressure is almost linear and it is expected to obtain a sensitivity of 0.93 pF/MPa.

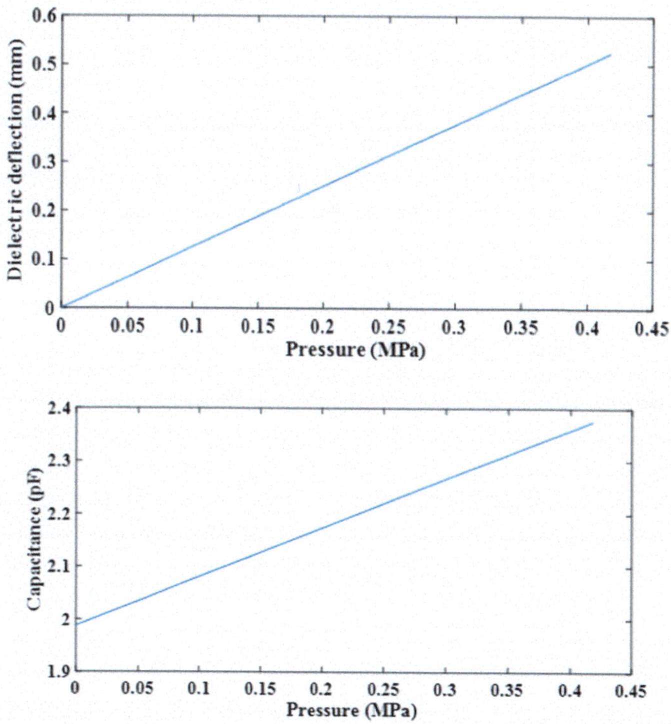

Fig. 2 Dielectric thickness deflection versus applied pressure plot, and capacitance versus applied pressure plot

Table 1 Dimensions of the fabricated sensors (r = radius, l = length, w = width, d = thickness/distance between electrodes)

Sensor	r (mm)	l (mm)	w (mm)	d (mm)	A (mm^2)
1	7	–	–	0.87	153.9
2	–	4	9	2.34	113.0
3	–	23	15	1.89	345
4	–	15	15	2.70	225

3 Fabrication

The sensor was fabricated with bottom electrode rectangular in shape but the top electrode is either circular or rectangular. Four different parallel plate capacitive sensors were fabricated with different electrode configurations. Geometrical parameters of the fabricated sensors are shown in Table 1. The sensors were fabricated on a metal clad glass substrate named as Fluorine-doped Tin Oxide (FTO). The metal film on the substrate was used as the bottom electrode of the sensors. Thin PDMS flexible film was deposited on the bottom electrode of each capacitor with thickness shown in Table 1. Top electrode with the area of cross section shown in Table 1 was made of either PEDOT: PSS or copper tape. The sensors are fabricated by sandwiching the dielectric material PDMS between two conducting substrates. Finally, connections were made on each electrode of the sensors using short-length cable. Figure 3 shows the photo of the four sensors. Top electrodes of sensors 1, 2, and 3 are made of PEDOT:PSS and sensor 4 was made of copper tape. Top electrodes of the first three sensors made of polymer material are circular but not of uniform shape. Table 2 presents the calculated and experimentally measured capacitance values in pF. There is some difference in the measured capacitance value from the theoretical value. Normally, the measured value is more than the theoretical value. This is due to lead capacitance effect. This is further to note that top electrode is nonuniform, there may be error to find the area of cross-section of the electrode.

4 Determination of Capacitance Response of the Sensors Due to Applied Pressure

Initial experiments were conducted by applying pressure of known values on each sensor one by one. The sensor was fixed on a fixed platform and a standard weight (gm) was applied on the top electrode. The sensors were tested by applying weights in the range of 0–100 gm. The equivalent pressure values in MPa obtained using known values of contact area on which weight was applied are given in Table 3. At each applied pressure, the capacitance value was measured by hand-held LCR meter interfaced to PC (Keysight, accuracy $<1\%$). Each experiment was repeated five times and the average capacitance value was observed corresponding to each applied

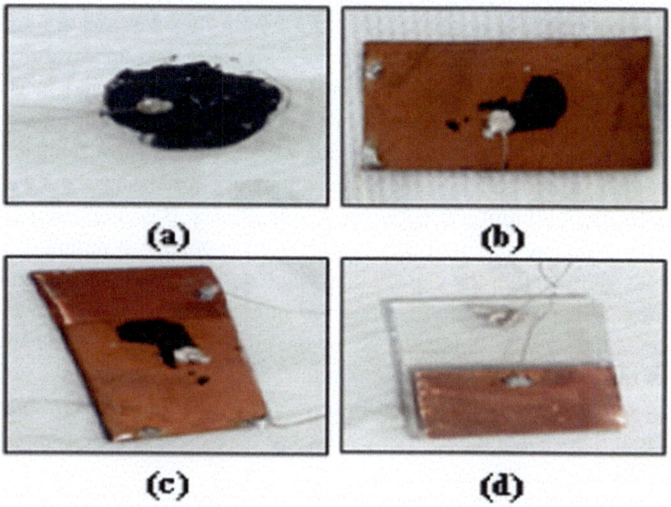

Fig. 3 Fabricated sensors, **a** Sensor 1, **b** Sensor 2, **c** Sensor 3, **d** Sensor 4

Table 2 The base capacitance value of the sensors

Sensor	Calculated value	Measured value	Error
1	4.36	5.89	1.52
2	1.20	1.72	0.52
3	3.82	3.25	0.57
4	2.07	2.97	0.90

pressure. The capacitance values were measured at three fixed excitation frequencies such as 1 kHz, 10 kHz, and 100 kHz.

Variation of the capacitance values of each sensor with the variation of applied pressure is shown in Fig. 4. It is observed that the capacitance value of all the sensors

Table 3 Pressure value ranges for all four sensors

Weights (*in gm*)	Sensor (MPa)	Sensor 2 (MPa)	Sensor 3 (MPa)	Sensor 4 (MPa)
0	0	0	0	0
5	0.033	0.044	0.015	0.022
10	0.065	0.088	0.029	0.044
15	0.097	0.133	0.043	0.067
20	0.130	0.177	0.058	0.089
57	0.370	0.504	0.165	0.253
94	0.611	0.832	0.272	0.418

increases with an increase in applied pressure. With the increase in applied pressure, the gap between the electrodes decreases, so the capacitance value increases. The capacitance variation of sensors 3 and 4 is linear as predicted in Eq. (7) but the capacitance variation of sensors 1 and 2 is nonlinear. This may be due to the nonuniformity of the cross-sectional area that leads to nonuniform electric field. The nonuniform electric field causes fringing capacitance. These issues can be eliminated by making electric field uniform between the electrodes. Base capacitance value of all the sensors is different. This is due to different geometrical parameters of each sensor. It is also shown that the capacitance value decreases with increase in excitation frequency. This phenomenon is observed in many polymer dielectric films. At high frequency, the sensitivity is reduced but the noise effect is also reduced. Among four sensors, the fourth sensor shows maximum sensitivity, the response is linear and the output is stable. Therefore, the fourth sensor was tested to measure hydration in concrete in real time. The characteristic response parameters of the four sensors are shown in Table 4.

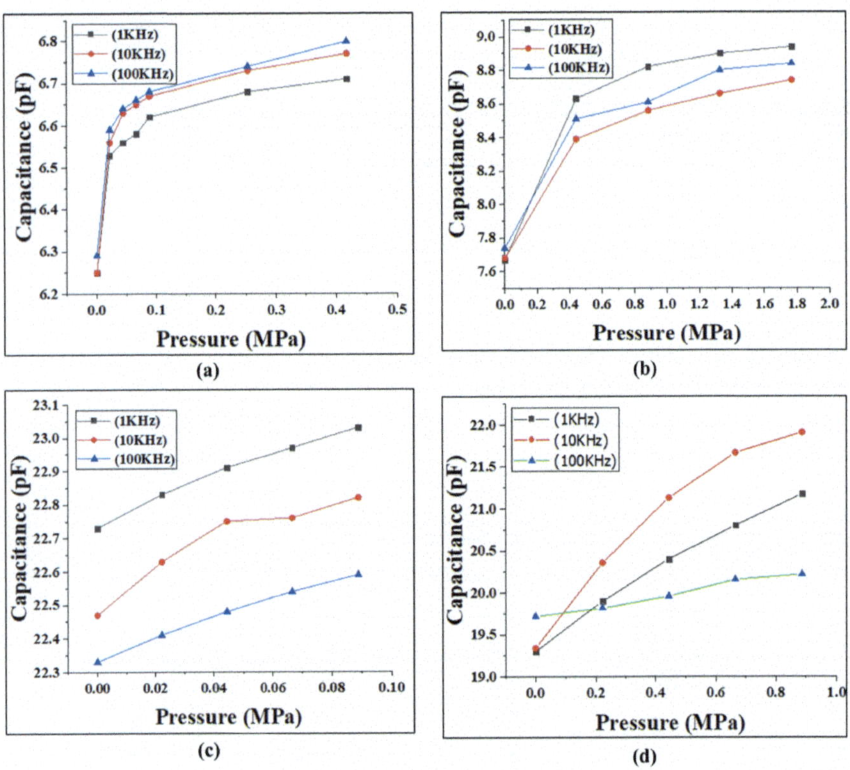

Fig. 4 Change in capacitance value with an increase in applied pressure

Table 4 Characteristic parameters of the sensor

Sensor	Sensitivity (pF/MPa)	Coefficient of correlation (R^2)	Nonlinearity (%)
1	1.17	0.947	2.94
2	5.39	0.783	17.92
3	5.77	0.893	10
4	28.8	0.946	9.14

5 Stress Measurement in Cement Concrete

Finally experiment was conducted with sensor 4 to determine stress during hydration in concrete. A cement concrete was made by mixing water in cement (Portland cement of 43 grade brand 147 Ultratech, India) and water (distilled water from Aquelix 5). Water is mixed with cement in a 0.3 ratio (78.9 gm water in 263.3 gm cement powder). The sensor was inserted into the cement paste to a height of 2.5 mm. The capacitance value of the sensor was measured by the Keysight GUI Data Logger for the handheld LCR meter. The measurements were taken continuously for 12 days. The LCR meter was connected to the PC. The photograph of the setup is shown in Fig. 5.

The experiment was performed with water to cement ratio "w/c" = 0.3, the hydration process starts just after the mixing and the reaction of cement with water takes place. The reaction involving water into the cement powder is called hydration. Due to hydration of cement power with water stress is developed in the concrete, the capacitance value tends to change. Figure 6 shows the variation of the capacitance due to hydration for a period of 12 days.

It is observed that the capacitance value initially decreases up to the fourth day, then it increases to a high value. The capacitance value then becomes stable for next few days and then slightly decreases. As the stress develops in the paste due to the hydration process, the chemical shrinkage forces in the paste produce tensile stress as the substrate resists deformation because the overall volume of the product formed is smaller than the volume of the reactants and the expansive forces in the coating results in the generation of compressive stress. A small tensile stress occurs during the early age of hydration followed by rapid development of compressive stress.

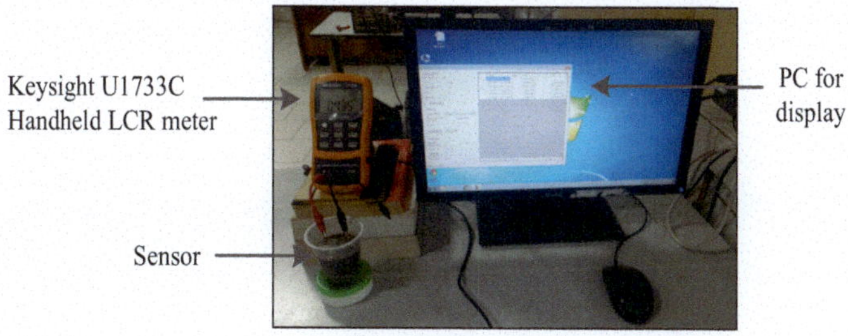

Fig. 5 Test setup for stress measurement using the sensor

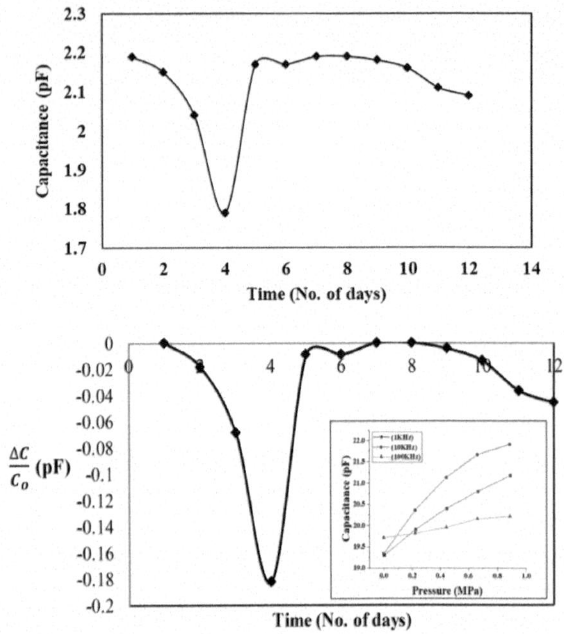

Fig. 6 Change in capacitance due to the development of stresses with time

Tensile stress reduces due to hydration of C_3A alone but compressive strength is continuously generated. Therefore, initially, the capacitance value tends to decrease due to the generation of tensile stress. With an increase in tensile stress, the flexible PDMS dielectric expands that leads to the increase in the gap between the electrodes of the sensor. The increase in gap leads to the decrease in capacitance value. This decrease in capacitance value is witnessed for 3 days. An increase in capacitance value is observed from the fourth day onward. This is because of the generation of compressive stress that leads to the decrease in gap "d" between the electrodes. The capacitance value then becomes almost constant or changes slightly. The pressure value with time is demonstrated in Fig. 7. The pressure value with time in days is predicted from the capacitance curve shown in Fig. 6. The curve shows that the stress generated during the hydration process, which at first is tensile for initial 3 days and then compressive up to day 7 and then further increases.

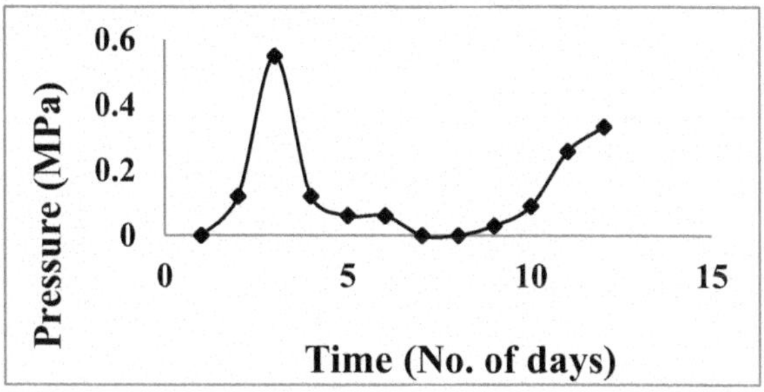

Fig. 7 Stress curve due to hydration of the cement sample

6 Conclusion

A capacitive sensor is proposed to measure stress developed in the hydration of concrete structures. Four capacitive sensors with flexible PMDS film were fabricated and initial experiments were conducted to measure stress in the range of 0–1 MPa. Sensor 4 with the best performance was finally used to measure the stress in a concrete made of 0.3 weight ratio of water and cement powder. It is observed that during the initial phase of hydration, tensile stress is developed, therefore the capacitance value decreases but later on due to the rapid development of compressive stress, the capacitance value increases. In concrete formation, the tensile stress reduces with aging but the compressive strength is continuously generated. The capacitive sensor truly monitors both the tensile and compressive stresses in the hydration process. Therefore, compared with the existing techniques, the proposed sensor is a potential low-cost alternative to measure the hydration stress. Further work is needed to measure the stress of hydration for concrete samples having different water-to-cement powder weight ratios.

References

1. Akram KJ, Ahmed A, Islam T (2020) Fringing field impedance sensor for hydration monitoring and setting time determination of concrete material. IEEE Trans Instrum Meas 69(5). https://doi.org/10.1109/TIM.2019.2918595
2. Pan HH, Huang MW (2020) Piezoelectric cement sensor-based electromechanical impedance technique for the strength monitoring of cement mortar. Constr Build Mater 254. https://doi.org/10.1016/j.conbuildmat.2020.119307
3. Santhakumar R (2015) Concrete technology. Oxford Higher Education
4. Akram KJ, Islam T, Ahmed A (2018) A simple method on transformer principle for early age hydration monitoring and setting time determination of concrete materials. IEEE Sens J 18(17). https://doi.org/10.1109/JSEN.2018.2853564
5. Ghafari E, Yuan Y, Wu C, Nantung T, Lu N (2018) Evaluation the compressive strength of the cement paste blended with supplementary cementitious materials using a piezoelectric-based sensor. Constr Build Mater 171. https://doi.org/10.1016/j.conbuildmat.2018.03.165

6. Salmaz U, Islam T, Sohail S (2020) A novel linear capacitive temperature sensor using polydimethylsiloxane. IEEE Trans Instrum Meas 69(10). https://doi.org/10.1109/TIM.2020.2986120
7. Yin X, Hutchins DA, Diamond GG, Purnell P (2010) Non-destructive evaluation of concrete using a capacitive imaging technique: Preliminary modelling and experiments. Cem Concr Res 40(12). https://doi.org/10.1016/j.cemconres.2010.08.015
8. Mishra M, Lourenço PB, Ramana GV (2022) Structural health monitoring of civil engineering structures by using the internet of things: a review. J Build Engin
9. Cheng Y, Gao F, Hanif A, Lu Z, Li Z (2017) Development of a capacitive sensor for concrete structure health monitoring. Constr Build Mater 149. https://doi.org/10.1016/j.conbuildmat.2017.05.150
10. Woodhead R, Stephenson P, Morrey D (2018) Digital construction: from point solutions to IoT ecosystem. Autom Constr 93. https://doi.org/10.1016/j.autcon.2018.05.004
11. Deschner F, Lothenbach B, Winnefeld F, Neubauer J (2013) Effect of temperature on the hydration of Portland cement blended with siliceous fly ash. Cem Concr Res 52. https://doi.org/10.1016/j.cemconres.2013.07.006
12. Noel AB, Abdaoui A, Elfouly T, Ahmed MH, Badawy A, Shehata MS (2017) Structural health monitoring using wireless sensor networks: a comprehensive survey. IEEE Commun Surv Tutor 19(3). https://doi.org/10.1109/COMST.2017.2691551
13. Grudzien K, Chaniecki Z, Romanowski A, Sankowski D, Nowakowski J, Niedostatkiewicz M (2016) Application of twin-plane ECT sensor for identification of the internal imperfections inside concrete beams. Conference record - IEEE instrumentation and measurement technology conference. https://doi.org/10.1109/I2MTC.2016.7520512
14. Myers A, Mahmud MA, Abdelgawad A, Yelamarthi K (2016) Towards integrating structural health monitoring internet of things (IoT). In:IEEE International; conference on eletro information technology (EIT), pp 438–441
15. Varghese SM, Sajeeb R, Philipose MC (2018) Concrete smart sensors for structural health monitoring – a review. In: International conference on circuits and systems in digital enterprise technology (ICCSDET)
16. Islam T (2017) Advanced interfacing techniques for the capacitive sensors. In: Smart sensors, measurement and instrumentation, vol 25. https://doi.org/10.1007/978-3-319-55369-6_2.

Application of OLCAO-GGA and OLCAO-MGGA Techniques for Investigating Properties of TiO_2

Neerja Dharmale, Saurabh Chaudhury, and Rupesh Mahamune

Abstract Here, the investigation of various properties of the rarest studied phase (Brookite) of titanium dioxide (TiO_2) is done. We have found these properties using two distinct approaches, namely the self-consistent Orthogonalized Linear Combination of Atomic Orbitals (OLCAO) with Generalized Gradient Approximation (GGA) and Meta-GGA (MGGA) under the framework of Density Functional Theory (DFT). Considering the optimal c value, the bandgap value is calculated using the MGGA approach. MGGA calculations are much closer to the experimental value than the GGA approach. However, the optical properties obtained using MGGA are lower than the GGA approach, especially the dielectric constant and refractive index. Absorption for brookite TiO_2 was observed in the U.V. region using both methods. E.M. spectrum shifted to a lower value by MGGA. A comparison of the two methods reveals that MGGA provides a more comprehensive description of the optical characteristics and electronic structure than GGA. Investigation done here is helpful to find applicability of brookite TiO_2 in Solar Cells.

Keywords Brookite TiO_2 · Density Functional Theory (DFT) · Electronic · Optical properties

1 Introduction

Nature makes anatase, rutile, and brookite [1]. Remaining forms of TiO_2 like columbite [2], baddeleyite[3], cotunnite [4], pyrite [5], fluorite [6], Tridymite [7], hollandite [8], and bronze [9, 10] are obtained using the four natural forms of TiO_2. It is advantageous to use TiO_2 in photocatalysis, solar cells, biogas sensors, white pigment, energy storage, etc. [11–20] Brookite TiO_2 was suitable for applications

N. Dharmale (✉) · R. Mahamune
Shri Sant Gajanan Maharaj College of Engineering Shegaon, Maharashtra 444203, India
e-mail: neerja.rupesh@gmail.com
URL: https://www.nits.ac.in/

S. Chaudhury
National Institute of Technology Silchar, Assam 788010, India

D.-I. Stroe et al. (eds.), *Emerging Electronics and Automation*, Lecture Notes in Electrical Engineering 1202, https://doi.org/10.1007/978-981-97-3090-2_6

such as DSSC, photovoltaic devices, etc. The performance of brookite as a photocatalyst is much better than any other polymorph of TiO_2, especially in CO_2 photo reduction [21, 22], photo reforming of oxygenates [23, 24], and ammonia oxidation [25]. As brookite TiO_2 absorption spectra lie in the U.V. region, it is suitable for light absorption in weak or diffused conditions. Light can travel more distances (in terms of wavelength (λ)) if brookite TiO_2 is doped with suitable material. Dima et al. [26] observed that doping brookite TiO_2 surfaces with ruthenium decreases its bandgap energy to a minimal value, causing a shift from U.V. to the visible region in the E.M. spectrum, thereby improving its performance in DSSCs and photocatalytic activities. Doping enhances photocatalytic activity. Hydrogen (H) and Nitrogen (N) doping reduce the bandgap of anatase and brookite TiO_2 [27]. Due to the large value of dielectric function and refractive index, TiO_2 is regarded as the more appropriate for harvesting of energy [28]. Numerous studies have examined the bandgap value from a theoretical [8, 9, 16, 29–32] and experimental perspective [33, 34]. The measured value falls between 3.2 and 3.4 eV, while the theoretically analyzed values thus far found are in the range of 1.86–3.78 eV. As a result, there is disagreement on the precise bandgap value. The properties of this material can be theoretically investigated using various methods, including LDA, GGA, and MGGA, within the DFT framework and various exchange-correlation functions. Different theoretical techniques are available to examine a material's electronic and optical characteristics, including G.W. methods, hybrid functionals, and the Bethe–Saltpeter Equation (BSE) [35–37]. For the crystalline phases of TiO_2, Zhu et al. [7] and Landmann et al. [28] discovered that HSE06, PBE G_0W_0 provides a reliable bandgap value. However, there are many disagreements over whether to use LDA or GGA. Yao et al. [38] discovered that LDA is superior to GGA for elastic properties calculations. Nonetheless, the GGA exchange-correlation potential more closely matches the measured value for the d orbital elements. Considering a hybrid functional like B3LYP (Becke 3-parameter Lee-Yang-Parr), the underestimation of Eg value induced by DFT is overthrown [30]. Plutonium Dicarbide (PbC_2) is a strongly correlated metal, and Yang et al. [39] applied the LSDA+U (Local Spin Density Approximation) method for investigating various properties. They discovered that the computed results align with the experimental findings. According to Samat et al. [31], investigation of the impact of Hubbard U parameter for 3d metals, DFT calculations align more with experimental findings than DFT+U calculations. When Shatendra Sharma et al. [40] used LDA, GGA, and MGGA to compute the electronic and optical characteristics of strontium sulfide (SrS), they found that MGGA produced findings that were more accurate to the experimental data than did LDA and GGA.

From the literature, it can be seen that there have been a few theoretical investigations into the electronic and optical characteristics of brookite TiO_2. Still, widespread agreement has not yet been on the choice of exchange-correlation functional or energy bandgap value. In-depth studies of brookite TiO_2 are rare because of the dearth of experimental data and the uncertainty in the theoretical results. Therefore, a thorough examination of the electronic and optical characteristics of brookite TiO_2 is motivated to discover its potential applications. In addition to relying on the density and kinetic energy Laplacian, the MGGA extends the GGA approximation. As a result, MGGA made significant progress toward improving electronic properties. In this paper, two different approaches have been used for the first principles calculations on electronic and optical properties of brookite TiO_2 and the similitude of their performances with each other as well as with previously available experimental [33, 34, 41] and theoretical data [7, 16, 29–32, 40].

In this paper, bandgap value using the MGGA approach is calculated by applying two methods: (i) considering the optimal c parameter and (ii) considering the c-parameter calculated by self-consistently (default c-parameter).

This is how the remaining sections are organized. The adopted methodology for studying the optical and electronic properties of brookite TiO_2 is described in Sect. 2. Results and a discussion of the acquired attributes are presented in Sect. 3. Section 4 presents the results that were drawn from this research.

2 Theoretical Approach

Under the framework of DFT, ground state calculations of brookite TiO_2 are performed with OLCAO [42]. The Limited Memory Broyden–Fletcher–Goldfarb–Shanno (LBFGS) algorithm is first used to optimize the structure [43–45]. Brookite TiO_2 represents the packing of the building block (TiO_6 octahedral), and its unit cell contains eight blocks [46]. This optimized structure is then used to do all theoretical calculations. Then, for the exploration of optical and electronic characteristics, GGA and MGGA with PBE [47] and Tran and Blaha (TBO9) [48, 49] are employed as exchange-correlation functionals. Atomistix Tool Kit (ATK) from Quantumwise is used to perform the calculations [50–52]. The sampling is carried out using the Monkhorst–Pack Scheme [53] and is set to 2 * 4 * 4, while the density mesh cut-off is taken to be 200 Hartree. State-of-the-art norm-conserving [54] pseudopotentials are utilized for titanium and oxygen. The damping factor is set to 0.5 using the FFT Poisson solution and without using SCF iterations. Two methods are taken into consideration while calculating electronic properties using the MGGA methodology.

Bandgap values Self may be consistently calculated in Quantumwise ATK by either using the optimal c-value (c_{opt}) or the default value of c, i.e., 1 (system-dependent parameter). Bandgaps for a variety of materials improved with a novel exchange potential developed by Tran and Blaha in 2009 utilizing a c-parameter. For all the solids under study, bandgap grows monotonically with respect to c when c is varied in Eq. (1) of [48]. A more significant value for c results in greater agreement with the experiment since the Eg calculated for c equal to 1 for all circumstances is so small in relation to experimental data. More precisely, c_{opt} (The c value which provides results consistent with experiment) is more prominent than 1.4 for solids with small bandgaps, while it lies within the range of 1.1–1.3 for solids with wide bandgaps. The bandgaps with various c-values are calculated in order to determine c_{opt}. A c-parameter versus bandgap value (Eg) plot is presented in Fig. 1 to obtain c_{opt}. c is investigated by Eq. (3) of [48].

Fig. 1 Graph to obtain c_{opt}

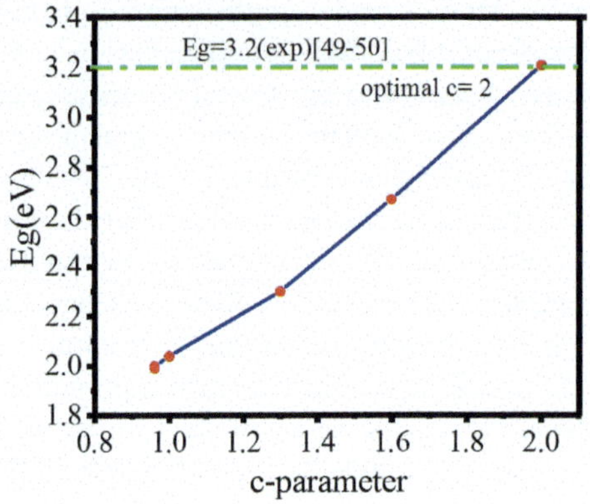

The c-value is self-consistently determined in the first technique. In the latter method, the bandgap value for various c-values is calculated in order to determine the best c-value. To determine the ideal c-value, the c-parameter versus bandgap value (Eg) plot is displayed in Fig. 1. The ideal c-parameter is found at the intersection of the experimental bandgap and the interpolated line of the estimated bandgaps. As seen in Fig. 1, at c = 2, Eg = 3.21, which is 0.31% greater than the practical value. In non-metals, bandgaps can be properly described by the semi-empirical functional TB09. It also expedites processing because it is quicker than LDA. TB09 meta-GGA functional is a useful tool to better show the electrical structure of semiconductors and insulators [55].

3 Results and Discussion

3.1 Energy Band Diargram and Density of State (DOS)

The energy bandgap is defined as the gap between the valence band's maximum and the conduction band's minimum. Here, two characteristics are thoroughly examined and analyzed: the band structure and the density of states (DOS).

The band structure is represented by the GGA technique in Fig. 2a, and the total density of states (TDOS) and partial density of states (PDOS) are represented by the GGA approach in Fig. 2 b. Once more, band structure is shown in Fig. 3a, and the DOS is shown in Fig. 3b using the MGGA method. Thus, for both calculations, the valence band's top and the conduction band's bottom lie along line G, implying a direct bandgap nature. The bandgap values are 2 eV and 2.71 eV, respectively. In both GGA and MGGA, the bands nearer to fermi levels are because of atomic mixing of O 2p and Ti 3d orbitals, with a small contribution from O 2 s and Ti 2p orbitals. The mixing of atomic orbitals Ti 3d and O 2p causes the greatest points to emerge at the top of the valence band and bottom of the conduction band, respectively, at −5.75,

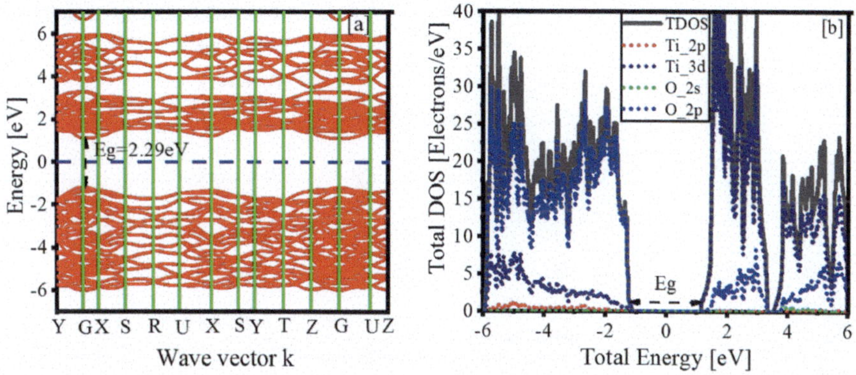

Fig. 2 **a** Energy band diagram of brookite TiO_2 using GGA. **b** TDOS and PDOS of brookite TiO_2 using GGA

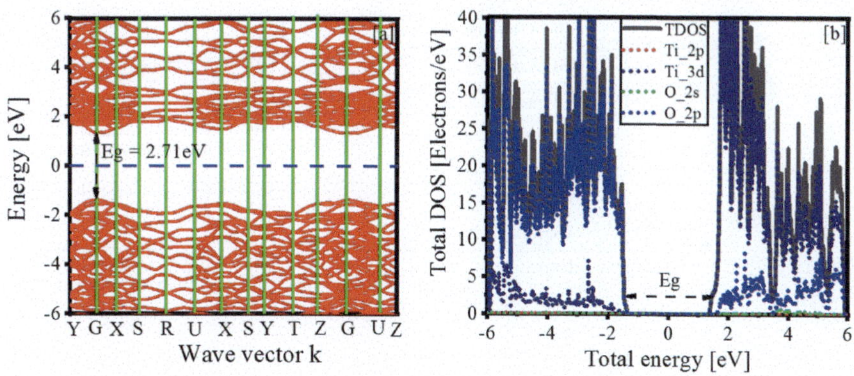

Fig. 3 **a** Energy band diagram of brookite TiO_2 using MGGA. **b** TDOS and PDOS of brookite TiO_2 using MGGA(c-parameter is calculated self-consistently)

−5.51 and 1.54, 2.41 eV, according to Fig. 2b. The energy levels that are engaged range from 1.128 to 6 eV and from −6 to −1.162 eV, respectively.

As seen in Fig. 3b, the mixing of the atomic orbitals of Ti 3d and O 2p causes the most significant points to arise at 5.88, 5.36 eV at the top of the valence band, and 1.79, 2.53 eV at the bottom of the conduction band, respectively. In the energy range of −6–6 eV, engaged energy states span from −6 to −1.38 eV and 1.35–6 eV. Ti 3d contributions provide significant conduction band peaks for both GGA and MGGA. The O 2s state contributes to the lowest valence band, which starts after 20 eV. Atomic mixing of O 1s and Ti 3d states produces the most occupied valence band in brookite TiO_2. Although Ti 2p and O 1s states contribute to developing the highest valence band, Ti 2p's contribution is significantly less. Significant peaks in the valence band are caused by the atomic mixing of the O 2p and Ti 3d states. Ti 2p state contributed very little to the creation of valence bonds. It is seen that

Table 1 Calculated bandgap and a comparison to results from earlier publications

Bandgap (eV)	Method	Reference
2.29(D)	OLCAO-PBE	This work
2.71(D)	OLCAO-MGGA(default c-parameter)	This work
3.21(D)	OLCAO-MGGA(optimal c-parameter)	This work
2.379(D)	GGA	This work
2.38(D)	GGA-PBE	[30]
1.86(D)	DFT-PBE	[29]
3.30(D)	PBE-G0 W0	[8]
3.68(D)	HSE06-G0 W0	[8]
3.45(D)	DFT-PBE	[8]
1.86(D)	DFT-PBE	[8]
1.86(D)	GGA-PBE	[8]
2.485(D)	GGA-RPBE	[9]
2.367(D)	GGA-PW91	[16]
2.323(D)	GGA-WC	[16]
2.34(D)	FP-PAW-EV-GGA-PBE	[16]
2.51(D)	FP-PAW-EV-GGA	[32]
3.14(D)	FP-PAW-mBJ-GGA	[32]
3.4(D)	Exp	[32]
3.2(D)	Exp	[33]
2.29(D)	OLCAO-PBE	[34, 41]

brookite TiO_2 has a direct bandgap, and the bandgap values using the GGA and MGGA approach with default c-parameter are 26.12 and 12.58 % less compared to the experimental value. But, Eg, using the optimal c-value is 3.21eV, which shows a minute variation with practical results [34, 41]. Moreover, the bandgap value is improved by 26 % using the MGGA approach with an optimal c-parameter than GGA. Table 1 represents a calculated bandgap and its comparison with previously reported experimental and theoretical results.

3.2 *Optical Properties*

All-optical properties are estimated up to 30 eV phonon energy using the GGA and MGGA approaches. Its dielectric function can be used to determine its optical characteristics. The complex number that describes the dielectric process is represented as follows:

$$\varepsilon(\omega) = \varepsilon_1(\omega) + i\varepsilon_2(\omega) \tag{1}$$

where $\varepsilon_1(\omega)$ is a fundamental component that represents the material's electronic polarizability, and $\varepsilon_2(\omega)$ is an imagined part that provides information about the material's absorption. When absorption starts, it stops being zero. In energy space, the material response to the incident light is given by $\varepsilon_2(\omega)$, whereas it is given by$\varepsilon_1(\omega)$ for incident light in actual space. Using GGA and MGGA, Fig. 4a and b depict $\varepsilon_1(\omega)$ and $\varepsilon_2(\omega)$, respectively. The peak in the absorption spectra is caused by a direct interband transition and is located at 2.31 eV for GGA and 2.68 eV for MGGA. The Kubo–Greenwood formalism [56, 57] has been used to evaluate dielectric function. It is provided as

$$X_{ij} = \frac{e^2h^4}{m^2\varepsilon_0 V\omega^2}\Sigma_{nm}\frac{f(E_m) - f(E_n)}{E_(nm) - h\omega - ih\Gamma}\Pi^i_{mn}\Pi^i_{mn} \tag{2}$$

The natural and imaginary parts are represented by the (3) and (4)

$$\varepsilon_1(\omega) = 1 + \frac{2}{\pi}P\int_0^\infty \frac{\varepsilon_2(\omega)\omega' d\omega'}{\omega'^2 - \omega^2} \tag{3}$$

where P is the dominant value of integral and

$$\epsilon_2(\omega) = \frac{2e^2\pi}{\omega\epsilon_0}\Sigma_{k,v,c}| < \psi_k^c|u, r|\psi_k^v > |^2\delta(E_k^c - E_k^v - E) \tag{4}$$

where e, ψ_k^c, ψ_k^v, and μ represent electronic charge, incident electric field, valence band, and conduction band wave function at k point, respectively. One can obtain $\varepsilon_1(\omega)$ using the Kramers–Kronig relation and the imaginary portion. The dielectric function for brookite TiO_2 utilizing the GGA and MGGA approaches is 4.98 and 4.38, respectively, according to Fig. 4a. For GGA and MGGA, the highest peak of $\varepsilon_1(\omega)$ occurs at 3.04 eV and 3.24 eV, respectively, while for GGA and MGGA, the highest peak of $\varepsilon_2(\omega)$ occurs at 3.55 and 4.57 eV, respectively. These peaks result from contributions from the titanium 3d and oxygen 2p states. To infer the direct bandgap nature of brookite TiO_2, the imaginary part of the dielectric constant begins at 2.29 for GGA and 2.71 for MGGA. The connection between dielectric function and refractive index can be given by

$$\eta(\omega) = \sqrt{\frac{|\varepsilon(\omega)| + \varepsilon_1(\omega)}{2}} \tag{5}$$

Calculated $\varepsilon(0)$ and $\eta(0)$ are shown in Table 2 and compared to previously published theoretical data. The refractive index, as provided by GGA and MGGA, is shown in Fig. 4c. The refractive index indicates the optical density of the medium. It is directly proportional to medium density and inversely proportional to the speed of light. Using GGA and MGGA, brookite has a refractive index of 2.22 and 2.09, respectively. We can infer from the observed data that the refractive index depends on

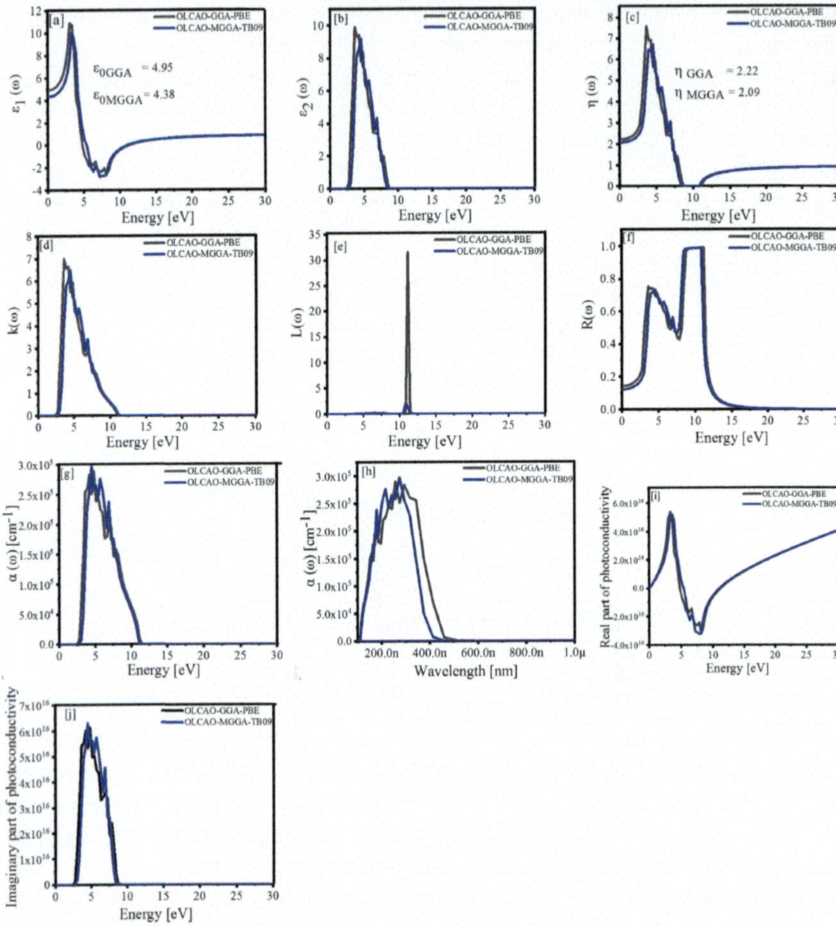

Fig. 4 **a** Real part of dielectric function **b** imaginary part of dielectric function **c** refractive index **d** Extinction coefficient **e** loss function **f** reflectivity **g** absorption versus energy **h** Absorption versus wavelength **i** Re (photoconductivity) **j** Im (photoconductivity)

the dielectric constant. Extinction coefficient $k(\omega)$, which represents the imaginary portion of the complex variable called refractive index, is written as

$$k(\omega) = \sqrt{\frac{|\varepsilon(\omega)| - \varepsilon_1(\omega)}{2}} \tag{6}$$

It provides information on light absorption. The k (ω) is depicted in Fig. 4d. Reflectivity $R(\omega)$ and the refractive index have the following relationship:

Table 2 Comparison of the calculated $\varepsilon_1(\omega)$ and $\eta(\omega)$ of brookite TiO_2 with the information already available

Method	$\varepsilon(0)$	$\eta(0)$	References
OLCAO-PBE	4.95	2.22	Present study
OLCAO-MGGA (Default c-parameter)	4.38	2.09	Present study
GGA	4.88	2.29	[40]
GGA+U	3.96	2.22	[40]
DFT-GGA	4.66	–	[7]
PBE-GGA	6.26	2.50	[32]
mBJ	4.90	2.21	[32]

$$R(\omega) = \frac{(\eta - 1)^2 + k^2}{(\eta + 1)^2 + k^2} \tag{7}$$

When an E.M. wave travels through a material, the absorption loss is represented by the symbol R(ω). At 11.08 and 10.83 eV, respectively, a dramatic reduction in the reflection spectra employing GGA and MGGA can be seen in Fig. 4f. Due to plasma resonance, R(ω) has decreased. The reflectance of brookite TiO_2 falls within the visible range of the U.V. spectrum, which is evident from R(ω), making it an appropriate coating material. The loss function utilizing the GGA and MGGA approaches is shown in Fig. 4e. Dielectric function loss function can be expressed as

$$L(\omega) = Im\left(\frac{-1}{\varepsilon(\omega}\right) = \frac{\varepsilon_2(\omega)}{\varepsilon_1^2(\omega) + \varepsilon_2^2(\omega)} \tag{8}$$

The electron's energy is lost when they pass through a homogeneous dielectric medium, and the loss function represents this. Here, L(ω) utilizing GGA and MGGA had energy peaks of 11.08 and 10.79 eV, respectively. A sharp decline in reflection spectra and Energy peaks of the loss function match with each other. Figure 4g and h represent absorption spectra versus phonon energy and absorption spectra versus wavelength, respectively. Absorption is expressed as

$$\alpha(\omega) = \frac{4 * \pi * k}{\lambda} \tag{9}$$

where c is the speed of light and k is the extinction coefficient. It refers to the amount of light of a specific wavelength that enters the substance before absorption. According to Fig. 4h, brookite TiO_2 exhibits optical absorption in the U.V. area at the absorption edge between 300 and 400 nm. For both GGA and MGGA, the energy maxima for the absorption spectra, loss function, and reflectivity occur in the same energy range. For GGA, these peaks vary from 2.31 to 11.36 eV, and MGGA ranges from 2.68 to 10.99 eV. Electrical conductivity is directly inversely correlated with phonon absorption. In light of this, brookite TiO_2 can offer electrical solid conductivity. The Interband Transition Region (IAR) is the most prominent peak in absorption spectra that results from the transition from the valence to the conduction band. For solar cell applications, brookite TiO_2 is favorable due to its high absorption value. Brookite TiO_2 exhibits transparency within the observed range of absorption. Photoconductivity produces electron–hole pairs in semiconductors due to the irradiation of photons with energy above the bandgap energy. Because photoconductivity involves transforming light energy into electrical power, it is crucial for solar cell applications. The following equation relates dielectric function and photoconductivity:

$$\sigma_r = \omega * \varepsilon_1(\omega) \tag{10}$$

$$\sigma_i = \omega * \varepsilon_2(\omega) \tag{11}$$

From Eqs. (10) and (11), photoconductivity is directly proportional to the dielectric function, and it is proved by Fig. 4a, b and blue Fig. 4i, j. Figure 4i and 4j represent real and imaginary parts of photoconductivity using the GGA and MGGA approach, respectively.

4 Conclusion

Here, utilizing GGA and MGGA, a computational investigation of the electrical and optical characteristics of brookite TiO_2 is reported. When utilizing MGGA with the proper c-parameter, the results are superior to those obtained through theoretical approaches [8, 9, 16, 32, 34, 47, 48] and are nearly in line with previous experimental findings [49, 50]. It is discovered that the bandgap value provided by the MGGA-based simulation is an approximation of the experimental results. GGA yielded higher refractive index and dielectric constant values than MGGA because the dielectric constant is inversely proportional to the bandgap value. It can be deduced from the analysis of the optical properties that brookite TiO_2 can offer electrical solid conductivity, making it an appropriate coating material for DSSC. Based on the analysis of electronic and optical properties using two distinct methodologies, it can be stated that OLCAO-MGGA-TB09 produces better results for electronic and optical properties than OLCAO-GGA-PBE. The experimental data on the optical properties of pure brookite TiO_2 are not accessible, hence the measured optical properties are compared with the previously reported theoretical data [7, 46, 47].

Further theoretical study is required to get accurate bandgap value and influence of exchange-correlation potential on crystals containing d electrons. In addition, it could be possible to reduce the deviation between the experimental and theoretical bandgap values using the Hubbard U parameter, HSE06, B3LYP.

As a result, the work described in this study provides a foundation for estimating the electronic and optical characteristics of brookite TiO_2, which will be helpful in both theoretical and experimental studies in the future.

References

1. Li QJ, Liu BB (2016) High pressure structural phase transitions of Tio_2 nanomaterials. Chin Phys B 25(7):076107
2. Arlt T, Bermejo M, Blanco M, Gerward L, Jiang J, Olsen JS, Recio J (2000) High-pressure polymorphs of anatase Tio_2. Phys Rev B 61(21):14414
3. Pischedda V, Hearne G, Dawe A, Lowther J (2006) Ultrastability and enhanced stiffness of 6 nm Tio_2 nanoanatase and eventual pressure-induced disorder on the nanometer scale. Phys Rev Lett 96(3):035509
4. Park SW, Jang JT, Cheon J, Lee HH, Lee DR, Lee Y (2008) Shape-dependent compressibility of Tio_2 anatase nanoparticles. J Phys Chem C 112(26):9627–9631
5. Swamy V, Kuznetsov A, Dubrovinsky LS, Caruso RA, Shchukin DG, Muddle BC (2005) Finite-size and pressure effects on the Raman spectrum of nanocrystalline anatase Tio_2. Phys Rev B 71(18):184302
6. Flank AM, Lagarde P, Itié JP, Polian A, Hearne G (2008) Pressure-induced amorphization and a possible polyamorphism transition in nanosized Tio_2: an X-ray absorption spectroscopy study. Phys Rev B 77(22):224112
7. Zhu T, Gao SP (2014) The stability, electronic structure, and optical property of Tio_2 polymorphs. J Phys Chem C 118(21):11385–11396
8. Li QJ, Liu BB (2016) High pressure structural phase transitions of Tio_2 nanomaterials. Chin Phys B 25(7):076107
9. Li Q, Cheng B, Yang X, Liu R, Liu B, Liu J, Chen Z, Zou B, Cui T, Liu B (2013) Morphology-tuned phase transitions of anatase Tio_2 nanowires under high pressure. J Phys Chem C 117(16):8516–8521
10. Dong Z, Song Y (2015) Size-and morphology-dependent structural transformations in anatase Tio_2 nanowires under high pressures. Can J Chem 93(2):165–172
11. Yang HG, Sun CH, Qiao SZ, Zou J, Liu G, Smith SC, Cheng HM, Lu GQ (2008) Anatase Tio_2 single crystals with a large percentage of reactive facets. Nature 453(7195):638–641
12. Etgar L, Zhang W, Gabriel S, Hickey SG, Nazeeruddin MK, Eychmüller A, Liu B, Grätzel M (2012) High efficiency quantum dot heterojunction solar cell using anatase (001) Tio_2 nanosheets. Adv Mater 24(16):2202–2206
13. Asahi R, Morikawa T, Ohwaki T, Aoki K, Taga Y (2001) Visible-light photocatalysis in nitrogen-doped titanium oxides. Science 293(5528):269–271
14. Chen X, Mao SS (2007) Titanium dioxide nanomaterials: synthesis, properties, modifications, and applications. Chem Rev 107(7):2891–2959
15. Chen X, Liu L, Huang F (2015) Black titanium dioxide (Tio_2) nanomaterials. Chem Soc Rev 44(7):1861–1885
16. Samat M, Adnan N, Taib M, Hassan O, Yahya M, Ali A. Structural, electronic and optical properties study of brookite Tio_2
17. Lagarec K, Desgreniers S (1995) Raman study of single crystal anatase Tio_2 up to 70 GPa. Solid State Commun 94(7):519–524
18. Gerward L, Staun Olsen J (1997) Post-rutile high-pressure phases in Tio_2. J App Crystallogr 30(3):259–264
19. Dubrovinsky LS, Dubrovinskaia NA, Swamy V, Muscat J, Harrison NM, Ahuja R, Holm B, Johansson B (2001) The hardest known oxide. Nature 410(6829):653–654
20. Swamy V, Muddle BC (2007) Ultrastiff cubic Tio_2 identified via first-principles calculations. Phys Rev Lett 98(3):035502
21. Ocaña IR, Beltram A, Jaén JJD, Adami G, Montini T, Fornasiero P (2015) Photocatalytic h2 production by ethanol photodehydrogenation: effect of anatase/brookite nanocomposites composition. Inorganica Chimica Acta 431:197–205
22. Kandiel TA, Feldhoff A, Robben L, Dillert R, Bahnemann DW (2010) Tailored titanium dioxide nanomaterials: anatase nanoparticles and brookite nanorods as highly active photocatalysts. Chem Mater 22(6):2050–2060

23. Altomare M, Dozzi MV, Chiarello GL, Di Paola A, Palmisano L, Selli E (2015) High activity of brookite Tio_2 nanoparticles in the photocatalytic abatement of ammonia in water. Catal Today 252:184–189
24. Magne C, Cassaignon S, Lancel G, Pauporté T (2011) Brookite Tio_2 nanoparticle films for dye-sensitized solar cells. ChemPhysChem 12(13):2461–2467
25. Xu J, Wu S, Jin J, Peng T (2016) Preparation of brookite Tio_2 nanoparticles with small sizes and the improved photovoltaic performance of brookite-based dye-sensitized solar cells. Nanoscale 8(44):18771–18781
26. Dima R, Maluta N, Maphanga R, Sankaran V (2017) Computational study of Tio_2 brookite (100), (010) and (210) surface doped with ruthenium for application in dye sensitised solar cells. In: Journal of physics: conference series, vol 905. IOP Publishing, p 012012
27. Xu H, Zhang L (2009) Controllable one-pot synthesis and enhanced photocatalytic activity of mixed-phase Tio_2 nanocrystals with tunable brookite/rutile ratios. J Phys Chem C 113(5):1785–1790
28. Landmann M, Rauls E, Schmidt W (2012) The electronic structure and optical response of rutile, anatase and brookite Tio_2. J Phys: Condensed Matter 24(19):195503
29. Zhang J, Zhou P, Liu J, Yu J (2014) New understanding of the difference of photocatalytic activity among anatase, rutile and brookite Tio_2. Phys Chem Chem Phys 16(38):20382–20386
30. Liu QJ, Zhang NC, Liu FS, Liu ZT (2014) Structural, elastic, electronic and optical properties of various mineral phases of Tio_2 from first-principles calculations. Physica Scripta 89(7):075703
31. Samat M, Ali A, Taib M, Hassan O, Yahya M (2016) Hubbard u calculations on optical properties of 3D transition metal oxide Tio_2. Res Phys 6:891–896
32. Mohamad M, Haq BU, Ahmed R, Shaari A, Ali N, Hussain R (2015) A density functional study of structural, electronic and optical properties of titanium dioxide: characterization of rutile, anatase and brookite polymorphs. Mater Sci Semicond Process 31:405–414
33. Yamauchi S, Yamamoto K, Hatakeyama S et al (2015) Enhanced photo-induced property of LPCVD-Tio_2 layer on PCVD-Tio_x initial layer. J Mater Sci Chem Eng 3(07):28
34. Wunderlich W, Oekermann T, Miao L, Hue NT, Tanemura S, Tanemura M (2004) Electronic properties of nano-porous Tio_2- and Zno thin films- comparison of simulations and experiments. J Ceram Process Res 5(4):343–354
35. Buckeridge J, Butler KT, Catlow CRA, Logsdail AJ, Scanlon DO, Shevlin SA, Woodley SM, Sokol AA, Walsh A (2015) Polymorph engineering of Tio_2: demonstrating how absolute reference potentials are determined by local coordination. Chem Mater 27(11):3844–3851
36. Di Paola A, Bellardita M, Palmisano L (2013) Brookite, the least known Tio_2 photocatalyst. Catalysts 3(1):36–73
37. Monai M, Montini T, Fornasiero P (2017) Brookite: nothing new under the sun? Catalysts 7(10):304
38. Yao H, Ouyang L, Ching WY (2007) Ab initio calculation of elastic constants of ceramic crystals. J Am Ceram Soc 90(10):3194–3204
39. Yang R, Tang B, Gao T, Ao BY (2016) First-principles study of structural, magnetic, electronic and elastic properties of PUC_2. Commun Theor Phys 66(4):447
40. Sharma S, Sharma J, Sharma Y (2016) DFT calculations of electronic and optical properties of SRS with LDA, GGA and MGGA functionals. In: AIP conference proceedings, vol 1728. AIP Publishing
41. www.jpo.go.jp/shiryou/sonota/youjungijutsu/hikarishokubai/06gaiyou.pdf
42. Broyden CG (1970) The convergence of a class of double-rank minimization algorithms 1. General considerations. IMA J Appl Math 6(1):76–90
43. Fletcher R (1970) A new approach to variable metric algorithms. Comput J 13(3):317–322
44. Goldfarb D (1970) A family of variable metric updates derived by variational means. Math Comput 24(109):23–26
45. Shanno DF (1970) Conditioning of quasi-newton methods for function minimization. Math Comput 24(111):647–656
46. Borisov S, Magarill S, Pervukhina N (2019) Crystallographic analysis of Tio_2 polymorphism (brookite, anatase, rutile). J Struct Chem 60:1783–1789

47. Perdew JP, Ruzsinszky A, Csonka GI, Vydrov OA, Scuseria GE, Constantin LA, Zhou X, Burke K (2008) Restoring the density-gradient expansion for exchange in solids and surfaces. Phys Rev Lett 100(13):136406
48. Tran F, Blaha P (2009) Accurate band gaps of semiconductors and insulators with a semilocal exchange-correlation potential. Phys Rev Lett 102(22):226401
49. Hartwigsen C, Gœdecker S, Hutter J (1998) Relativistic separable dual-space Gaussian pseudopotentials from H to Rn. Phys Rev B 58(7):3641
50. Smidstrup S, Markussen T, Vancraeyveld P, Wellendorff J, Schneider J, Gunst T, Verstichel B, Stradi D, Khomyakov PA, Vej-Hansen UG et al (2019) Quantumatk: an integrated platform of electronic and atomic-scale modelling tools. J Phys: Condensed Matter 32(1):015901
51. Brandbyge M, Mozos JL, Ordejón P, Taylor J, Stokbro K (2002) Density-functional method for nonequilibrium electron transport. Phys Rev B 65(16):165401
52. Soler JM, Artacho E, Gale JD, García A, Junquera J, Ordejón P, Sánchez-Portal D (2002) The siesta method for ab initio order-n materials simulation. J Phys: Condensed Matter 14(11):2745
53. Monkhorst HJ, Pack JD (1976) Special points for Brillouin-zone integrations. Phys Rev B 13(12):5188
54. Schlipf M, Gygi F (2015) Optimization algorithm for the generation of ONCV pseudopotentials. Comput Phys Commun 196:36–44
55. https://docs.quantumwise.com/tutorials/optical/optical.html
56. Dash D, Chaudhury S, Tripathy SK (2018) First principle investigation of structural and optical properties of cubic titanium dioxide. In: AIP Conference Proceedings, vol 1953. AIP Publishing
57. Desjarlais MP (2005) Density functional calculations of the reflectivity of shocked xenon with ionization based gap corrections. Contrib Plasma Phys 45(3–4):300–304

Design and Implementation of Multifunctional Logic Circuits Using SRAM Cells Utilizing In-Memory Computing Concept

Nazida Ansari, Umayia Mushtaq, Seema Eram, and Md. Waseem Akram

Abstract In-memory computing (IMC) is a cutting-edge technology that has disrupted the conventional approach of computing, instead of data is stored separately and processed, in-memory computing eliminates delays, reduces energy consumption, and minimizes resource wastage by addressing the memory bottleneck issue during computation. Leveraging the speed and efficiency of RAM (Random Access Memory), IMC offers fast data access and computation, making it ideal for handling large-scale data-intensive tasks and real-time analytics. This paper consists of design and simulation of 6T SRAM cell array and Multifunctional in memory computing logic circuit to implement the basic two input digital logic gates such as AND, NOR, NAND, OR, XOR and XNOR. Additionally same architecture can perform Half Adder operation and 2-bit binary to gray conversion. The proposed circuit is designed using PTM32nm (Predictive Technology Model) technology node with 1 V operating voltage and implemented in LTSpice Software. The circuit performs bitwise operations, enabling simultaneous read and perform each logic operation parallelly.

Keywords Memory bottleneck · In memory computing · 6T SRAM cell · Bitwise computing · Parallel processing

---

N. Ansari (✉) · U. Mushtaq · S. Eram · Md. Waseem Akram
Department of Electronics and Communication Engineering, Jamia Millia Islamia, New Delhi 110025, India
e-mail: nazidaansari2072@gmail.com

U. Mushtaq
e-mail: umayia1909870@st.jmi.ac.in

S. Eram
e-mail: seema.eram@gmail.com

Md. Waseem Akram
e-mail: makram1@jmi.ac.in

D.-I. Stroe et al. (eds.), *Emerging Electronics and Automation*, Lecture Notes in Electrical Engineering 1202, https://doi.org/10.1007/978-981-97-3090-2_7

1 Introduction

In the rapidly evolving technology, innovations like big data and artificial intelligence (AI) have given rise to intelligent applications, notably edge computing which demands swift memory access during event handling. This poses challenges for traditional widely used Von Neumann computing architecture with its separated memory and computing units [1, 2]. These architectures struggle with extensive data exchanges between memory and processing components, leading to energy inefficiencies, reduced memory bandwidth, and data processing delays [3]. Despite impressive leaps in processor performance due to Moore's Law, memory technology has lagged, resulting in a "memory wall" scenario where processor speed outpaces data transfer capabilities, causing resource wastage [4].

To address these issues, the concept of "In-Memory Computing" has emerged, performing computations directly within memory to minimize data transmission, reduce delays, and overcome Von Neumann architecture limitations [5]. Utilizing Static Random-Access Memory (SRAM), SRAM-IMC shows promise in revolutionizing various computing systems, especially in data-intensive applications like neural networks and pattern matching [3, 6, 7]. The need for faster and more efficient data processing becomes critical. IMC, fast data access and computation make it an attractive solution for handling large-scale data-intensive tasks and real-time analytics.

The 6T SRAM cell faces a shared read–write path issue, causing a reduction in its read noise margin [8–10]. The read operation is the most challenging for a 6T SRAM cell due to the need for strong transistors to discharge the pre-charged bit-line through pull down transistor without altering the stored value [11–13]. To address this, alternatives like 8 or 10T SRAM cells are available. The 8T SRAM, for instance, separates the read and write paths, enabling independent optimization [14]. However, it results in a 30% larger area compared to the 6T SRAM due to its single read BL, suitable only for the AND operation, requiring unique architectures for other logic functions [8, 15, 16]. A novel 9-transistor 3-D computing-in-memory SRAM cell has been developed, featuring two layers. This design incorporates a traditional 6T SRAM cell along with three additional transistors on both the bottom and top layers, respectively [17, 18]. The 10T SRAM resolves some issues but incurs a larger area penalty [19, 20]. While DRAM (Dynamic Random Access Memory) is an option for IMC, it poses challenges with destructive reads and data copying [21, 22]. RRAM (Resistive Random Access Memory) devices have limitations like low endurance and high write latency/energy usage [23, 24].

The paper is structured as follows: Sect. 1 consists introduction of In-Memory Computing (IMC) and also necessity of IMC concept, accompanied by a literature review. Section 2 elaborates on the methodology, implementation, and operation of a multifunctional IMC circuit using LTSpice on a 32 nm technology node. This multifunctional circuit is designed to execute two-bit logic gates and also an area-efficient implementation of the Half Adder based on 6T SRAM cells instead of 8T SRAM cells, which typically occupy more space. Additionally, Binary-to-Gray conversion is accomplished within the SRAM array, eliminating the necessity for additional multiplexer circuitry embedded in the memory in conventional method [25–27]. In Sect. 3, discussions on simulation outcomes presented with respect to voltage levels. Section 4 concludes the paper by summarizing contributions, discussing implications, and highlighting research findings. It also outlines potential avenues for further research and improvement of the proposed model.

2 Methodology: Implementation of Logic Gates Using 6T SRAM Cells Utilizing in Memory Computing Concept

In-Memory Computing (IMC) multifunctional circuit has been implemented in open source LTSpice software and its outcomes are represented by comparing voltage levels with the reference voltage of the comparator. Simulated on the 32 nm technology node in LTSpice, this choice leverages increased transistor density without advanced node challenges, establishing a standardized platform for research and development.

In Fig. 1, we have implemented two SRAM cells arranged in a column configuration. These cells, distinguished as "Cell 1" and "Cell 2, connected to common bit lines (BL and BLB). Both the cells have different word lines Word1 and Word2 to "ON" the access the transistors. By employing different word lines for different cells, the array gains the ability to enable or disable access to specific cells as required as shown in Fig. 2. This selective access prevents inadvertent data alteration and ensures that only the desired cell is involved in read or write operations. Another advantage lies in facilitating parallel operations. Separate word lines permit simultaneous access to distinct cells, enabling efficient parallel processing of data. This proves particularly advantageous in scenarios where multiple memory operations need to be executed concurrently.

2.1 Logic Gates Implementation Using SRAM Cells Utilizing IMC

The functioning of SRAM-based in-memory computing relies on a concept known as bit-wise computing [8]. In this approach, the outcomes of computations are represented through the voltage levels of Q_1 and Q_2 compared with the reference voltage of Sense Amplifier/Rail Comparator. In the existing literature [25, 26], AND gate and NOR gate computations rely on comparing the bit line voltage with the reference voltage of the sense amplifier. In contrast, our approach involves computation through the access of data Q_1 and Q_2 from two SRAM cells, followed by a comparison with the comparator reference voltage instead of the sense amplifier. Furthermore, while Binary-to-Gray conversion is conventionally achieved using a multiplexer circuit

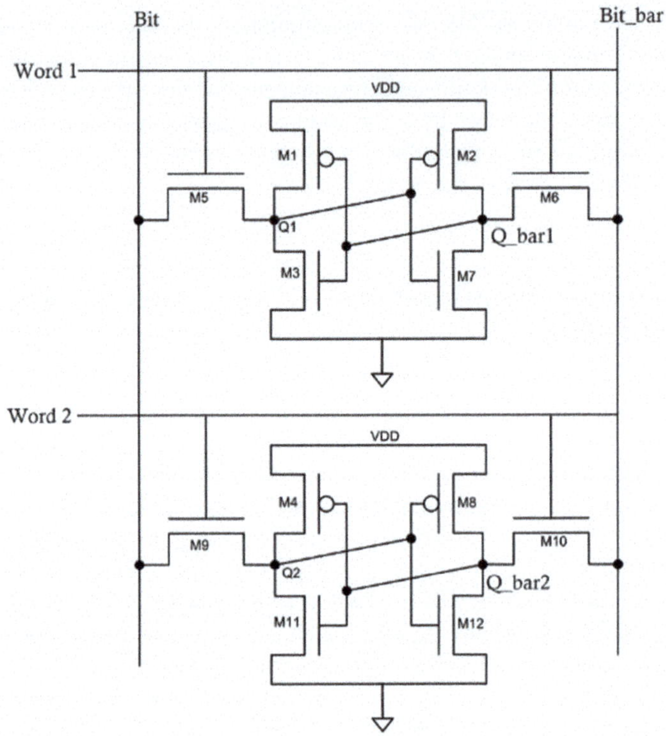

Fig. 1 Schematic of 2-bit SRAM cell array

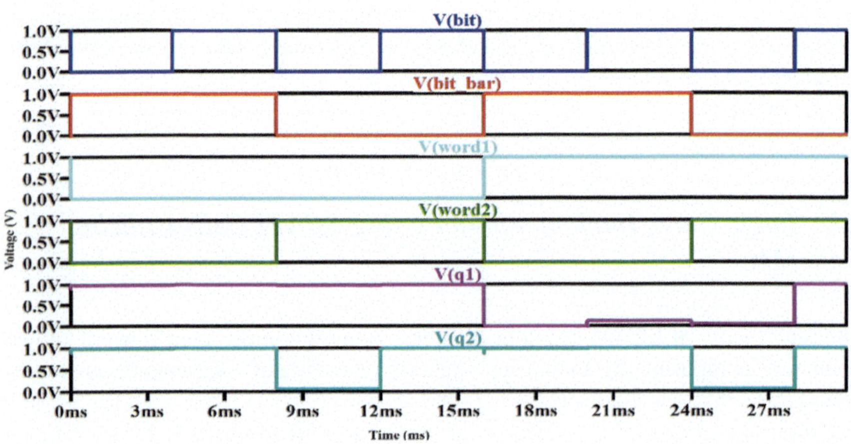

Fig. 2 Transient behaviour of 2-bit SRAM cell array

embedded in the memory [26], our method accomplishes this directly within the SRAM array itself, eliminating the need for any additional circuitry within the SRAM memory. Notably, the IMC Circuit's implementation of the Half Adder [27] relies on the 8T SRAM cell, whereas our proposed circuit employs the more area-efficient 6T SRAM Cell for the Half Adder IMC implementation. The comprehensive operation details are outlined in Table 1.

In Fig. 3a, initially both the BL (Bit-Line) and BLB (Bit-Line Bar) are pre-charged to the v_{DD} voltage level. Then, the Word Line (WL) corresponding to the Cell is activated. Then subsequent step involves the discharge of the two Bit-Lines based on the data stored. If Q_1Q_2 equals '11', the BL remains unaltered, maintaining its charge. For Q_1Q_2 values of '01' or '10', the BL discharges to a certain level. Finally, if Q_1Q_2 equals '00', the BL undergoes maximum discharge. The Sense Amplifier (SA) plays a pivotal role in realizing the AND operation, setting voltage reference (V_{ref}) to v_{DD}. It then compares the voltages of BL and $V_{ref} = 1\text{V}$ to generate the output. When Q_1Q_2 equals '11', the undischarged BL results in a higher output voltage, interpreted as logic '1'. For Q_1Q_2 values of '01' or '10', the discharged BL yields an output voltage below V_{ref}, representing logic '0'. Meanwhile, in the case of Q_1Q_2 equal to '00', the BL's maximum discharge leads to a significantly lower output voltage than V_{ref}, signifying logic '0', as shown in Fig. 3b. For NOR operation, when Q_1Q_2 is '11', BLB discharges to its maximum, while for Q_1Q_2 as '01' or '10', it discharges partially. If Q_1Q_2 is '00', BLB remains unchanged. The Sense Amplifier (SA) uses V_{ref} to perform NOR operations by comparing BLB and V_{ref}. When Q_1Q_2 is '00', the undischarged BLB results in a high output voltage (Logic '1'). For Q_1Q_2 as '01' or '10', the discharged BLB yields a low output voltage (Logic '0'). When Q_1Q_2 is '11', maximum BLB discharge leads to a low output voltage (Logic '0') as shown in Fig. 3b [3, 5].

In Fig. 3c, proposed circuit consists of two 6T SRAM cells 'Cell 1' and 'Cell 2'are arranged in a column configuration, accompanied by two comparators C_1 and C_2. . Each cell is capable of storing 1 bit of data. To execute two input logic gate operations, we read the 2- bit data from the cells and give as the input to the comparator

Table 1 Different operations performed by proposed circuit

Operations	Equations
AND	$Y = Q_1Q_2$
NOR	$Y = \overline{Q_1 + Q_2}$
NAND	$Y = \overline{Q_1Q_2}$
OR	$Y = Q_1 + Q_2$
XOR	$Y = \overline{Q}_1Q_2 + Q_1\overline{Q}_2$
XNOR	$Y = \overline{Q}_1\overline{Q}_2 + Q_1Q_2$
Half Adder	$Sum = \overline{Q}_1Q_2 + Q_1\overline{Q}_2$ $carry = Q_1Q_2$
2-bit binary to gray code converter	$G_1 = Q_1, G_0 = \overline{Q}_1Q_2 + Q_1\overline{Q}_2$

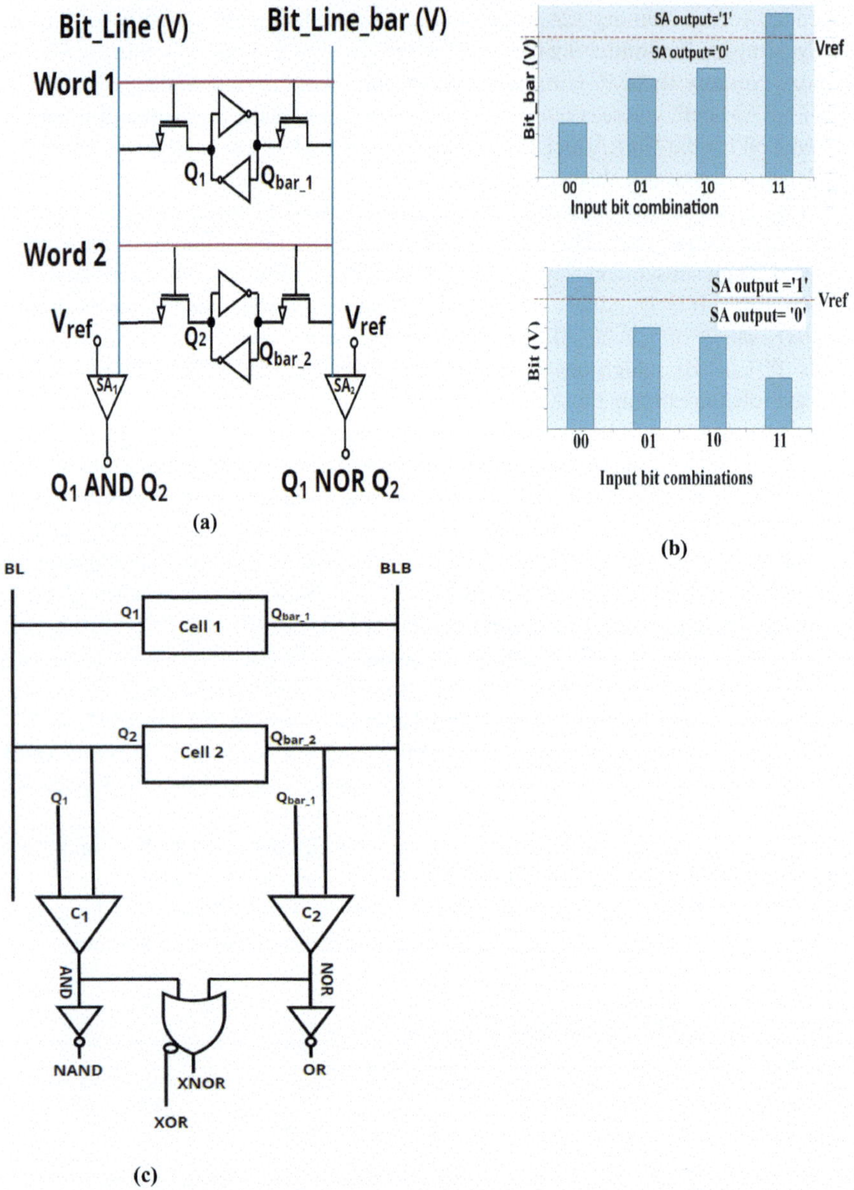

Fig. 3 **a** AND gate and NOR gate bit-line computing in SRAM Array [3, 25, 28]. **b** Logic AND and NOR operation Voltage Histogram [25]. **c** Implementation of IMC circuit to perform Multifunctional Logic Operations

in computing mode. The reference voltage (V_{ref}) of the comparator is set to 1 V. Two variable data from cells are then compared by the V_{ref} of the comparator to yield the output. When the two data points exhibit similar voltage levels, the rail-to-rail comparator generates an output exceeding the reference voltage. Conversely, if there is a substantial voltage difference between the data points, the output registers as a positive voltage lower than the V_{ref} of the comparator. Here the voltage which is above the V_{ref} of rail-to-rail comparator is interpreted as logic '1', and the positive voltage less than the V_{ref} interpreted as the logic '0'.

3 Results and Discussions

In Fig. 4, If both Q_1 and Q_2 are high ('11'), the output voltage is 1.24 V which is higher than the V_{ref}, signifying an AND gate's logic '1' operation. Conversely, if one the Q_1 or Q_2 is low ('10/01'), the output is positive 0.7 V voltage lower than V_{ref}, indicating logic '0' result. If both the Q_1 and Q_2 are low ('00') results in the minimum positive voltage of 0.2 V, compared to V_{ref} interpreted as logic '0' result. This setup implements a two-input AND gate function. Simultaneously, NOR gate is evaluated by receiving input from Q_{bar_1} and Q_{bar_2} to the comparator 2. In Fig. 5, If both Q_{bar_1} and Q_{bar_2} are high ('11'), the output voltage is 0.28 V which is lower than the V_{ref}, signifying an NOR gate's logic '0' operation. Conversely, if one the Q_{bar_1} or Q_{bar_2} is low ('10/01'), the output is positive 0.29 V voltage lower than V_{ref}, indicating logic '0' result. If both the Q_{bar_1} and Q_{bar_2} are low ('00') results in maximum positive voltage of 1.24 V, compared to V_{ref} interpreted as logic '1' result.

Apply the NOT operation (inversion) to the output of an AND gate a NAND (NOT AND) gate is effectively generated as shown in Fig. 6. Similarly, apply the NOT operation (inversion) to the output of an NOR gate an OR gate is effectively generated as shown in Fig. 7. By utilizing an OR gate to merge the outputs of an AND

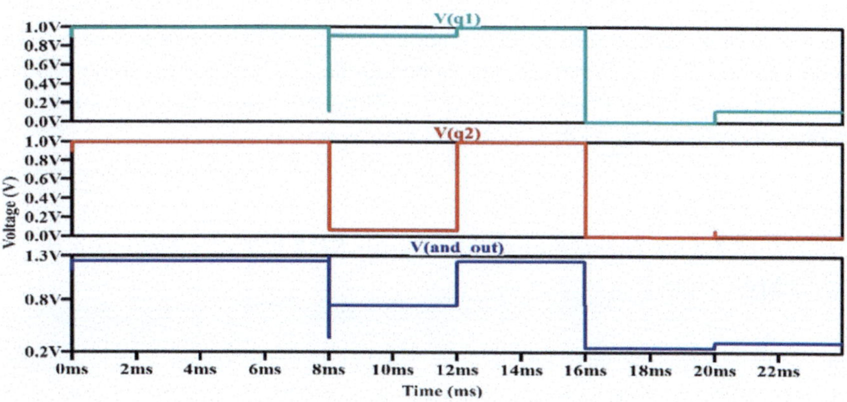

Fig. 4 Transient waveform of proposed IMC AND gate

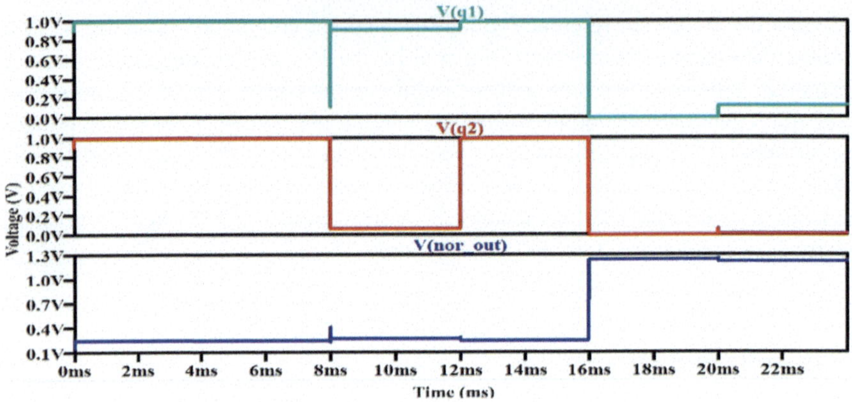

Fig. 5 Transient waveform of proposed IMC NOR gate

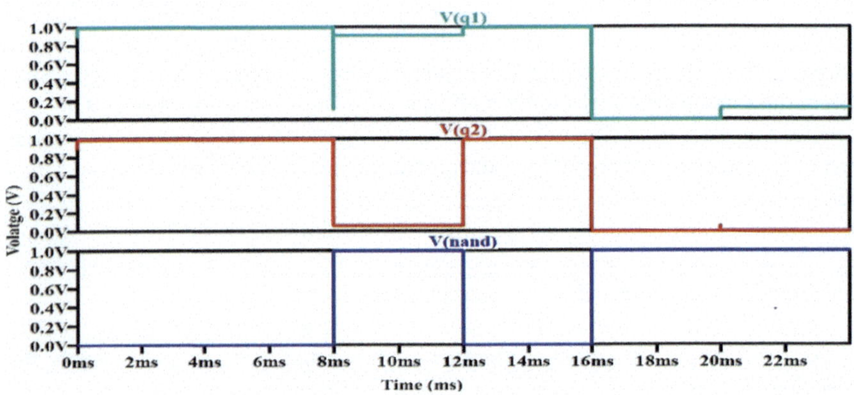

Fig. 6 Transient waveform of proposed IMC NAND gate

gate and a NOR gate, results in the creation of a circuit that performs the XNOR logical operation as shown in Fig. 8. Applying the NOT operation to the XNOR process yields a logical inversion of the XNOR output, results in the creation of a circuit that performs the XOR logical operation as shown in Fig. 9.

3.1 Half Adder Implementation Using SRAM Cells Utilizing IMC

Half Adder is a fundamental digital circuit used in digital electronics and computer systems to perform binary addition of two single-digit binary numbers. It's called a

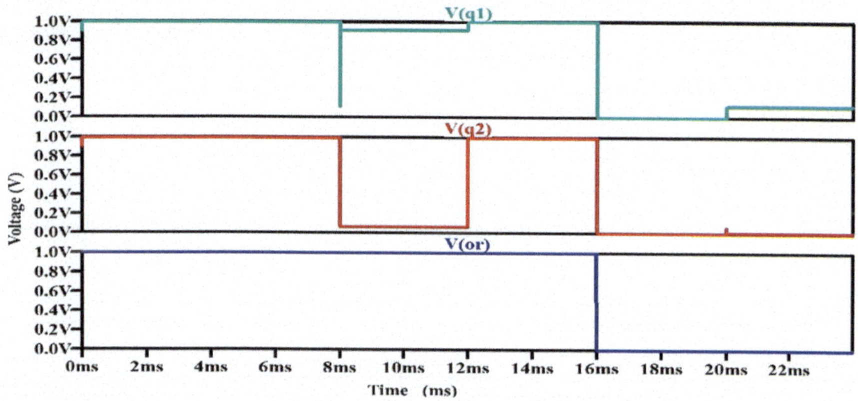

Fig. 7 Transient waveform of proposed IMC OR gate

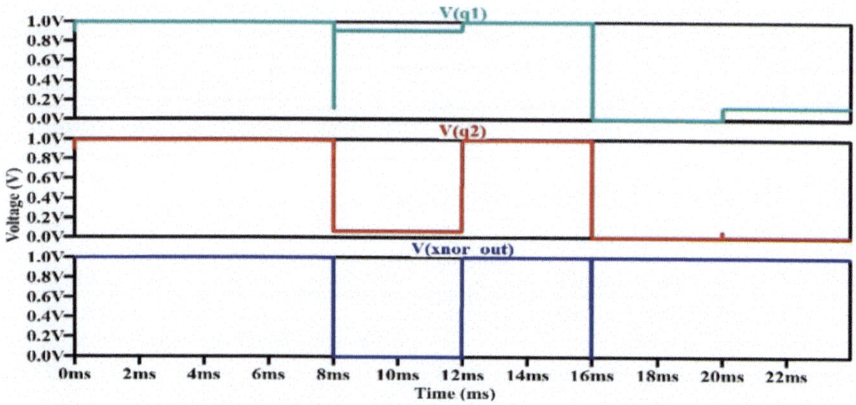

Fig. 8 Transient waveform of proposed IMC XNOR gate

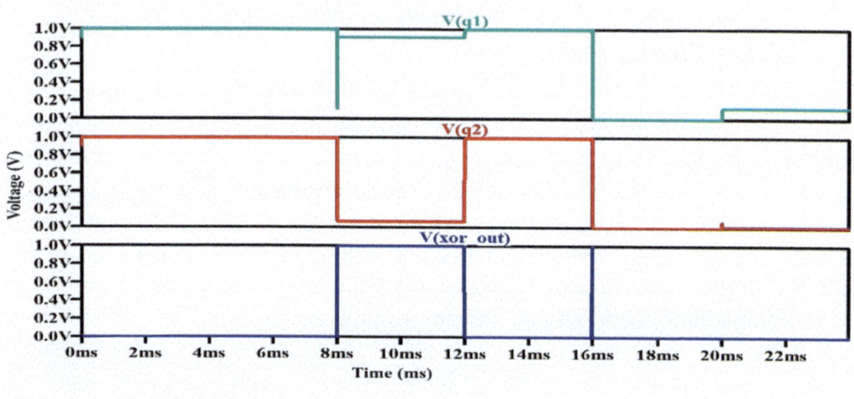

Fig. 9 Transient waveform of proposed IMC XOR gate

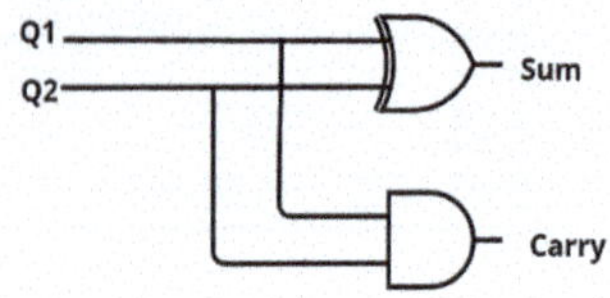

Fig. 10 Half adder circuit [27]

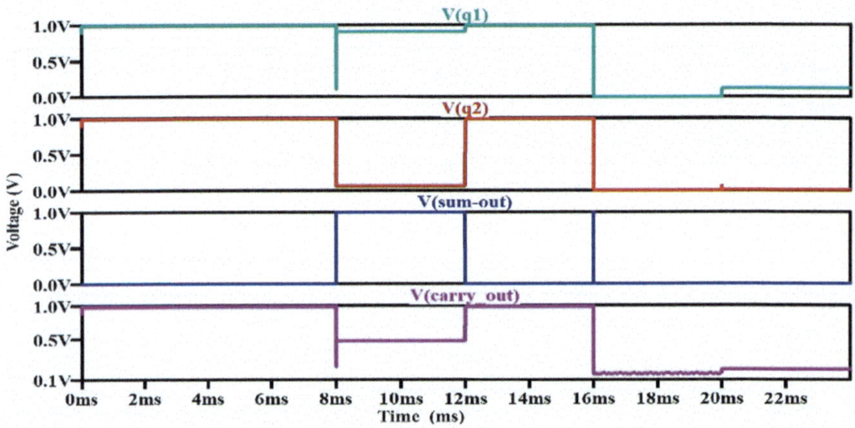

Fig. 11 Transient waveform of proposed IMC half adder

"half" adder because it can only add two bits and doesn't account for any carry from a previous addition as shown in Fig. 10 [29].

In Fig. 11, carry has been generated by performing AND operation for the data taken from Q_1 and Q_2. Sum output will be same as the XOR operation of Q_1 and Q_2.

3.2 *Binary to Gray Converter Implementation Using SRAM Cells Utilizing IMC*

The presence of numerous codes for representing the same discrete information elements leads to varying code usage among different digital systems. When two systems employ different codes for the same information, a conversion circuit becomes necessary to bridge the gap. This circuit, known as a code converter, is a logical construct with input bit patterns representing numbers or characters in one code and output corresponding representations in another code. Its purpose is to ensure compatibility between two systems, even when they employ distinct binary codes [28].

Gray code, also known as reflected binary code, is a binary numeral system in which adjacent values differ in only one bit position. A 2-bit Binary to Gray code converter is a digital circuit that transforms a 2-bit binary input into its equivalent

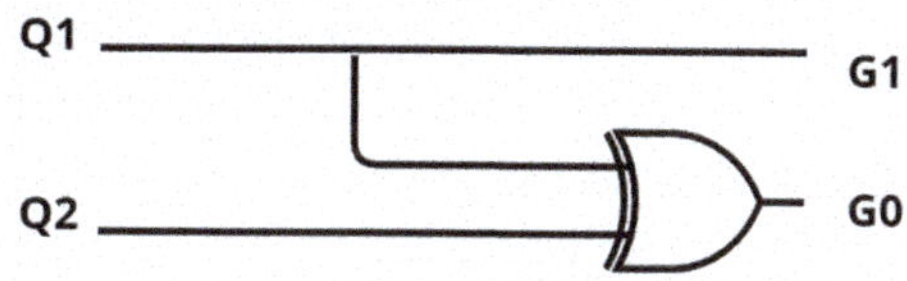

Fig. 12 2-bit Binary to Gray Code Converter [29]

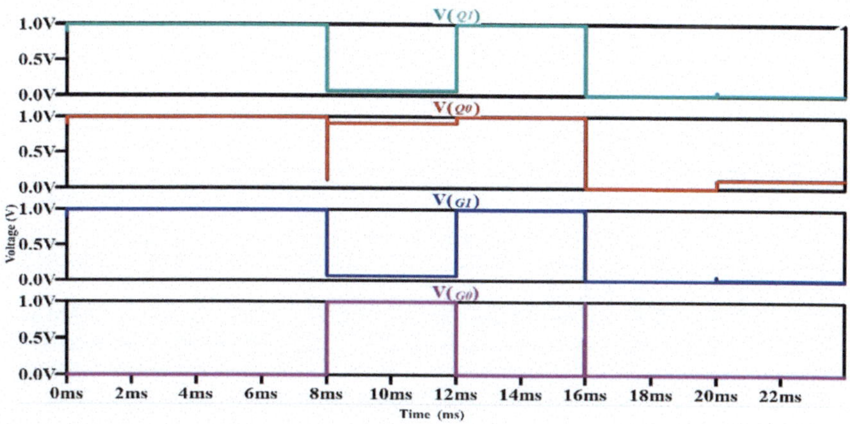

Fig. 13 Transient waveform of proposed IMC 2-bit binary to gray converter

2-bit Gray code output. It typically uses a combination of XOR gates to perform this conversion as shown in Fig. 12.

In Fig. 13, G_1 has been generated by reading the Q_1 bit, and G_0 is the XOR operation of Sum output will be same as the XOR operation of Q_1 and Q_2.

4 Conclusion

This research, commence with the application of robust SRAM cell technology, initiated a thorough functional validation process for two SRAM cell in column configuration. Our proposed circuit highlights its versatility by successfully implementing various logic gates, encompassing AND, NOR, NAND, OR, XNOR, and XOR. Comprehensive functional verification has been conducted to confirm the circuit's effective operation, emphasizing its flexibility and resilience. This proposed circuit can also function as a half-adder circuit and 2-bit Binary-to-Gray code converter, both validated through functional verification. The proposed approach to AND gate and NOR gate implementation uses data from two SRAM cells and compares it with the comparator reference voltage, eliminating the need for additional circuitry for the implementation of Binary to Gray code converter. It also employs a more area-efficient 6T SRAM cell for the Half Adder IMC implementation, unlike the existing 8T SRAM cell implementation. In conclusion, this work contributes to advancing innovative computing approaches within the SRAM cell framework, with the potential to revolutionize digital circuitry and memory systems, opening doors to unparalleled data processing efficiency and real-time analytics. The integration of In-Memory Computing and SRAM cell technology opens doors to unparalleled data processing efficiency, real-time analytics, and transformative innovations in memory-centric computing. This research not only enhances our understanding of innovative computing paradigms but also establishes the groundwork for potential advancement in the digital circuitry and the memory system in the years ahead.

References

1. Si X et al (2019) A dual-split 6T SRAM-based computing-in-memory unit-macro with fully parallel product-sum operation for binarized DNN edge processors. IEEE Trans Circ Syst I: Reg Pap 66(11):4172–4185
2. Khwa W-S et al (2018) A 65nm 4Kb algorithm-dependent computing-in-memory SRAM unit-macro with 2.3 ns and 55.8 TOPS/W fully parallel product-sum operation for binary DNN edge processors. In: 2018 IEEE international solid-state circuits conference-(ISSCC). IEEE
3. Lin Z et al (2022) A review on SRAM-based computing in-memory: circuits, functions, and applications. J Semicond 43(3):031401
4. Challenges and opportunities: from near-memory computing to in-memory computing. In: Proceedings of the 2017 ACM on international symposium on physical design
5. Jaiswal A et al (2019) 8T SRAM cell as a multibit dot-product engine for beyond von Neumann computing. IEEE Trans Very Large-Scale Integr (VLSI) Syst 27(11):2556–2567
6. Jhang C-J et al (2021) Challenges and trends of SRAM-based computing-in-memory for AI edge devices. IEEE Trans Circ Syst I: Reg Pap 68(5):1773–1786
7. Simon W et al (2019) A fast, reliable and wide-voltage-range in-memory computing architecture. In: Proceedings of the 56th annual design automation conference
8. Jaiswal A, Agrawal A, Ali MF, Sharmin S, Roy K (2020) I-SRAM: interleaved wordlines for vector boolean operations using SRAMs. IEEE Trans Circ Syst I Regul Pap 67(12):4651–4659. https://doi.org/10.1109/TCSI.2020.3005783
9. Mushtaq U, Sharma VK (2019) Design of 6T FinFET SRAM cell at 7nm. In: 2019 international conference on communication and electronics systems (ICCES). IEEE
10. Mittal S et al (2021) A survey of SRAM-based in-memory computing techniques and applications. J Syst Archit 119:102276
11. Sil A et al (2008) A novel high write speed, low power, read-SNM-free 6T SRAM cell. In: 2008 51st midwest symposium on circuits and systems. IEEE
12. Alorda B, Torrens G, Bota S (2018) 6T CMOS SRAM stability in nanoelectronic era: from metrics to built-in monitoring. In: Complementary metal oxide semiconductor
13. Mushtaq U, Sharma VK (2020) Design and analysis of INDEP FinFET SRAM cell at 7-nm technology. Int J Num Model: Electron Netw Dev Fields 33 (5):e2730
14. Mittal S, Wang R, Vetter J (2017) DESTINY: A comprehensive tool with 3D and multi-level cell memory modeling capability. J Low Power Electron Appl 7(3):23
15. Lin Z et al (2020) In-memory computing with double word lines and three read ports for four operands. In: IEEE transactions on very large scale integration (VLSI) systems, vol 28, no 5, pp 1316–1320, May 2020. https://doi.org/10.1109/TVLSI.2020.2976099
16. Dong Q et al (2018) A 4 + 2T SRAM for searching and in-memory computing with 0.3-V VDDmin. IEEE J Solid-State Circ 53(4):1006–1015. https://doi.org/10.1109/JSSC.2017.2776309

17. Hsueh F-K et al (2017) TSV-free FinFET-based Monolithic 3D+-IC with computing-in-memory SRAM cell for intelligent IoT devices. In: 2017 IEEE international electron devices meeting (IEDM). IEEE
18. Lin Z et al (2021) Two-direction in-memory computing based on 10T SRAM with horizontal and vertical decoupled read ports. IEEE J Solid-State Circ 56(9):2832–2844
19. Zhang Y et al (2017) Recryptor: a reconfigurable in-memory cryptographic Cortex-M0 processor for IoT. In: 2017 symposium on VLSI circuits. IEEE
20. Chang IKJ et al (2009) A 32 kb 10T sub-threshold SRAM array with bit-interleaving and differential read scheme in 90 nm CMOS. IEEE J Solid-State Circ 44(2):650–658
21. Parveen F, He Z, Angizi S, Fan D (2018) HielM: highly flexible in-memory computing using STT MRAM. In: 2018 23rd Asia and South Pacific design automation conference (ASP-DAC), Jeju, Korea (South), pp 361–366. https://doi.org/10.1109/ASPDAC.2018.8297350
22. Wang Y et al (2021) An in-memory computing architecture based on two-dimensional semiconductors for multiply-accumulate operations. Nat Commun 12(1):3347
23. Mittal S (2018) A survey of ReRAM-based architectures for processing-in-memory and neural networks. Mach Learn Knowl Extract 1(1):75–114
24. Halawani Y et al (2019) ReRAM-based in-memory computing for search engine and neural network applications. IEEE J Emerg Sel Top Circ Syst 9(2):388–397
25. Sebastian A et al (2020) Memory devices and applications for in-memory computing. Nat Nanotechnol 15(7):529–544
26. Dhakad NS et al (2023) In-memory computing with 6T SRAM for multi-operator logic design. Circ Syst Signal Process 1–15
27. Izhar Md, Ahmad S, Rahman SA (2022) Logic circuit implementation for enabling SRAM based in-memory computing. In: 2022 5th international conference on multimedia, signal processing and communication technologies (IMPACT). IEEE
28. Simon WA, Qureshi YM, Rios M, Levisse A, Zapater M, Atienza D (2020) BLADE: an in-cache computing architecture for edge devices. In: IEEE transactions on computers, vol 69, no 9, pp 1349–1363, 1 September 2020. https://doi.org/10.1109/TC.2020.3373332
29. Kumar AA (2014) Switching theory and logic design. PHI Learning Pvt. Ltd.

A Capacitive Inkjet Printed Humidity Sensor on a Paper Substrate

Md Shakir, Rubina, and Tarikul Islam

Abstract This paper presents a parallel plate capacitive humidity sensor using the paper substrate. The sensor was fabricated by inkjet printing silver electrodes on each side of a 205 μm thickness paper substrate. The top electrode was mesh type. Experiments were conducted to determine the static and dynamic response parameters of the sensor with variation of humidity from 0 to 50% RH. The electrical response parameters are sensitivity 2.40 pF/% RH, rise time 5 min. The device is simple in fabrication, inexpensive, and biodegradable, and multiple identical sensors can be made in large quantities in less time. However, due to the thick sensing film, the response time and hysteresis error are high. These parameters can be improved by proper selection of the optimum pore size of the mesh, the dimension of the electrodes, and adding an extra sensing film. The output is precise, repeatable, and suitable for many day-to-day common applications.

Keywords Capacitive sensor · Humidity measurement · Biodegradable · Inkjet printing

1 Introduction

The dynamic range of humidity measurement in solid, liquid, or gas for various needs is extremely wide ranging from few traces in ppm to high %RH [1, 2]. Application needs of humidity are also very wide. Therefore, it is difficult to measure such a wide range of humidity by a single sensor. An important full-scale range of humidity measurement for some applications, including textiles, is 0 to 50%RH [3]. Among

Md Shakir (✉) · Rubina · T. Islam
Department of Electrical Engineering, Jamia Millia Islamia, New Delhi 110025, India
e-mail: mdshkr123@gmail.com

Rubina
e-mail: rubina2008655@st.jmi.ac.in

T. Islam
e-mail: tislam@jmi.ac.in

D.-I. Stroe et al. (eds.), *Emerging Electronics and Automation*, Lecture Notes in Electrical Engineering 1202, https://doi.org/10.1007/978-981-97-3090-2_8

various types, capacitive sensors are extensively used in fabricating inexpensive, low-power-consuming, linear sensors [4]. For these reasons, several commercial humidity sensors are of capacitive type.

For fabricating the sensor, the substrate must be carefully chosen considering various factors like porosity, surface roughness, thickness, permittivity, and unit cost. More precisely, paper has evolved as a suitable material of choice for inkjet printing due to its cheaper cost and adaptability to roll-to-roll and reel-to-reel printing [5–8]. A paper substrate possessing characteristics such as low cost, water absorptive, and biodegradability, brings new possibilities to fabricate low-cost, mass-producible, environment-friendly sensors. One unique feature of a paper-based sensor is the fabrication of identical sensors, using the inkjet printing technique. Therefore, in recent years, increased interest in the growth of paper-based devices and technology for flexible and green electronics has been observed. However, most of the paper-based humidity sensors are made using an interdigital electrode (IDE) capacitive structure that suffers from low sensitivity, high nonlinearity, high parasitic capacitance effects, and a complex approximate analytical expression [8, 9]. An IDE sensor is fabricated on the paper substrate in [5] but this utilizes the manual screen printing procedure for fabricating the sensor.

In this work, efforts have been made to design and fabricate a parallel plate capacitive humidity sensor. The novel approach in fabricating this sensor is that here just a paper is used both as the substrate and the sensing material. It does not utilize any extra material to be deposited to respond to humidity changes. The sensor utilizes the paper substrate which has a nanoporous coating on its top surface and serves the purpose of an active sensing layer for the moisture range of 0–50% RH. The device was fabricated by sandwiching a paper substrate, which serves the purpose of a hydrophilic dielectric layer between two metal plates. The electrodes were printed on paper by providing the electrode design to the inkjet printer that utilizes silver electrodes. The key advantage of these sensors is that the electrodes are conductive just after printing without the need for heating which means ready-to-use sensors are fabricated. The dimensions of the paper substrate and the electrodes are properly selected to minimize the sensor dimension and at the same time to maintain an exact expression of the parallel plate capacitor. Also, by this inkjet printing large quantities of identical sensors can be fabricated at low cost. The critical electrical parameters of the sensor were determined with humidity variation from 0 to 50%RH. The lower value of humidity is not exactly 0% RH but is close to it. This is a novel approach for fabricating low-cost disposable sensors.

1.1 Design and Fabrication

The sensor was fabricated using the pe_smart type 2 (FS2) paper having a nanoporous coating on the top layer. The substrate was procured from Felix Schoeller [10]. As per specifications, these papers are characterized by high dimensional stability even at relatively high temperatures, and high chemical purity free from heavy metal

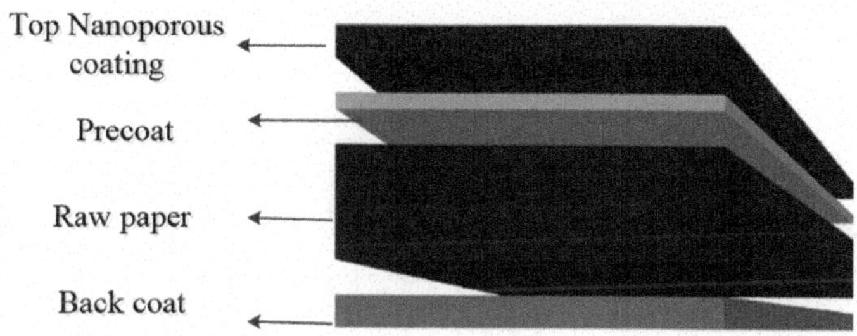

Fig. 1 The structural layout of the pe_smart type2 paper [10]

impurities. Hence, it leads to significantly less waste when printing organic circuits compared to conventional substrates (e.g., PET films, and polyimide films). The structure of the paper is shown in Fig. 1. It consists of a precoat and back coat each side of a RAW paper substrate. On the top layer, there is a nanoporous coating of 100 μm thickness. The contact angle of FS2 with DI water was found to be 76.14∘. The porous film has an average porosity of 15 nm [11]. The overall thickness of the substrate is 205 μm. It can withstand temperatures up to 250 °C which means this can also be used for high-temperature applications. It was cleaned to remove impurities from its surface. Epson Eco-Tank Monochrome M1170 printer was used for printing the electrodes using silver nanoparticle ink (25%w/w Ag) acquired from Novacentrix [12]. The resistivity of electrically conductive ink varies from 1.5 × 10–5 Ω-cm to 3.0 × 10–6 Ω-cm. The most desirable advantage is that the electrodes printed are conductive just after printing without the need for heating as in earlier printers. Multiple identical sensors can be printed at a single time ensuring uniform electrode thickness and identical structures. Hence errors due to these inabilities that occur in manual screen printing can be minimized. In addition, paper being a biodegradable material minimizes environmental pollution due to its disposal.

The idea behind the sensor design is that it utilizes the paper itself as the substrate as a humidity-sensing dielectric material and does not need any other sensing film for moisture adsorption. Ceramic film sometimes cracks during high-temperature sintering so errors due to cracked film can be avoided.

A parallel plate capacitor was fabricated by printing electrodes on each side of the substrate. The top and bottom electrodes are shown in Fig. 2. The dimension of the top electrode is 10 mm × 10 mm whereas the dimension of the bottom electrode is 15 × 15 mm. Connection to the electrodes was made from the contact pad of the electrode. The dimensions of the electrodes were taken to be larger than the dielectric film thickness to minimize the fringe effect (distorted E-field at the edges). Using this configuration, the distorted field at the edge of the electrodes moves out of the area of interest and does not add up significant parasitic capacitance. The top electrode is made of mesh type since we want humidity to purge from the top surface to the sandwiched paper substrate. This can be made possible if the top electrode has pores in it to allow the humidity to purge through it. So, the top electrode is designed as a mesh having pores of size 0.5 mm × 0.5 mm distributed uniformly. Thus, water

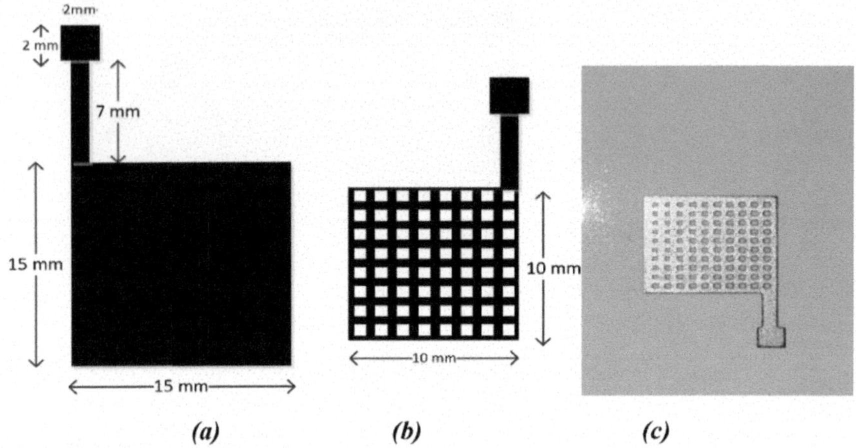

Fig. 2 Sensor schematic **a** bottom electrode **b** top electrode **c** photograph of top electrode

molecules in the air can purge through the top electrode and reach the nano-porous top surface of the paper.

Since the thickness of the dielectric layer is much smaller than the dimension of the electrodes, the capacitance value can be given by

$$C = \varepsilon_0 \varepsilon_r A/t \tag{1}$$

$$\varepsilon_r^{1/3} = \left[\gamma\left(\varepsilon_w^{1/3} - \varepsilon_s^{1/3}\right) + \varepsilon_s^{1/3}\right] \tag{2}$$

where A is the effective overlap area which is 100 mm^2, t is the thickness of the dielectric medium which is 205 μm, ε_o is the relative permittivity of the vacuum and is a constant and ε_r is the effective relative permittivity of the dielectric medium in between the opposite two electrodes, which is the paper that we have used. As described above, the paper used for the sensor fabrication has a nanoporous coating on its top surface. So, when the sensor is exposed to increasing/decreasing humidity, moisture purges in/out through the top electrode and gets adsorbed/desorbed respectively in the nano-order pores of the top surface of the paper. When moisture gets adsorbed during increasing humidity in the nano order pores of the paper the effective relative permittivity of the paper increases as per the empirical relation Eq. (2) where γ is the fractional volume of water adsorbed, ε_w is the relative permittivity of water, ε_s is the relative permittivity of the paper in dry state (0% RH) and ε_r is the effective relative permittivity of paper with adsorption/desorption of humidity [10]. Thus, capacitance increases as per Eq. (1), and similarly when the moisture gets desorbed from the nano order pores during decreasing humidity its effective relative permittivity decreases leading to a decrease in capacitance. This phenomenon is modeled through various equations [10, 11].

1.2 Sensor Electrical Characterization

The schematic diagram of the setup for the sensor calibration is shown in Fig. 3. The sensor was installed in a 100-cc SS-sealed test chamber having inlet and outlet pipes. The sensing chamber is connected in series to the commercial RH meter (Honeywell). The accuracy of the meter is ±3.5% RH. The shielded cable of short length was used to minimize the effect of parasitic capacitance. The electrical response due to the variation of the capacitance of the sensor with the humidity was measured with a Keysight hand-held LCR meter (U1733C). The accuracy of the LCR meter was ±0.2% for the capacitance range from 0 to 20 nF and response time of approximately 1 s/DUT (Device Under Test) [12]. Initially, the sensor was refreshed using dry nitrogen for 30 min to remove settled humidity in the nanoporous layer. The sensor response was taken at 0%RH. Then, moist nitrogen gas with humidity varying from 0 to 50%RH was prepared by mixing water vapor with the dry N_2 gas. Humidity level in the moist gas was measured by the commercial hygrometer which is connected in series with the sensor chamber. The length of the PTFE pipe carrying humidity from the commercial hygrometer to the sensor chamber was kept minimum to minimize the humidity gradient and avoid inaccuracies in the final result.

The capacitance response was determined for increasing and decreasing humidity. Necessary precautions were taken to record the capacitance values and to reduce the lead effect.

Static Response

Static response such as variation of capacitance value with humidity was determined. The sensor was exposed to static humidity, the capacitance value was measured at three excitation frequencies of 1, 10, and 100 kHz respectively. Humidity level was then increased in steps and when it reached the stable value, the capacitance was

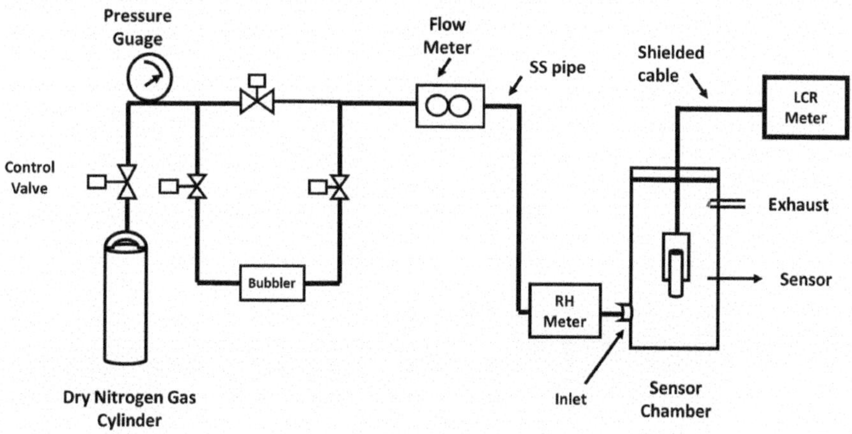

Fig. 3 Schematic diagram of the calibration setup of the humidity sensor

noted. The sensor inside the chamber was given sufficient time (10 min) to reach the equilibrium vapor condition. The capacitance value at each humidity level was recorded at least five times and it was then averaged and plotted at different humidity levels. It can be inferred from the response curve that the capacitance value increases with an increase in humidity. But at lower signal frequencies, the sensor shows maximum sensitivity as compared to other higher frequencies. It is also reported in various works that the capacitance change is more significant at low frequency and is suitable for high sensitivity but the response is more nonlinear. A detailed investigation of the frequency effect of a capacitive humidity sensor is reported [13]. Due to the cyclic variation of the excitation signal, the polar water molecules try to align with the excitation signal. At lower frequencies, water molecules easily follow the slowing varying electric field but at higher frequencies due its inertia (mass of water molecules), they are unable to follow the fast-changing electric field. Hence sensitivity reduces with an increase in signal frequency.

The sensitivity of the sensor which in general is defined as the ratio of change in capacitance value for the full-scale humidity range to the change in %RH. The sensitivity value was found to be 2.40 pF/%RH which is quite significant (Fig. 4).

Dynamic Response

The experiment was then conducted to determine the transient response due to step changes in humidity. This response curve helps to understand the behavior of the sensor due to sudden changes in humidity. Response and recovery time are two important characteristic parameters of the sensor that can be determined from the dynamic response. The response time can be defined as the time taken by the sensor for a 10–90% change in capacitance value for a given 0–50% step change in RH, while the recovery time denotes the time taken by the sensor for the change in

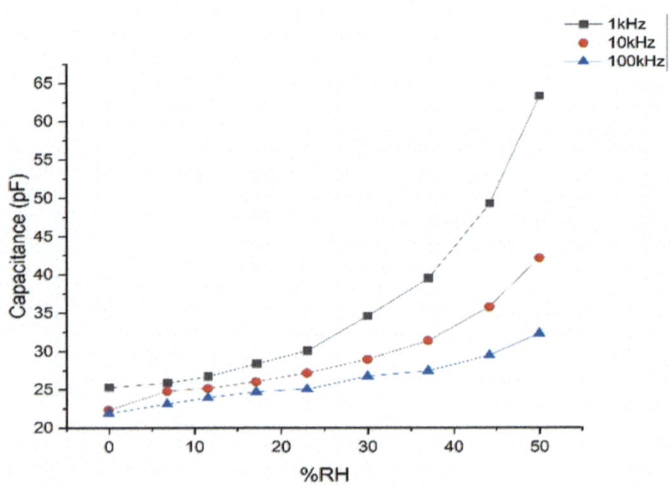

Fig. 4 Variation of the capacitance of the sensor with change in humidity

capacitance from the maximum to 10% of the final value when humidity undergoes the step change from 50%RH to 0%RH. To determine these parameters, the sensor was initially refreshed to 0%RH, and it was then suddenly exposed to humidity of 50% RH. For refreshing the sensor, the moist path of the setup is almost closed, and the dry gas path is opened to pass the gas at a fixed flow rate. Then, the flow path of dry gas is closed, and the wet flow path is opened suddenly to a fixed flow rate. The flow rate of the wet path for the desired RH of 50% is calibrated with the help of the reference RH meter. The dynamic response curve for two identical step cycles is shown in Fig. 5. Due to the steep increase in humidity, the capacitance value rises from 25 to 97 pF and reaches a saturation value. When the sensor was refreshed by dry N_2 gas, water vapor desorbs from the nano order pores and the capacitance reached the initial value. The response and recovery times are 9.64 and 4.90 min respectively. The time constant of the device is close to 5 min (63% step change). The response parameters are quite high and undesirable for certain applications. However, we have considered the worst situation of maximum step humidity change to determine the response times. Such a value of step change rarely occurs in practice. Also, this issue may arise due to the large size of the sensor chamber (360 cc) as a longer time is required for the water molecules to reach the equilibrium vapor condition. The response time can be further improved by reducing the thickness of the paper substrate and controlling the flow rate properly. Also, as in these sensors, only the paper hydrophilicity is used for sensing the humidity, so to increase the sensitivity and the response time parameters of the sensor can be enhanced by adding an extra moisture-sensitive layer on the paper substrate [5]. Further, it is observed that the response of the sensor is repeatable as the capacitance change in two cycles is identical. Then the sensor response to increasing and decreasing humidity was noted. Since adsorption and desorption of water molecules do not take place at a similar rate from nano-order pores there is a possibility of hysteresis error because water molecules take time to adsorb and desorb water vapor [14].

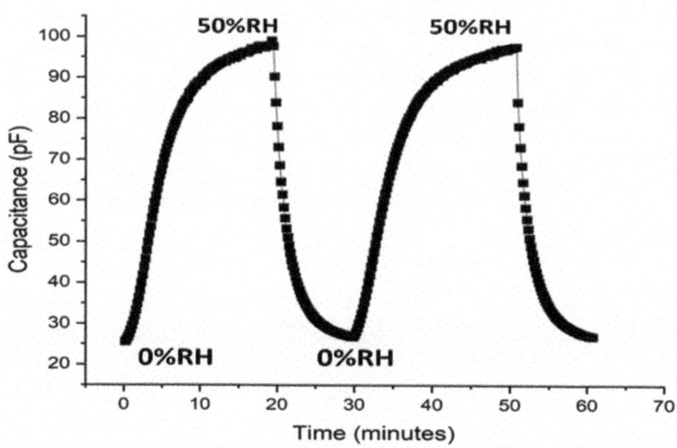

Fig. 5 Dynamic response of the sensor with a step change in humidity

This error occurs due to differences in the rate of adsorption and desorption of water molecules during increasing and decreasing humidity concentrations. The hysteresis effect is studied by observing the capacitive response of the sensor at a fixed frequency for the full-scale increase and decrease in RH. First, the capacitance value is noted with the increase in RH from 0 to 50%, and then, the experiment is repeated with the reverse order of RH from 50 to 0%. It was calculated by taking the change in capacitance at a particular RH that results in maximum deviation when approached in both directions. The hysteresis error plot is shown in Fig. 6. As it can be seen that appreciable hysteresis error is observed. This is dependent on various factors like the type of paper substrate, and porosity of the paper surface which affects the rate of adsorption/desorption of water molecules by the substrate and determines the response/recovery rate of the sensor, etc. The error can be reduced by reducing film thickness or by calibrating the humidity response taking the average value of increase and decreases in capacitance value at a fixed humidity (reduced by almost 50%). The hysteresis error for the average value was noted to be 1.07% and has been plotted in Fig. 6.

In Table 1 comparison of our work has been done with the existing works that are based on interdigital fringing field capacitors. It can be observed that most of the work has been done using manual screen-printing methods for electrode fabrication which in itself is time-consuming, complex, and inhibiting large-scale production in a short time. Also, most of the sensor utilize additional sensing film for humidity sensing which require additional steps for sensor fabrication, but in our sensor, only the pores already present on the paper is utilized for sensing purpose means less/ minimal expertise requirement. The large response and recovery time for the existing

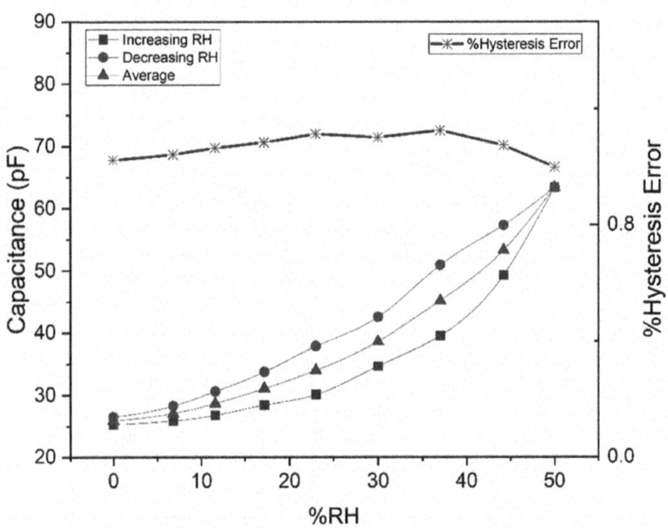

Fig. 6 Hysteresis error of the sensor for increasing and decreasing humidity

Table 1 Comparison with existing literature

Substrate	Sensitive material	Output	Response/recovery time (s)	References
Paper	NFC/CNTS	Resistive	330/377	[15]
Paper	GO	Capacitive	180/-	[16]
Paper	GO	Capacitive	174/40	[5]
Paper	None	Capacitive	9.64/4.9 min	Our work

work can be mitigated by optimizing the mesh pore size, thickness of the film and will be part of future work.

1.3 Conclusion

This paper presents a paper substrate-based capacitive sensor for midrange humidity measurement. The sensor was fabricated by sandwiching a paper substrate between silver electrodes. The silver electrodes were printed on each side of the substrate by injecting printing of silver ink. The top electrode was mesh type. Experiments were conducted to determine the static and dynamic response parameters of the sensor with a humidity range of 0–50% RH. The device electrical parameters observed are sensitivity 2.40 pF/% RH, rise time 5 min. The device is simple in fabrication, inexpensive, biodegradable and the identical sensors can be made in bulk quantity. However, its response time and hysteresis error are high which can be minimized by reducing film thickness and processing the sensor output signal.

References

1. Ma Z, Fei T, Zhang T (2023) An overview: sensors for low humidity detection. Sens Actuat, B Chem 376:133039. https://doi.org/10.1016/j.snb.2022.133039
2. Traversa E (1995) Ceramic sensors for humidity detection: the state-of-the-art and future developments. Sens Actuat, B Chem 23(2–3):135–156. https://doi.org/10.1016/0925-4005(94)01268-M
3. Mathew AP, Dufresne A (2002) Plasticized waxy maize starch: effect of polyols and relative humidity on material properties. Biomacromol 3(5):1101–1108. https://doi.org/10.1021/bm020065p
4. Mahboob MdR, Zargar ZH, Islam T (2016) A sensitive and highly linear capacitive thin film sensor for trace moisture measurement in gases. Sens Actuat, B Chem 228:658–664. https://doi.org/10.1016/j.snb.2016.01.088
5. Song Z, Li S, Hou B, Cheng Z, Xue Y, Chen B (2023) High-sensitivity paper-based capacitive humidity sensors for respiratory monitoring. IEEE Sens J 23(3):2291–2302. https://doi.org/10.1109/JSEN.2022.3228540
6. Korotcenkov G (2023) Paper-based humidity sensors as promising flexible devices: state of the art: Part 1. General consideration. Nanomaterials 13(6):1110. https://doi.org/10.3390/nano13061110

7. Jia W et al (2022) A flexible capacitive paper-based pressure sensor fabricated using 3D printing. Chemosensors 10(10). https://doi.org/10.3390/chemosensors10100432
8. Guan X et al (2021) Flexible humidity sensor based on modified cellulose paper. Sens Actuat, B Chem 339:129879. https://doi.org/10.1016/j.snb.2021.129879
9. Gugliandolo G et al (2022) Design and test of an inkjet-printed microwave interdigital capacitor on flexible Kapton substrate. In: 25th IMEKO TC-4 international symposium on measurement of electrical quantities, IMEKO TC-4 2022 and 23rd international workshop on ADC and DAC modelling and testing, IWADC 2022, pp 346–351
10. Special paper p_e:smart® | Felix Schoeller [Online]. Available: https://www.felix-schoeller.com/en/products/other-specialty-papers/p-esmart. Accessed: Oct. 30, 2024
11. Islam T, Khan AU, Akhtar J, Rahman MZU (2014) A digital hygrometer for trace moisture measurement. IEEE Trans Industr Electron 61(10):5599–5605. https://doi.org/10.1109/TIE.2013.2297295
12. U1733C 100Hz/120Hz/1kHz/10kHz/100kHz Handheld LCR Meter | Keysight. Accessed October 12, 2023. https://www.keysight.com/zz/en/product/U1733C/handheld-lcr-meter-100-hz-120-hz-1-khz-10-khz100-khz.html
13. Islam T, Ur Rahman MZ (2016) Investigation of the electrical characteristics on measurement frequency of a thin-film ceramic humidity sensor. IEEE Trans Instrum Meas 65(3):694–702. https://doi.org/10.1109/TIM.2015.2506302
14. Siddiqui A, Mahboob R, Islam T (2017) A passive wireless tag with digital readout unit for wide range humidity measurement. IEEE Trans Instrum Meas 66(5):1013–1020. https://doi.org/10.1109/TIM.2016.2647478
15. Zhu P et al (2019) Flexible and highly sensitive humidity sensor based on cellulose nanofibers and carbon nanotube composite film. Langmuir 35(14):4834–4842. https://doi.org/10.1021/acs.langmuir.8b04259
16. Alrammouz R et al (2019) Highly porous and flexible capacitive humidity sensor based on self-assembled graphene oxide sheets on a paper substrate. Sens Actuat, B Chem 298:126892. https://doi.org/10.1016/j.snb.2019.126892

Delay Assessment of Parallel Multiplier Using Ultra-Scale FPGA Devices

Shivangi Awasthi, Chandrasekhar Aditya, Parveen Kumar, and Balwinder Singh

Abstract A chip designer mainly focuses on reducing area, power consumption, and delay when designing Integrated Circuit (IC). In many Digital Signal Processing (DSP) systems, multipliers play a vital role. If the delay of multipliers gets reduced, it will change the whole circuit delay, so researchers are going to reduce the area and delay of the circuit. This paper basically gives information about the comparison of all types of parallel multipliers such as an array, Baugh-Wooley, braun, and a tree multiplier that is the daddamultiplieronultra-scale Field Programmable Gate Array (FPGA) devices. Which gives analysis about the capability and performance of multipliers. This simulation work is done on the Xilinx using the ISM simulator, different ultrascale FPGA devices (Znyque, Kintex 7, Virtex 7, and Artix 7). In the last, a comparative table gives information about the different parameters like delay and area (Number of Look-up tables) used by the particular multiplier. The abstract should summarize the contents of the paper in short terms, i.e. 150–250 words.

Keywords Baugh-Wooley · Dadda multiplier · Znyque · Xilinx · ISM simulator

1 Introduction

Very Large Scale Integration (VLSI) gives knowledge about the integration of thousands of transistors on a single chip of IC that comes under the VLSI [1, 2]. According to Moore's law, "the number of transistors gets doubled in every 18 months" [3].

S. Awasthi (✉) · C. Aditya · P. Kumar · B. Singh
Centre for Development of Advanced Computing, Mohali, India
e-mail: shiv18gyan@gmail.com

C. Aditya
e-mail: imeraditya@gmail.com

P. Kumar
e-mail: parveen.eng@gmail.com

B. Singh
e-mail: balwinder@cdac.in

D.-I. Stroe et al. (eds.), *Emerging Electronics and Automation*, Lecture Notes in Electrical Engineering 1202, https://doi.org/10.1007/978-981-97-3090-2_9

Therefore, many semiconductor companies are trying to decrease the size of the IC to decrease power consumption and area. The reduction of area, delay, and power consumption has a great contribution to the manufacturing and high performance of DSP systems [4–6]. Multipliers play a vital role in the design of many DSP systems [7]. This multiplier performs multiplication, a fundamental operation of arithmetic operations that is used in most computational operations [8–11]. Multiplication is a series of repeated additions so that it requires more execution time and high speed of the DSP system. Various types of algorithms are analyzed for reducing the cycle time of instructions, the area of multipliers [12] and the high-speed performance of multipliers. Many multipliers are designed for low power, reduced area, high speed, and better performance by researchers and chip designers.

FPGA works as a bridge between software and hardware. The FPGA is the IC, which is the combination of various configurable logic blocks that gives us freedom to program and reprogram. Its name, "field programmable," means it is adjustable and gives you the freedom to reprogram the device according to your needs. In this work, we have proposed a comparative analysis of different parallel multipliers [13]. These multipliers consume less voltage and provide high speed for signal processing. In which we discussed the array multiplier [14] and the types of array multipliers that are braun multipliers and Baugh-Wooley multipliers. Another type of parallel multiplier we have discussed about the tree multiplier is the Dadda multiplier, and the analysis is based on the different types of ultrascale FPGA devices using Xilinx ISE [13, 14].

These VLSI circuits consist of low-power and energy-efficient ICs used for designing high-performance and energy-efficient DSP systems, mobile phones, and various biomedical electronic systems.

2 Methodology

2.1 Types of Multiplier

In a general way, multipliers are classified into three types on the basis of the initialization of operands, which are serial, parallel, and serial-parallel multipliers (Fig. 1).

In serial multipliers, the operands are initialized or inputs are given serially, which requires less chip area. As chip area is smaller, it requires a smaller number of hardware, and as hardware is smaller, it takes less area, but operands are entered serially because of that speed, and performance is not good.

In digital signal processing applications such as convolution and filtering, which require high computational speed and high mathematical operations to perform all these things, parallel multipliers are proposed by the designers. Power consumption, chip area, and complexity are also higher, but they are useful in Reduced Instruction Set Computers (RISC). Parallel multipliers are further classified into array and tree

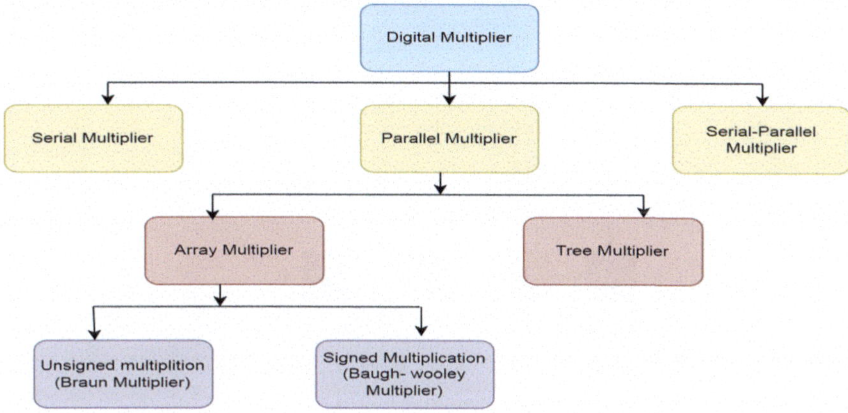

Fig. 1 Types of multipliers [13]

multipliers. Examples of array multipliers are booth and high-radius multipliers. Wallace multipliers and binary multipliers are examples of tree multipliers. Tree multipliers are used for performing complex computations.

Serial-parallel multipliers are useful for performing high-speed tasks, but at the same time, they consume less area. Operands entered serially in the multiplier get stored in parallel and have a fixed number of bits. This multiplier is used for performing high speeds, which require less chip area.

A brief introduction of multipliers is shown in the above block diagram with different types of subparts of parallel multipliers, which are going to be discussed in this paper.

Array Multiplier

The algorithm of the array multiplier is based on the basic multiplication. This multiplier is mainly used for the small multiplication [14]. Its simplicity and structural regularity makes it preferable for the VLSI implementation. This multiplier provides a high degree of uniformity and regularity. As its name suggests basically its an array of AND Gate, half adders, and full adders. In this multiplier, the partial product of the multiplicand and multiplier is generated by the array of AND gates [15]. Which provides reduced area with high speed of working (Figs. 2 and 3).

In general, the multiplication of two N bits binary numbers requires N^2 AND gates, N number of half adders and $N(N-2)$ full adders. It is the basic method which is used for the designing of array multipliers. Verilog code for half adder is given in this block.

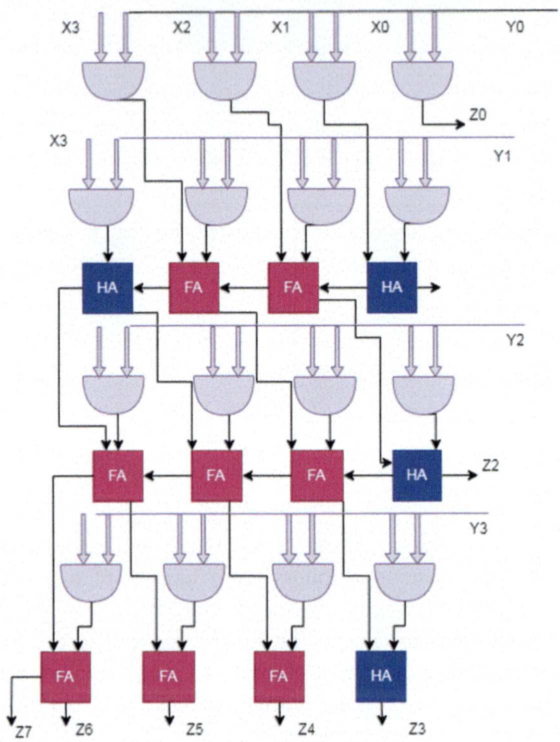

Fig. 2 Simplified block diagram of array multiplier [14]

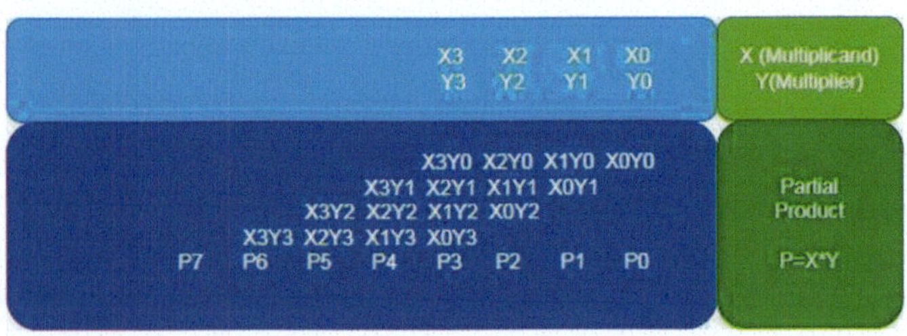

Fig. 3 Algorithm structure of array multiplier (simple multiplication) [13]

Braun Multiplier

This multiplier is proposed by Braun Edward Louis and after that many amendments are done by the researchers for improving the performance. This multiplier can only perform multiplication of two unsigned numbers [13]. Arrays of AND gates and full adders are used for the hardware part of this multiplier. If we have an nXn multiplier then $n(n-1)$ full adders and n^2 AND gates will be used for implementation of the

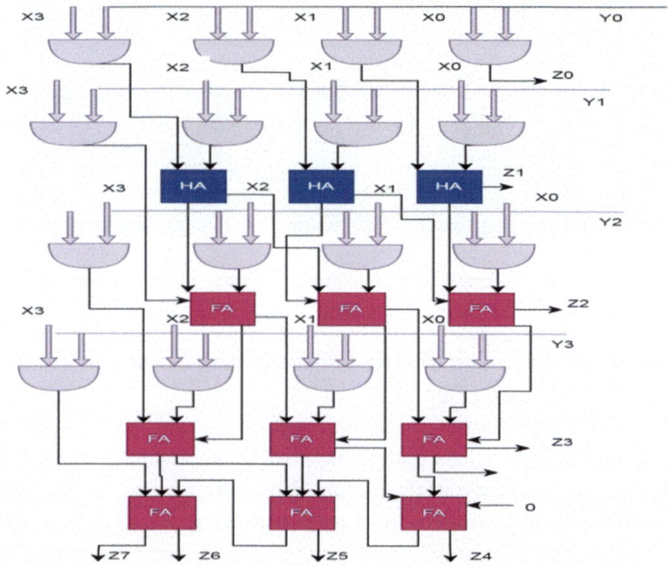

Fig. 4 Simplified block diagram of Braun multiplier [13]

Braun Multiplier. This multiplier is good in performing unsigned multiplication in the field of VLSI and DSP.

As the carry and sum is critical operation due to that delay of the multiplier depends on the full adder. The worst case time taken by the multiplier is

$$t_{\text{Braun}} = (n-1)t_{\text{carry-save}} + t_{\text{AND gate}} + (n-1)t_{\text{ripple-carry}}$$

In that $= t_{\text{carry-save}}$ time taken by carry save adder for generating carry and sum
$t_{\text{riple-carry}}=$ time taken by Ripple carry adder for generating carry out and sum
$t_{\text{AND gate}} =$ delay due to AND gate
$t_{\text{Braun}} =$ time taken by the Braun multiplier.

As shown in the figure the product of multiplicands is generated firstly from AND gates. This product is a partial product and added to the previously given partial product by the given rows of the full adder [16]. Carry is generated by the different adder it gets added diagonally. Ripple Carry Adder (RCA) is used for getting the final result or output. Verilog code for ripple carry adder is given in the block (Fig. 4).

Baugh-Wooley

Baugh-Wooley multiplier is designed for the multiplication of unsigned and signed binary numbers [16]. It is basically an enhanced version of Braun Multiplier output gets in the 2's complement number system [17]. The architecture is designed in such

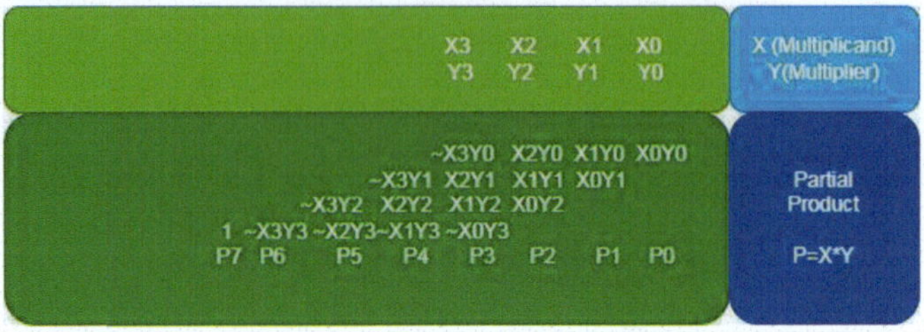

Fig. 5 Algorithm structure of Baugh-Wooley multiplier (signed multiplication) [13]

a way due to that negative sign transfer to the last step. Consistency and regularity of the array gets maximized (Fig. 5).

When the multiplier operates for signed multiplication then it is necessary that signed operands should be in 2's complement form. The table shows the algorithm which works for the generation of the output. Where eight vertical columns are there of partial products, these partial products are generated by the XOR gate. Terms with bar sign shows inverted output, which is generated using the NAND gates.

In the circuit of the multiplier each column's partial products are added by the full adders and half adders. When the regularity and locality of any silicon based circuit will increases, it will decrease the power consumption in the circuit because area and power consumption of a multiplier vary with the number of bit operands and layout strategies. Although it is the enhanced version of Braun Multiplier, it gives better performance in comparison to previous elaborated circuits (Fig. 6).

Dadda Multiplier (Tree Multiplier)

Luigi Dadda was the inventor of this binary multiplier which consists of many half adders and full adders. Multiplication process in this multiplier by forming the tree structure of partial products [18]. Height of the tree structure varies according to the number of products or we can say that it depends on the number of bits of multiplier and multiplicand. The maximum height sequence (dj) of the multiplier is controlled by this value of dj. Where the value of dj can be dj = 2, 3, 4, 6, 9, 13 etc. The value of dj depends on the size of the multiplier and multiplicand.

$$Dj < \min(N, m)$$
$$N = \text{size of bits of multiplier}$$
$$m = \text{size of bits of multiplicand}$$

In our model we have taken the size of multiplier and multiplicand is 4. The minimum of 4 is 3 so that the value of the height sequence is 3. Due to less number of hardware being used in this multiplier, less power is consumed by the multiplier,

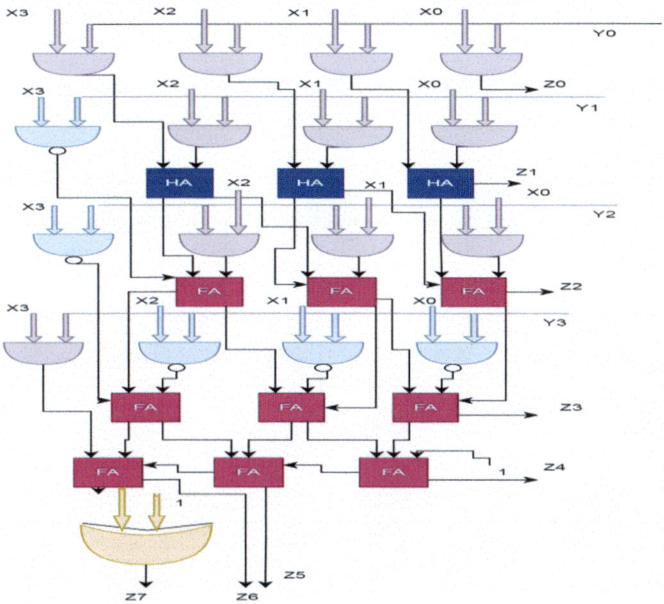

Fig. 6 Block diagram of Baugh-Wooley Multiplier [14]

but this multiplier is not suitable for the higher size of multiplicand and multiplier (Fig. 7).

2.2 *Ultra Scale FPGA Device*

The next-generation architecture is the Xilinx ultrascale architecture, which is known for providing ultra-high capacity to designers. These ultrascale FPGAs were first introduced by the Virtexultrascale FPGA and Kintexultrascale FPGA, which are made up of 20 nm node-span planners through 16 nm FinFET technology. This design provides improvements in communication, clocking, and where massive data flow is required because hundreds of gigabits are processed within 1 s. This next-generation architecture becomes the most demanding FPGA device application because of its massive data flow, high performance, and memory bandwidth [20].

Massive data flow in the terabits and teraflops is a requirement of the designers, which is the enhancement of the ultrascale device that is ultrascale+. The FPGA devices of this family that are used in this project are:

Kintex: This device is good for the highest signal processing, and with this, it provides cost-effectiveness, high performance, and low-cost packaging.

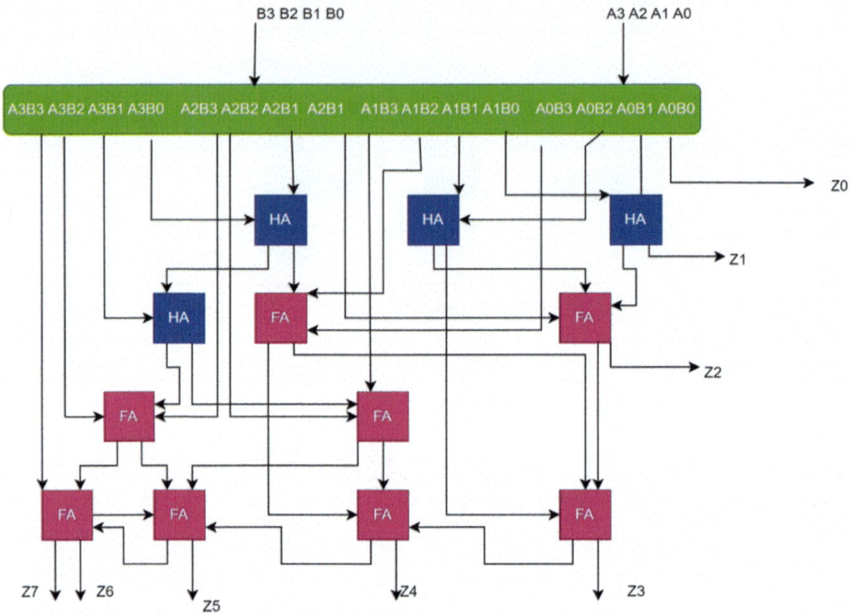

Fig. 7 Block diagram of Dadda multiplier [19]

Virtex: The device has good logic capacity, integration, and good performance at 20 nm. This family provides serial I/O bandwidth useful for the application of 400 G networking to a large range of ASIC prototyping.

Artix: Artix offers 16 Gb/s transceivers for the advanced protocols. This quality of device is useful for networking, video, and machine vision. This device provides enhanced packaging with an ultra-compact form factor and compute density.

Znyque: It is a Multi-Purpose System on Chip (MPSoc) device with 64-bit processor scalability. This is the multi-processing device that is used for the application of smarter systems, for example, smarter control, smarter vision, and a smarter network.

3 Result and Analysis

The result and analysis have been done by synthesizing the HDL code on Xilinx. A design summary table comes along, which gives information about the number of LUTs and IOBs. The number of LUTs is about the blocks used on the FPGA or the area used by the particular device. The number of IOBs is an input/output buffer; the number of pins is used on the FPGA device. On the basis of that design summary, a result table about the number of LUTs, the number of IOBs, and the delay of each multiplier is described below in the given table of comparative analysis of multipliers

Table 1 Comparative analysis of multipliers on ultra-scale FPGA

Multipliers	Parameters	Znyque	Virtex7	Airtex7	Kintex7
Array	No. of slice LUTs	20/17600	20/20400	20/63400	20/41000
	No. of full used LUT-FF pair	0/20	0/20	0/20	0/20
	No. of bonded IOBs	16/100	16/100	16/210	16/300
	Delay	2.705 ns	2.705 ns	3.395 ns	2.705 ns
Braun	No. of slice LUTs	20/17600	20/20400	20/63400	20/41000
	No. of full used LUT-FF pair	0/21	0/21	0/21	0/21
	No. of bonded IOBs	16/100	16/100	16/210	16/300
	Delay	2.582 ns	2.582 ns	3.309 ns	2.582 ns
Baugh-Wooley	No. of slice LUTs	20/17600	20/20400	20/63400	20/41000
	No. of full used LUT-FF pair	0/22	0/22	0/22	0/22
	No. of bonded IOBs	16/100	16/100	16/100	16/100
	Delay	2.569 ns	2.569 ns	3.29 ns	2.569 ns
Dadda	No. of slice LUTs	20/17600	20/20400	20/63400	20/41000
	No. of full used LUT-FF pair	0/22	0/22	0/22	0/22
	No. of bonded IOBs	16/100	16/100	16/210	16/300
	Delay	2.658 ns	2.658 ns	3.325 ns	2.658 ns

on ultra-scale FPGA devices. In which Artix 7 has more delay in comparison to other FPGA devices and Baugh-Wooley multiplier takes less delay in comparison to other parallel multipliers (Table 1).

4 Conclusion

In this paper, a comparative analysis between different types of parallel multipliers using different ultra-scale FPGA devices was conducted. Implementation of multipliers has been done by adders like the full adder, half adder, and ripple carry adder. All the multipliers described have their own properties and disadvantages, depending on their execution time. The analysis of delay can be performed as both signed and unsigned multiplication. It is observed that the Baugh-Wooley multiplier consumes less delay than all the other multipliers. In the future, designing of different Filters like FIR filter and IIR filter can be done using these multipliers.

References

1. Semiconductor Industry Association (1997) The national technology roadmap for semiconductors: technology needs. Sematech
2. Moore M (2020) International roadmap for devices and systems. Accessed January 2020
3. Colinge J-P (2014) Multigate transistors: pushing Moore's law to the limit. In: 2014 international conference on simulation of semiconductor processes and devices (SISPAD). IEEE
4. Jaya E, Rao K (2016) Power, area and delay comparison of different multipliers. Int J Sci Eng Technol Res 5(6):2093–2100
5. Kumar VS (2015) Analysis of energy efficient PTL based full adders using different nanometer technologies. In: 2015 2nd international conference on electronics and communication systems (ICECS). IEEE, pp 310–315
6. Dubey A, Akashe S, Dubey S (2013) A novel high-performance CMOS 1 bit full-adder cell. In: 2013 7th international conference on intelligent systems and control (ISCO). IEEE
7. Pucknell DA, Eshraghian K (1994) Basic VLSI design. Prentice-Hall, Inc.
8. Badawi A et al (2016) FPGA realization and performance evaluation of fixed-width modified Baugh-Wooley multiplier. In: 2015 third international conference on technological advances in electrical, electronics and computer engineering (TAEECE). IEEE
9. Hassan H, Ibrahim S, Kim HyungWon (2018) Power-gating sub-threshold source-coupled logic (PG-STSCL) circuits for ultra-low-power applications. Microelectron J 74:127–140
10. Johannah JJ, Korah R, Kalavathy M (2017) Standby and dynamic power minimization using enhanced hybrid power gating structure for deep-submicron CMOS VLSI. Microelectron J 62:137–145
11. Wang N et al (2018) Power-gating-aware scheduling with effective hardware resources optimization. Integr 61:167–177
12. Kavand N, Darjani A, Rai S, Kumar A (2023) Design of energy-efficient RFET-based exact and approximate 4:2 compressors and multipliers. IEEE Trans Circuits Syst II Express Briefs 70:3644
13. Yeo, KS, Roy K (2009) Low voltage, low power VLSI subsystems. McGraw Hill Edition, India
14. Shanmuganathan R, Brindhadevi K (2019) Comparative analysis of various types of multipliers for effective low power. Microelectron Eng 214:28–37
15. Biradar VB et al (2017) Design and performance analysis of modified unsigned Braun and signed Baugh-Wooley multiplier. In: 2017 international conference on electrical, electronics, communication, computer, and optimization techniques (ICEECCOT). IEEE
16. Kishore P, Sirimalla R, Sushma KS, Reddy RS (2023) Implementation of Braun and Baugh-Wooley multipliers using QCA. In: 2nd international conference for innovation in technology (INOCON). Bangalore, India, pp 1–4
17. https://docs.xilinx.com/v/u/en-US/wp435-Xilinx-UltraScale
18. Afzali-Kusha H, Vaeztourshizi M, Kamal M, Pedram M (2020) Design exploration of energy-efficient accuracy-configurable Dadda multipliers with improved lifetime based on voltage overscaling. IEEE Trans Very Large Scale Integr (VLSI) Syst 28(5):1207–1220
19. Padmanabhan KK, Seerengasamy U, Ponraj AS (2022) High-speed grouping and decomposition multiplier for binary multiplication. Electronics 11(24):4202
20. Khosla R, Singh B (2013) Implementation of efficient image processing algorithm on FPGA. In: IEEE international conference on machine intelligence research and advancement (ICMIRA), SMVDU, Jammu, pp 335–339

Design and Evaluation of 14NM N-MOSFET with HFO2/WSI2 for Substrate Noise Analysis Using Physical Simulation Models

Sanjay Sharma, R. P. Yadav, and Vijay Janyani

Abstract Silicon technologies advance at a quicker rate every year as the size of semiconductor devices reduces in response to the growing demand for high performance, such as higher operating frequency or faster-operating speed devices based on FETs and other transistors. There is a minimum of a single common substrate interface for every gate in a circuit that is digital. As a result, there will be a huge number of substrate interactions in the digital design. Consequently, the digital circuit achieves an extremely low impedance to the substrate surface throughout its area. A low impedance can be produced especially when the substrate comes in contact from the analog circuit to the ground, will result in substrate noise within the analog region. In order to solve this problem and prevent low impedance performance, the analog circuit needs to have a high enough PSRR. For PSSR, the capacitive coupling-based shielding technique is used. In this designed model, a 14 nm N-MOSFET is designed on SILVACO ATHENA with HfO2/WSi2 dielectric medium for evaluating the substrate noise in this designed model. Using the simulator, it's possible to manually inject noise models such as Drain Current Noise, Poisson's Equation (shot noise) and Energy Balance etc. Subsequently, the noise is reduced through the substrate noise coupling technique. Shielding is the most popular technique for reducing noise in capacitive coupling. For evaluating the performance of the model, the I-V characteristics, transfer characteristics and substrate noise evaluation are considered. The attained I-V characteristics, On/Off ratio and minimum noise figure values of the proposed model are 0.2 mA for 1.2 (Vd), 10-4-10-3 and 0–1.5. Thus, an excellent output of noise reduction is done by using capacitive coupling.

S. Sharma (✉) · R. P. Yadav · V. Janyani
Department of ECE, Malaviya National Institute of Technology Jaipur (MNIT), Jaipur 302017, Rajasthan, India
e-mail: sanjay77mnit@gmail.com; lncs@springer.com

R. P. Yadav
e-mail: lncs@springer.com

V. Janyani
e-mail: lncs@springer.com

D.-I. Stroe et al. (eds.), *Emerging Electronics and Automation*, Lecture Notes in Electrical Engineering 1202, https://doi.org/10.1007/978-981-97-3090-2_10

Keywords N-MOSFET · Capacitive coupling · Substrate noise · SILVACO ATHENA · I–V characteristics

1 Introduction

Cost-cutting is the primary driver of the consumer electronics sector, and this is made possible by the ongoing transistor size' reduction. It is possible to combine an increasing amount of functionality on a single die. So, single-chip radio implementations, also known as Systems-on-Chip (SoC), become more and more common [1]. This type of SoC has multiple sensors, cameras, GPS, touchscreen technology, wireless internet access, and phone capabilities. On the same die, these devices combine digital and analog/RF capability. Unfortunately, technological issues also arise about by the merging of analog and digital circuitry, particularly in deep submicron CMOS technology [2]. In addition to being expensive and challenging to design, analog circuits are susceptible to noise coupling of any kind since they take advantage of the low-level physics of the fabrication process. One of the most crucial parts of any digital system, such as a mobile device or microprocessor, is a digital electronic circuit. Transistors, including MOS transistors, make up the majority of these digital systems [3]. This is the fundamental building block of every digital system. Studying the optimal MOS transistor behavior is a crucial task in VLSI circuit design because the performance of the digital system depends heavily on the choice of transistors [4]. TCAD software tools can produce valuable findings by modelling the device structure under various steady-state and transient situations.

SILVACO ATHENA is a commercial simulator process used to find out the electrical characteristics and electro-thermal parameters such as gate leakage current, electric field, output current, lattice temperature rise, built-in potential, localized heating, etc. [5]. Dissimilarity of such parameters commonly leads to the device's structural failure. A 14 nm N-MOSFET is designed with the SILVACO ATHENA simulator for evaluation and reduction of substrate noise. At the substrate [6], transitory signals are the source of substrate noise. This occurs because the substrate impedance is not zero. The substrate remains uniform in an ideal state, flawlessly conducting a medium that works as a voltage source. In a restricted parasitic, interconnect resistance and capacitance, substrate resistivity, bond wire conductance, and the printed circuit board generated finite source impedance [7]. Furthermore, the power supply impedance changes at every position on the substrate [8], which results fluctuation in substrate's voltage.

In the primary transistor production process, substrate noise coupling was utilised for power supply and switching noise. The P-MOS and N-MOS transistors can both switch to the ON position throughout a duration of time when the Digital Inverter fluctuates among the maximum and minimum or vice versa [9, 10]. At this time, the flow of current will be moving from the power supply to the ground source. Noise is injected into the substrate through the P + substrate [11] aperture by the instantaneous current flowing through the ground/power lines and package inductance. By utilizing capacitive coupling based on the shielding technique, the injected noise also referred to as noise from the supply can be suppressed. Major contributions of this work are:

- Design and Evaluation of 14 nm N-MOSFET for reducing the substrate noise analysis using physical simulation models.
- Modelling of substrate for N-MOSFET using silicone by doping it with phosphorous.
- Injection of noise into the substrate with various models for evaluation.
- Reduction of noise from the substrate using the capacitive coupling technique.

The next section of the paper is structured as follows, while Sect. 2 discusses specific research papers pertaining to the methods currently in use for object detection. A brief summary of the suggested methodology is provided in Sect. 3. The outcomes and performance measures of the suggested framework are described in Sect. 4. Section 5 brings the entire research project with a conclusion.

2 Literature Review

To minimize the noise in the substrate, many researchers have constructed MOSFET devices using a variety of approaches. Several of the designed models are discussed here.

Substrate Current Assessment for lightly and heavily doped MOSFETs was created by Sharma et al. [12]. The manufacture of PMOS and NMOS devices is simulated using the ATHENA process simulator. The simulated devices incorporate the hot carrier effects. The MOS devices are built on weakly and extensively doped substrates, and the substrate current is measured and compared with the aid of the ATLAS device simulator. Substrates that are lightly doped have better substrate currents than those that are heavily doped. Moreover, compared to heavily doped substrates, the drain current is better in substrates that have been mildly doped. Silvaco TCAD Tool is used for simulation and virtual fabrication. The ATLAS device is used for device characterization, while the ATHENA process simulator is used for virtual fabrication.

A doped pocket region was used by Saha and Goswami [13] to improve MOSFET performance. The device with a 20 nm channel length is subjected to a simulation-based study. The identical device is made with a channel length of 30–40 nm and a typical MOSFET of 30–40 nm for study and comparison. The device's characteristics are identical to those of a normal MOSFET, according to the simulation results. The modified structure outperforms the conventional one when the drain current of the modified MOSFET is contrasted with that of a conventional MOSFET.

Using SILVACO ATLAS, Bhuyan et al. [14] created a 100 nm n-MOSFET. After describing the design processes for an n-MOSFET, the electrical characterization of an n-MOSFET is simulated at 100 nm using the SILVACO ATLAS software, a tool for process and device simulation. All simulations are run using the ATLAS software, and all graphs are plotted using TONYPLOT in SILVACO. The MOS device is essentially built-in to SILVACO using ATHENA. The mesh structure of the n-MOSFET device, as well as its transfer and output properties, doping and carrier concentration plot, and other information, are then determined from the simulation results.

For 28 nm stacked high-k n-MOSFETs, Chen et al. [15] conducted hot carrier stress sensing bulk current. This largely focuses on the bulk current degradation degree for 28 nm stacked high-k (HK) n-channel MOSFETs, detected and stressed by the channel-hot-carrier and the drain-avalanche-hot-carrier test, and then applies a lifetime model to estimate the lifetime of the tested devices. The results demonstrate that as IB approaches its maximum, the ratio of VGS/VDS values in the investigated devices gradually shifts from long channel to short channel. The possibility of ratiocination, wherein the short-channel devices offer bigger Ids, is brought on by the On-current (IDS). As a result, the drain junction is affected and the surface channel next to the drain site produces extra hole carriers. The slope parameter m of the lifetime model is ultimately extracted by the device lifespan using Berkley's model.

A lumped-parameter equivalent circuit model for a commercial 90-nm MOSFET has been presented by Wu et al. [16]. A more effective substrate coupling network has been developed for an RF CMOS model. The proposed model focuses on characterizing the nonlinear phase of S12 when a transistor is at zero bias. The method for extracting the matching parameters for the model is also created. A set of transistors made using a commercial 90-nm CMOS process are tested under multi-bias settings to confirm this theory.

3 Proposed Methodology

Noise resistance is becoming a key challenge in several production domains while developing integrated circuits (IC) and very large-scale integrated circuits (VLSI). The chip is often faster and more sophisticated, which makes the circuit more sensitive to both internal and external noise. The decreased transmittance results inclusion of a large number of transistors on the device. The information technology revolution is expanding quickly as a result of the rapid scaling of MOSFETs, which increases CPU performance. In this designed model, a 14 nm N-MOSFET is created with the HfO2/WSi2 for analyzing the substrate noise in this designed model. The developed model is manually given the noises for evaluation. The substrate noise coupling technique is used to decrease these injected noises.

Figure 1 illustrates the design of 14 nm HfO2/WSi2 N-MOSFET. In this designed model, the silicon p-substrate is doped with phosphorus for a concentration of 3.75 × 1012 cm-3. The HfO2 (Hafnium Dioxide)/WSi2 (Tungsten Silicide) is used as the High k dielectric medium and mid-metal electrode for the N-MOSFET. The silicon nitride is used as a gate metal, which is implanted with the indium at $3 \times 1012\ \text{cm}^{-3}$,

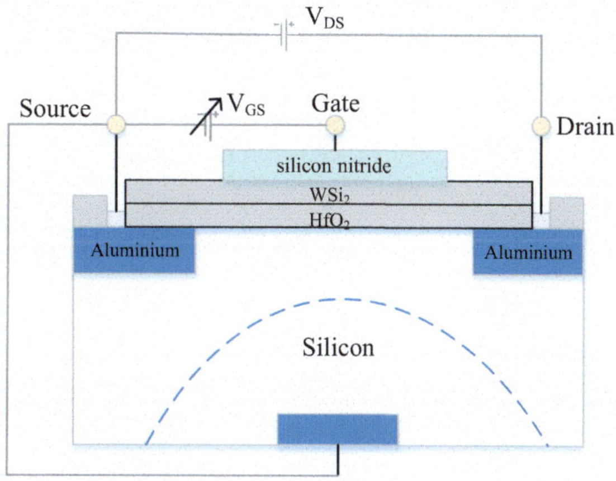

Fig. 1 Design of 14 nm HfO2/WSi2 N-MOSFET

and the aluminium is implanted with antimony at 1.0 × 1014 cm^{-3} for source/drain metal.

3.1 *Substrate Modelling*

Substrates are modelled within the circuits of the same chip to relate it with the substrate device. Substrate modelling is done until all the different layers are included with the different levels of doping or the geometrics of different doping levels [17]. So, the substrate is sliced into a number of smaller elements, and each element is required to have a constant level of doping determined by using Maxwell's equation. Each and every element will have a constant permittivity and resistivity. The model of an element is achieved by solving the equation.

$$\varepsilon \frac{\partial}{\partial t}(\nabla \cdot E) + \frac{1}{p}\nabla \cdot E = 0 \tag{1}$$

In this Eq. (1), E, p and ε are denoted as electric field, resistivity and permittivity of the element. According to Gauss law, an electrical field's divergence at a given location equals a constant value. As a result, the cube provides the separation of node I, which can be expressed as

$$\nabla \cdot E = l \tag{2}$$

Here, ∇ is the divergence operator and $\nabla \cdot E$ is formed by the cube, which is rewritten as

$$\int_V \nabla \cdot E dV = \int_V k dV = 8d^3 l \tag{3}$$

For Eq. (3), the divergence theorem is applied and expressed in Eq. (4)

$$\int_V \nabla \cdot E dv = \int_S E ds \tag{4}$$

Substituting Eq. (3) in (4)

$$\frac{1}{8d^3} \int_s E dS = l \tag{5}$$

Equation (2) is substituted in (5)

$$\nabla \cdot E = \frac{1}{8d^3} \int_s E dS \tag{6}$$

Substituting $EdS = \sum_{j=1}^{6} E_{ij} 4d^2$ and $E_{ij} = \frac{V_i - V_j}{d/2}$ in Eq. (6).

$$\nabla \cdot E = \frac{1}{8d^3} \sum_{j=1}^{6} \frac{V_i - V_j}{d/2} 4d^2 = \sum_{j=1}^{6} \frac{V_i - V_j}{d^2} \tag{7}$$

$$\sum_{j=1}^{6} \left[\frac{(V_i - V_j)}{R} + C\left(\frac{dV_i}{dt} - \frac{dV_j}{dt} \right) \right] = 0 \tag{8}$$

The resulting substrate element model is shown in Fig. 2.

To achieve related outcomes from the system, the substrate is divided into several elements to obtain a capacitor and resistor mesh. This method is not suitable for hand calculation because it consists of a large number of nodes. So, a simulator is used for the calculations. The connection between various substrate regions is examined using the circuit simulator.

(I) Substrate noise injection

For designing the 14 nm MOSFET with the substrate noise evaluation, some substrate noises are injected into the designed model to represent the noise at the substrate. The substrate noise that is injected into the MOSFET are Lucky Electron Hot carrier injection model, Drain Current Noise, Energy Balance Transport Model, Poisson's Equation (shot noise), Fermi- Dirac, Energy Balance Equations and Boltzmann statistics.

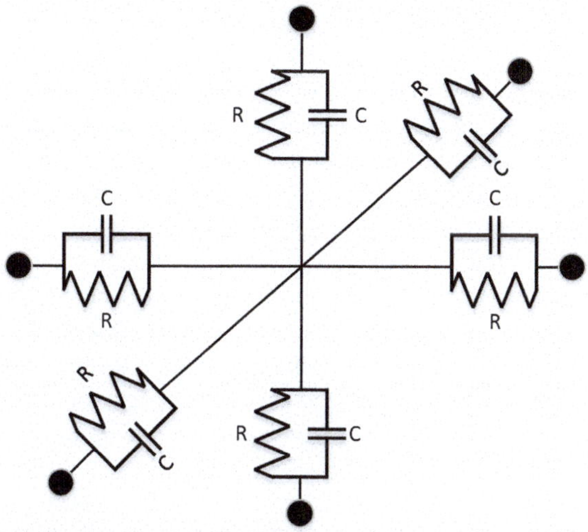

Fig. 2 Model of substrate Element

3.2 *Substrate Noise Reduction*

From the above seven models, the noise are manually injected into the substrate for evaluating the noise rejection by capacitive coupling of the p-substrate. The transistor channel is surrounded by an evenly doped NMOS transistor substrate. The substrate is directly coupled to the NMOS transistor's biased contacts. In certain situations, it is conceivable to shift the substrate noise to lower or higher frequencies [18]. Alternatively, the frequency spectrum of the substrate noise can be modified during the noise is being induced. However, the frequency components of the substrate noise are outside the band of signal, and the noise effects may still be presented in the band of signal due to the intermodulation. Figure 3 illustrates the Capacitive coupling for the p-substrate.

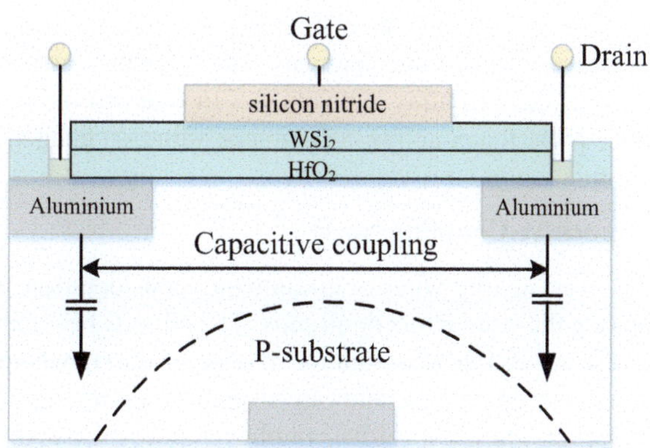

Fig. 3 Capacitive coupling for p-substrate

A common substrate contact is present in every gate of a digital circuit thus, the digital design would therefore have a large number of substrate connections. Consequently, there will be very minimal resistance between the substrate surface and the digital circuit within its boundaries. Additionally, there will be fluctuations in voltage at the substrate area of the digital circuit. In digital ICs, the injection of noise mechanism typically predominates (Integrated circuits). The substrate noise at analog area are characterized with the analog ground due to low impedance. To solve this problem, an analogue circuit with a high enough PSRR (Power Supply Rejection Ratio) is required to prevent poor performance. Capacitive coupling is utilized for PSSR. Shielding is the most used technique for reducing noise in capacitive coupling. Shielding is connected to the ground at a location apart from the connection to the signal source and is placed between capacitively linked circuits. The noise currents travel directly into the earth via the ground connection.

$$\mathrm{C} = \sqrt{\frac{q\varepsilon}{2(\varnothing_0 + V_{cs})}\left(\frac{N_c * N_s}{N_c + N_s}\right)} \tag{9}$$

where, VCS the collector-to-substrate bias voltage, $\varnothing_0$ is the built-in junction potential, NC and NS are the collectors and the substrate doping levels, q is the electron charge, and H is the substrate dielectric permittivity. Thus, the capacitive coupling substrate reduces the noise injected by the models, and this method improves the performance of the N-MOSFET. A detailed description of capacity coupling noise reduction is given as follows.

3.3 Capacitive Coupling Noise Reduction

Voltage changes in a circuit have an impact on coupling capacitance and noise from the capacitive coupling. Dependent on the distance between two circuits is the coupling capacitance. Shielding is the method used most frequently to reduce capacitive coupling noise. Shielding equipment known as Faraday cages is employed to reduce coupled interferences. The coupling capacitance value rises as a result of the existence of stray capacitance. In order to reduce capacitive coupling noise in the circuit, the coupling capacitance value is changed as a result of the reduction or elimination of stray capacitance.

Capacitive Coupling Noise

Take two nearby wires or circuits into consideration. There is a built-in capacitance among the conductors or circuits. The circuits become noisy as a result of the capacitance's creation of a conductive conduit for signals to mingle between lines. These capacitive coupling sounds are also known as near-field interferences or electric field coupling. These capacitive coupling sounds are also known as electric field coupling or near-field interferences. The coupling capacitance value and the variations in the circuit voltage have an impact on the capacitive coupling noise. The separation of the two circuits affects the coupling capacitance. In both cases, the coupling capacitance increases as the circuits are coupled closer together. Mutual capacitance is another name for coupling capacitance.

Capacitive Coupling Noise Reduction Methods

Lowering coupling capacity is the most effective method for reducing capacitive coupling noise. Increasing the distance between two circuits is a good approach since the coupling capacity of two circuits is inversely proportional to the distance between them. Nevertheless, this is a non-workable approach due to circuit shrinkage. Take a look at the options below.

Shielding

The most popular technique for lowering capacitive coupling noise is shielding. Shielding tools called Faraday cages are employed to reduce coupled interferences. The shielding is positioned somewhere other than where the connection to the signal source is made, between the capacitively coupled circuits, and is connected to the ground. The ground connection is used to direct the noise currents directly into the ground. It is vital to ground the shield when deploying a Faraday cage. Leaving it open-circuited or floating causes it to generate more capacitive coupling noise in the system.

4 Simulation Results

An N-channel MOSFET device with a 14 nm gate length, 0.01 mm gate oxide thickness and 0.01 mm surface net doping is employed to execute the simulation. In order to transfer current through the buried layer towards the drain layer, the buried layer is attached to the sinker. The ATHENA simulator is used for MOSFET simulation, whereas the SILVACO process simulator is used for device fabrication. For the simulator in intricate patterns that have been realized by industrial procedures, ATHENA is highly anticipated. ATHENA combines all of the fabrication processes into a single structure. In this designed model, the silicon p-substrate is doped with phosphorus for the concentration of 3.75×10^{12} cm^{-3}. The HfO_2 (Hafnium Dioxide)/ WSi_2 (Tungsten Silicide) is used as the High k dielectric medium and mid-metal electrode for the N-MOSFET. The silicon nitride is used as a gate metal, which is implanted with the indium at 3×10^{12} cm^{-3}, and the Aluminum is implanted with antimony at 1.0×10^{14} cm^{-3} for source/drain metal.

Figure 4 illustrates the ATHENA structure of 14 nm N-MOSFET. The materials used in the ATHENA structure are silicone substrate that is doped with phosphorus, polysilicon, SiO_2 and Aluminium. These materials are doped with the other materials, which are illustrated in Table 1.

Table 1 illustrates the materials that are used in N-MOSFET, consisting of initial substrate silicon doped with phosphorous at 3.75×1012 cm^{-3}, gate oxide thickness

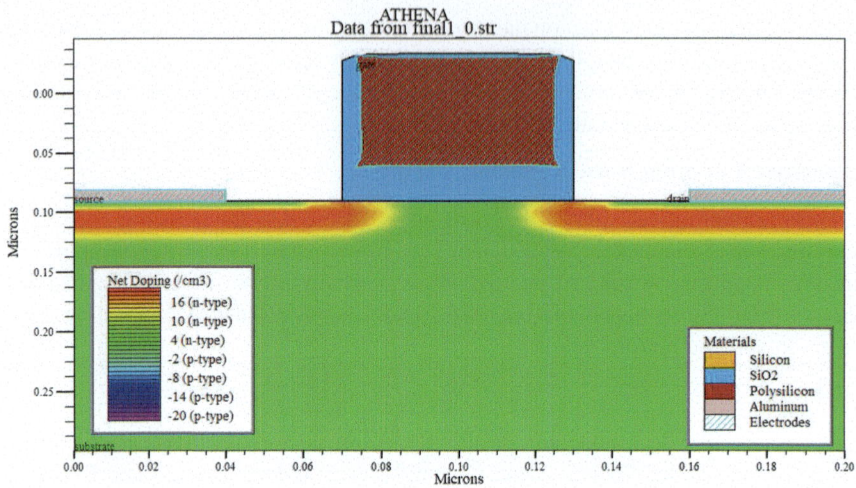

Fig. 4 ATHENA structure of 14 nm N-Channel MOSFET

Table 1 Material properties of N-MOSFET

Simulation process	Lightly doped NMOS
Substrate	Phosphorous = 3.75 × 1012 cm^{-3}, Orientation = <100>
Gate oxide thickness	0.01 μm
Poly deposition	0.01 μm
V_t implant	1.75 × 1011 cm-3 Boron difluoride 5 keV implant energy
LDD implantation	3 × 1012 cm-3 Indium
High-K/Metal gate deposition	0.002 um HfO2, 0.05 um WSi2
Source drain implant	1.0 × 1014 cm^{-3} antimony 12 keV implant energy
Final rapid thermal	800° C/s
Metal 1	0.04 um aluminium
Metal 2	0.12 um aluminium
Sidewall spacer deposition	0.008 um Si3N4

for 0.01 μm, poly deposition for 0.01 μm, Vt implant doped with boron difluoride at 1.75 × 1011 cm^{-3}, LDD implant doped with indium at 3 × 1012 cm^{-3}, source/drain implant doped with Antinomy at 1.0 × 1014 cm-3 and High-k/metal gate deposition with 0.002 um HfO2 and 0.05 um WSi2. These parameters are used to design the N-MOSFET.

The simulated outcome of the drain current and drain bias voltage is illustrated at Fig. 5. The breakdown voltage of the created model is shown by this simulated graph. The drain-source voltage (V_{DS}), also known as the breakdown voltage (B_{VDS}), of a FET transistor is the voltage at which the transistor enters the breakdown zone. In this region, the transistor experiences excessive voltage across its drain-source

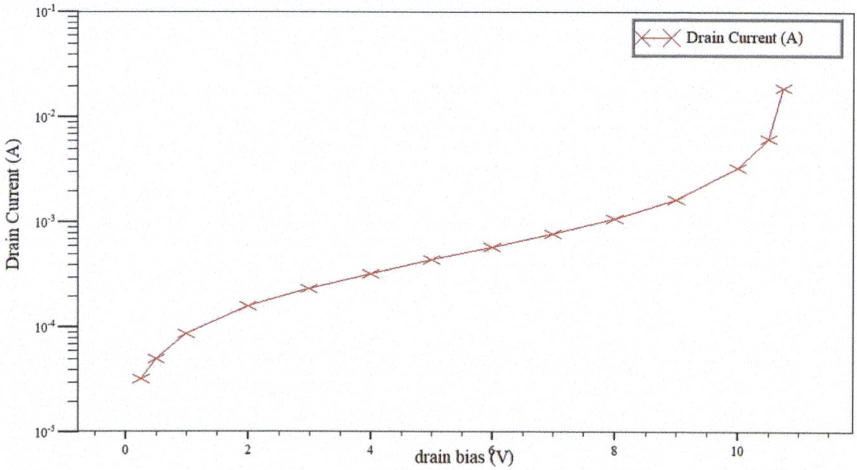

Fig. 5 Simulation of drain bias (V) versus drain current (A) for N-MOSFET

terminal, which damages the terminal and significantly raises the drain current ID. According to Fig. 5, the planned model's B_{VDS} breakdown voltage is 10 V.

The V–I characteristics of N-MOSFETs are shown in Fig. 6. The cut-off, ohmic, and saturation areas are the N-three MOSFET's operational regions. In the cut-off zone, the MOSFET will be OFF because there won't be any current passing through it. When they are needed to act as electronic switches, MOSFETs are chosen because they behave in this region like an open switch. The zone known as the ohmic or linear region is one in which the current I_{DS} rises as the value of V_{DS} rises. The planned model's ohmic range is 0 to 0.2 V. If MOSFETs are made to function in this range, they can be employed as amplifiers. The MOSFETs' IDS remains constant in the saturation area even though their VDS increases once it surpasses the pinch-off voltage VP. The designed model's saturation range is between 0.2 and 1.2 V. When this happens, the device acts like a closed switch that allows information to flow through an IDS value that has been saturated with it. As a result, this operational area is chosen whenever MOSFETs are required to perform switching operations.

Figure 7 illustrates the result of gate bias versus drain current, and the drain current will linearly increase when the gate bias voltage increases with a constant drain voltage. The Frist linear line varies gate voltage from 0.2 to 1.2v, which increases the drain current from 0.000421–0.000424 with $V_D = 0$. Similarly, for $V_D = 2$ and 4, the drain current increases from 0.000435–0.000437 to 0.000447–0.000448 for gate voltage 0.2–1.2 v. Figure 8 illustrates the gate bias versus substrate current simulation result with noise. The leakage of channel current is assumed as substrate current, and for that leakage current, the injection of noise modules is done. For the different gate voltages, the evaluation of substrate noise has been done three times. The first, second and third substrate current due to injection of noise varies from 0.160–0.165, 0.162–0.170 and 0.169–0.178 for 0–5 v.

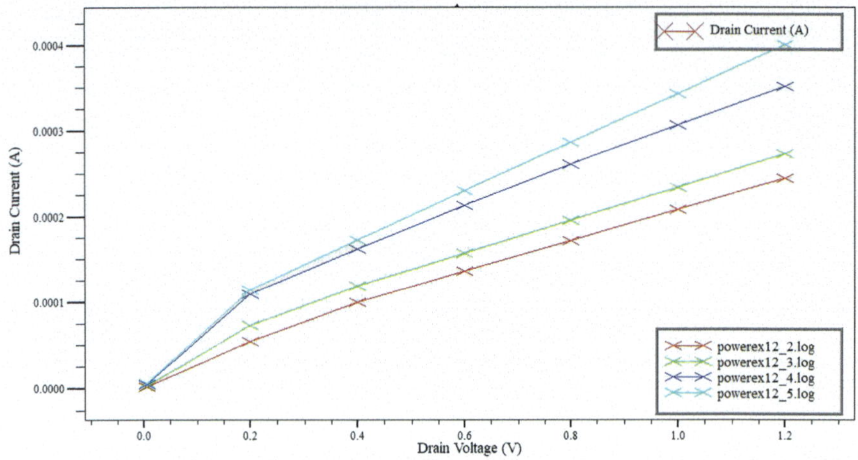

Fig. 6 Simulated result of V–I characteristic for N-MOSFET

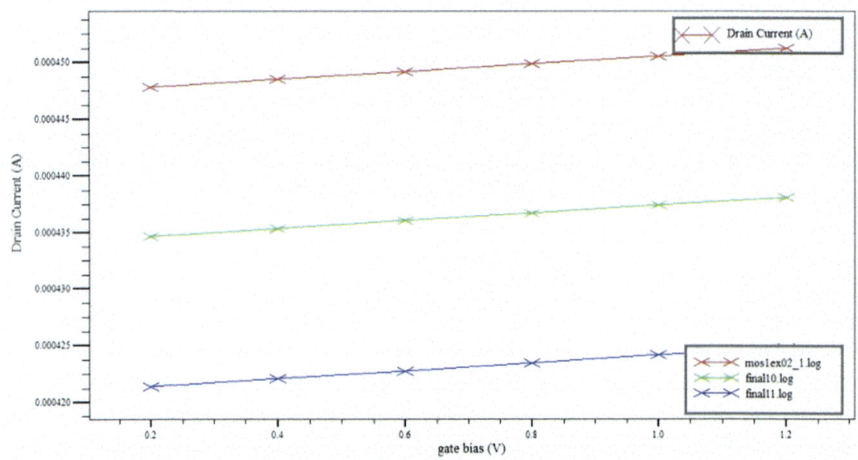

Fig. 7 Simulated result for gate bias (V) versus drain current (A)

Figure 9 and 10 illustrates the simulated result for injected and reduced noise figure. The simulation of this graph is taken by injecting noise into the designed model with 7 different noise-injecting models such as Drain Current Noise, Poisson's Equation, Energy Balance Transport Model, Energy Balance Equations, Fermi-Dirac, Boltzmann statistics, Lucky Electron Hot carrier injection model. The noise-injected designed model is treated with capacitive coupling techniques to reduce noise. From extremely low (100 MHz) to high (1 GHz) coupling levels, there will be a rise. With increased capacitance to the substrate, lines exhibit higher coupling values.

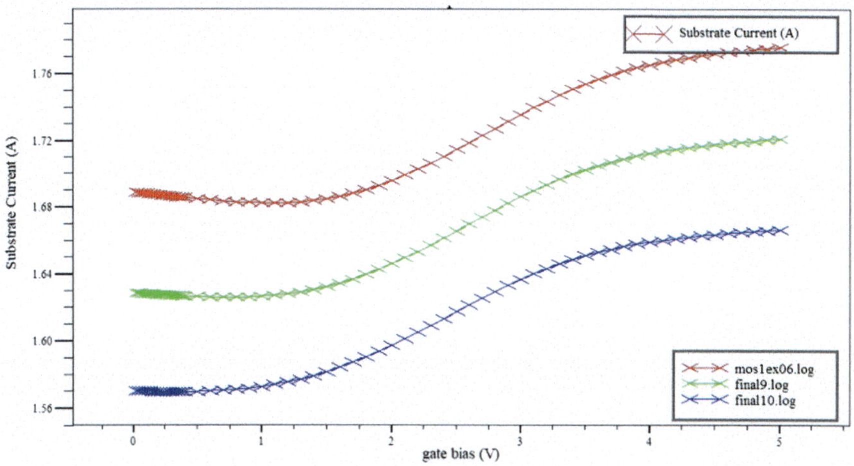

Fig. 8 Simulated result for gate bias (V) versus substrate noise current (A)

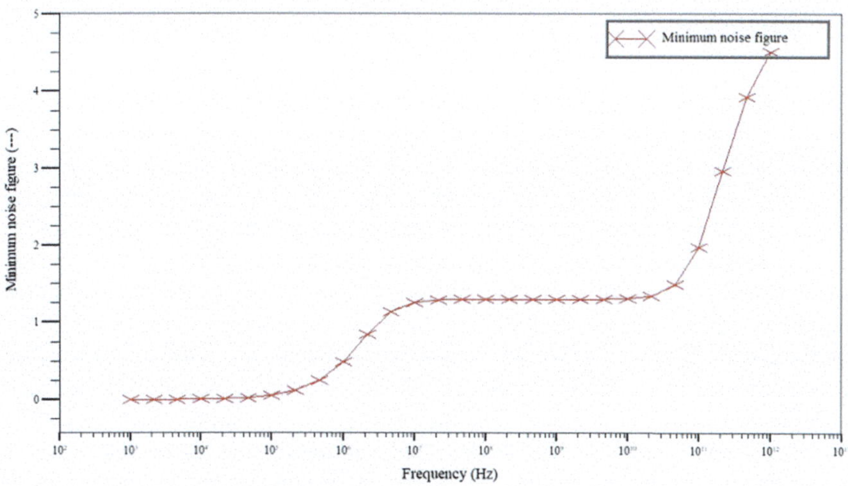

Fig. 9 Simulated result for injected noise figure

The overall result explains that the frequencies above GHz interconnect to couple a significant fraction of voltage to the substrate.

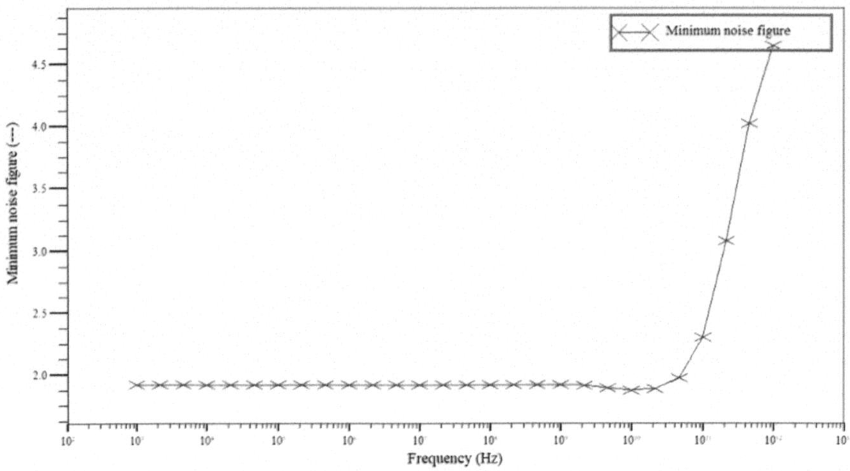

Fig. 10 Simulated result for reduced noise figure

5 Conclusion

In a VLSI circuit, each gate has a minimum of one common substrate contact. Therefore, there would be a lot of substrate connections in the digital design. As a result, the digital circuit's impedance to the substrate surface inside its boundaries is extremely low. Substrate noise can be produced in the analog area if the substrate is in contact from the analog circuit to the analog ground due to low impedance. To solve this problem and prevent poor performance, the analog circuit must have a high enough PSRR (Power Supply Rejection Ratio). Thus, in this model, the capacitive coupling is used to enhance the PSRR. In this designed model, a 14 nm N-MOSFET is designed on SILVACO ATHENA by inducing noise into the substrate to evaluate the performance of the designed model. These noises are manually injected into the designed model and reduced using the substrate noise coupling technique. The attained I-V characteristics, On/Off ratio and minimum noise figure values of the proposed model are 0.2 mA for 1.2 (V_d), 10^{-4}–10^{-3} and 0–1.5. Thus, an excellent output of noise reduction is done by using capacitive coupling in this designed 14 nm N-MOSFET.

References

1. Dasgupta S, Mondal C, Biswas A (2019) Role of grooving angle of 14-nm-InAs channel quantum well MOSFETs in improving analogue/RF and linearity performance. IET Circ Dev Syst 13(8):1292–1298
2. Honda T, Yano H (2021) Simple method to estimate the shallow interface trap density near the conduction band edge of MOSFETs using hall effect measurements. Jpn J Appl Phys 60(1):016505
3. Nguyen-Gia Q, Kang M, Jeon J, Shin H (2019) Characteristic length of macaroni channel MOSFET. IEEE Electron Device Lett 40(11):1720–1723
4. Mendiratta N, Tripathi SL (2022) 18nm n-channel and p-channel dopingless asymmetrical Junctionless DG-MOSFET: low power CMOS based digital and memory applications. SILICON 14(11):6435–6446
5. Araga Y, Kikuchi K, Aoyagi M (2018) Evaluation of substrate noise suppression method to mitigate crosstalk among trough-silicon vias. Jpn J Appl Phys 57(4S):04FC07
6. Ashraf NS (2022) Low substrate temperature modeling outlook of scaled N-MOSFET. Springer Nature
7. Carangelo G, Reggiani S, Consentino G, Crupi F, Meneghesso G (2021) TCAD modeling of bias temperature instabilities in SiC MOSFETs. Solid-State Electron 185:108067
8. Sudarsanan A, Nayak K (2021) TCAD-based investigation of statistical variability immunity in U-Channel FDSOI n-MOSFET for sub-7-nm technology. IEEE Trans Electron Devices 68(6):2611–2617
9. Li C, Chye BC, Yang Y, Yao E, Fujishima M (2019) MOSFET small-signal model considering hot-carrier effect for millimeter-wave frequencies. J Infrared Millimeter Terahertz Waves 40:419–428
10. Mahapatra S, Sharma U (2020) A review of hot carrier degradation in n-channel MOSFETs—Part I: Physical mechanism. IEEE Trans Electron Devices 67(7):2660–2671
11. Kang SC, Lim D, Kang SJ, Lee SK, Choi C, Lee DS, Lee BH (2019) Hot-carrier degradation estimation of a silicon-on-insulator tunneling FET using ambipolar characteristics. IEEE Electron Device Lett 40(11):1716–1719
12. Sharma S, Yadav RP, Janyani V (2016) Substrate current evaluation for lightly and heavily doped MOSFETs at 45 nm process using physical models. Bull Electr Eng Inform 5(1):120–125
13. Saha P, Goswami B (2019) Implementation of a doped pocket region in order to enhance the device performance of MOSFET. In: Advances in communication, devices and networking: proceedings of ICCDN 2018. Springer, Singapore, pp 29–36
14. Bhuyan MH, Islam MT (2022) Study of an n-MOSFET by designing at 100 nm and dimulating using SILVACO ATLAS dimulator. IOSR J VLSI Signal Process 12(1):7–15
15. Chen CW, Wang MC, Chang CHT, Chu WL, Sung SP, Lan WH (2020) Hot carrier stress sensing bulk current for 28 nm stacked high-k nMOSFETs. Electronics 9(12):2095
16. Wu Y, Xu Q, Zhao C, Liu J, Yu Y, Yin W, Kang K (2018) An improved RF MOSFET model accounting substrate coupling among terminals. IEEE Microwave Wirel Compon Lett 28(2):138–140
17. Zhang H, Niu G, Liang Q, Imura K (2018) Extraction of drain current thermal noise in a 28 nm high-${k} $/metal gate RF CMOS technology. IEEE Trans Electron Devices 65(6):2393–2399
18. da Silva EM, Doria RT, Doria RT (2021) Effect of substrate bias and temperature variation in the capacitive coupling of SOI UTBB MOSFETs. J Integr Circ Syst 16(2):1–7

Spline Interpolation of Atrial Repolarization Wave as a Promising Indicator for Atrial Arrhythmias: A Novel Approach

Arya Bhardwaj, R. Pradeep Kumar, J. Sivaraman, Kunal Pal, and Bala Chakravarthy Neelapu

Abstract Atrial repolarization wave (Ta wave) is seldom used as an indicator of atrial arrhythmia due to the lack of perceiving it in standard 12-lead Electrocardiogram (ECG), where the QRS complex obscures the low-voltage Ta wave. Hence, this study proposes spline model to interpolate the hidden Ta wave within the QRS complex using the data points from the PR and ST segments. Fifty Sinus Rhythm (SR) and fifty Sinus Tachycardia (ST) 10 s denoised ECGs were taken from the Chapman University and Shaoxing People's Hospital database. Each ECG was segmented into single-beat ECGs after creating the window surrounding the R peak and was further averaged. PR and ST segment data points from averaged ECG were taken and an arc interpolation was plotted in ST. P and Ta wave characteristics were acquired and further analyzed. Ta peak location lies within the QRS complex however near the PR segment in ST condition. This study also introduces a new segment in SR called QRS inter-segment duration, i.e., end of the P wave to beginning of the T wave, which corresponds to the Ta wave duration in ST as the Ta wave does not manifest in SR. This segment will help to understand the probable Ta wave duration variations in SR. Moreover, through the correlation plot between the Ta and P wave with HR, it is noted that Ta correlates better ($r = 0.5$). Age-wise alterations in Ta wave were examined and found that Ta wave duration and amplitude increase with the age in SR and ST; however, the increase in the Ta wave duration with age in ST condition is not as frequent to that of the SR condition. To conclude, Ta wave is interpolated in ST condition and its features were extracted to analyze the wave variations between SR and ST. Further insightful studies are needed to study Ta wave characteristics in different heart conditions.

A. Bhardwaj (✉) · J. Sivaraman · K. Pal · B. C. Neelapu
Department of Biotechnology and Medical Engineering, National Institute of Technology Rourkela, Odisha 769008, India
e-mail: 521BM1003@nitrkl.ac.in

R. Pradeep Kumar
Institute of Cardiac Care, MIOT International, Chennai, Tamil Nadu 600089, India

D.-I. Stroe et al. (eds.), *Emerging Electronics and Automation*, Lecture Notes in Electrical Engineering 1202, https://doi.org/10.1007/978-981-97-3090-2_11

Keywords Atrial arrhythmia · Atrial repolarization wave · Electrocardiogram · Spline interpolation · Sinus tachycardia

1 Introduction

Electrocardiogram (ECG) represents the electrical activity in the atrial and ventricular chambers of heart. Standard 12-lead ECG in Sinus Rhythm (SR) comprises atrial depolarization, P wave, i.e., depolarization impulse spreading in both right and left atria; ventricle depolarization, QRS complex, and repolarization wave, T wave. However, atrial repolarization wave, Ta wave is not witnessed in surface ECG. It has long been in a debate within scientific community about whether the Ta wave characteristics can be useful to improve the current clinical decisions. Various studies have been performed mentioning the different methods to manifest the Ta wave either in normal heart using signal averaging techniques [1–4], Modified Limb Lead (MLL) placement [5], or studying the Ta wave in diseased conditions such as Atrioventricular Block (AVB) conditions [6, 7]. However, the study on Ta wave grasped further consideration when the cases of false positive diagnosis of Myocardial Infarction (MI) increased among the patients [8–10]. As the most accepted criterion to detect MI was ST segment depression of more than 0.1 mV. Later, it came to notice that Ta wave has an effect on ST segment of exercise stress test ECG recording leading to its depression and false diagnosis of MI. Effect of false diagnosis is the unrequired drug intake which may further lead to side effects on patients. Therefore, study on Ta wave is timely as still its characteristics remain a known dilemma due to its limited manifestation and low amplitude. Hence, Sinus Tachycardia (ST) ECG manifesting Ta wave in PR and ST segment became a handy model to study Ta wave.

Previous studies [11–13] have stated the arc-like structure of Ta wave due to its low amplitude and longer width extending from PR segment up to ST segment. Few studies [6, 14] have shown that Ta wave duration may extend even up to T wave in AVB conditions. Studies by Sivaraman et al. proposed MLL system [5, 15, 16], in which the electrode position of limb leads replaced to the right side of the chest to enhance the atrial activity waves, which gave a great insight into Ta wave along with enhanced P wave in SR subjects. However, the Ta wave could not be observed in the QRS complex. Nevertheless, Ta wave characteristics have been proven to be a significant indicator for prognosis of acute atrial infarctions and Atrial Fibrillation (AF). The complete morphology of Ta wave is not retrieved due to its low amplitude and obscured by the ventricular depolarization complex. Therefore, it has less likelihood to acquire its activity's characteristics. Hence, this study proposes the method to interpolate the Ta wave morphologies hidden within the QRS complex using the data points from PR and ST segments.

2 Methodology

2.1 Database Used

The Chapman University and Shaoxing People's Hospital database [17] contains denoised standard 12-lead ECGs of 10,646 patients for a sample size of 5000 samples (10 s). ECGs are recorded with 500 Hz sampling rate which features 11 common rhythms and 67 additional cardiovascular conditions, labeled by professional experts.

2.2 Study Population

Lead-II ECGs of 50 SR (mean age 47 years) and 50 ST (mean age 45 years) were taken, with a 1:1 male–female ratio, from the database recorded at 500 Hz sampling frequency. ST ECG of heart rate ≥100 bpm were taken in the study. ECGs with cardiovascular co-morbidity such as first-degree atrioventricular block, right axis shift, atrial premature beats, early repolarization of the ventricles, MI, T wave change, and left ventricle high-voltage beats were excluded and only none-beat ECGs of SR and ST were analyzed. Figure 1a and b show SR and ST ECGs.

2.3 Spline Interpolation

SR and ST ECGs were plotted in MATLAB R2021a version. R peak was located for 10 s ECG using Pan–Tompkins algorithm. A window of R-250 ms and R+350 was created to segment ECG into beat signal and averaged. Using the ST averaged ECGs, the PR and ST segment data points were taken to interpolate the Ta wave within QRS complex. Further, this averaged interpolated ECG was used to acquire features.

2.4 Data Acquisition

Lead-II ECG were plotted for P and Ta wave features acquisition. Figure 1c shows the features allocation on single-beat ST ECG.

- Points in ECG marked in this study (refer Figs. 1 and 2):

  P_{on} is the point when the P wave crossed the isoelectric line for the first positive deflection,

  P_{off} is the point when P wave ends and combines with the isoelectric line,

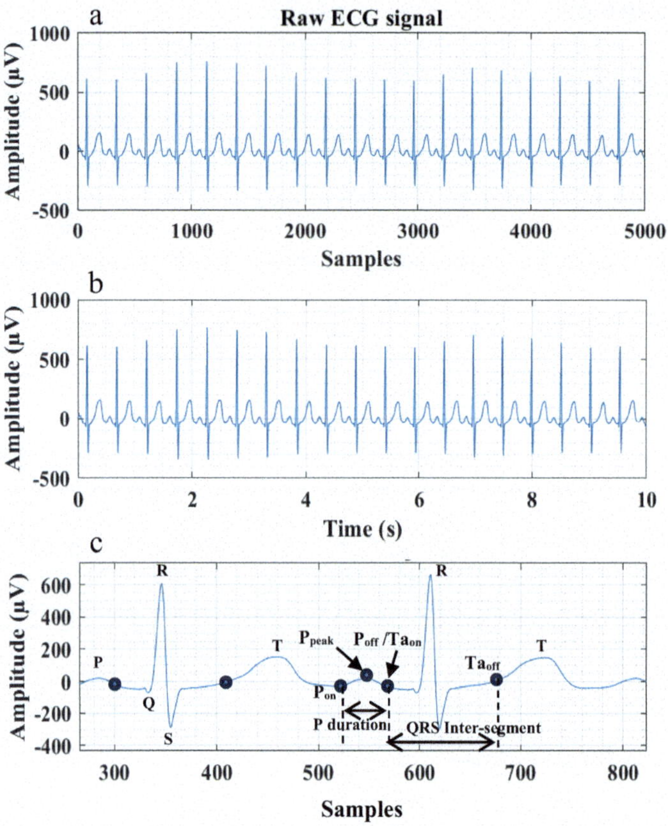

Fig. 1 10 s ECG of **a** SR and **b** ST volunteer **c** Single-beat ECG displaying feature annotation

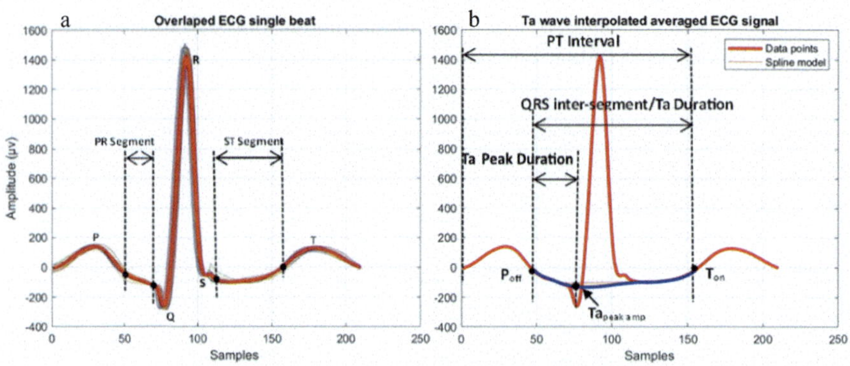

Fig. 2 **a** Averaged beat ECG of ST volunteer **b** Ta wave interpolated ST signal with Ta features

P_{peak} amp (μV) is the point when P wave reaches its peak (maximum) amplitude,

Ta_{peak} amp (μV) is the point when Ta wave reaches its maximum amplitude, and

Ta_{on} is the point when Ta wave begins, crossing the isoelectric line for positive deflection.

- Durations were further derived using the points mentioned:

  P duration (ms) = Time taken from P_{on} to $P_{off} = P_{off} - P_{on}$

  P_{peak} duration = Time taken from P_{on} to P_{peak} amp

  Ta duration = Time taken from P_{off} to $T_{on} = T_{on} - P_{off}$

  Ta_{peak} duration = Time taken from P_{off} to Ta_{peak} amp

  QRS inter-segment duration = Duration of (PR segment + QRS complex + ST segment) = PTI – P duration

  PTI = PT Interval; Time taken from P_{on} to T_{on}.

Here, QRS inter-segment is equivalent to Ta duration in ECG; however, Ta duration can only be calculated in ECG with Ta wave.

2.5 *Statistical Analysis*

The feature values are mentioned as mean. To check the normality distribution of each feature Shapiro–Wilk test was performed. To check the correlation between the features r value (Pearson's correlation) was calculated and to determine the significant differences, 2-tailed Student's t-test and ANOVA test were calculated for parametric and non-parametric data, respectively. The level of significance is considered as $p <$ 0.05.

3 Results

Table 1 shows the mean values of P and Ta wave features in SR and ST. A significant difference between the two groups in duration and amplitude aspects is noted.

Figure 2 represents the (a) averaged waveform of ST ECG and (b) interpolated Ta wave within QRS complex using spline interpolation method and annotation of the Ta peak amplitude, Ta duration and Ta peak duration. The interpolated Ta wave forms an arc like structure using the data points from PR and ST segments. Further, the correlation between the heart rate versus wave amplitude (P and Ta) was plotted to determine the relation (see Fig. 3). It was noted that Ta wave correlated well with heart rate compared to P wave which signifies the prospective utility of Ta wave features in diagnosis/classification of atrial arrhythmias. The outcome of the

Table 1 P and Ta wave feature measurements in Mean ± SD for SR and ST volunteers

	Sinus Rhythm ($n = 50$)	Sinus Tachycardia ($n = 50$)	p value
Heart rate (bpm)	77.02 ± 10	109 ± 9	<0.05
$P_{peak\ amp}$ (μV)	85.42 ± 41	95.66 ± 42.76	<0.05
P_{peak} duration (ms)	47.92 ± 13	49.92 ± 13.97	<0.05
P duration (ms)	87.92 ± 14	79.33 ± 17.01	<0.05
$Ta_{peak\ amp}$ (μV)	–	59.64 ± 21	–
Ta_{peak} duration (ms)	–	175.04 ± 33	–
Ta duration/QRS inter-segment duration (ms)	255.6 ± 33	243.2 ± 21	<0.05
PT Interval (ms)	325.68 ± 27.71	343.52 ± 37.49	<0.05

scatter plot was used for determining statistical correlation between the heart rate and wave amplitude which illustrates the stronger association of Ta wave with heart rate followed by P wave. The positive relation between heart rate and Ta wave explicates the dependence of increase or decrease in Ta wave amplitude with higher or lower heart rate, respectively which is in agreement with a study [14, 18].

Tables 2 and 3 show the age-wise distribution of Ta wave amplitude and duration in SR and ST volunteers, respectively. The bar charts for age-wise distribution of Ta wave features (see Figs. 4 and 5). Bar chart distribution displays an increasing trend of Ta wave amplitude with advancing age. A study by Macfarlane [19] also noted an increase in amplitude with age which is in agreement with this study. Moreover, (in Fig. 5.) intra-SR condition Ta duration is also noted to increase with age; however, the increase is not complimentary to the increase in ST condition.

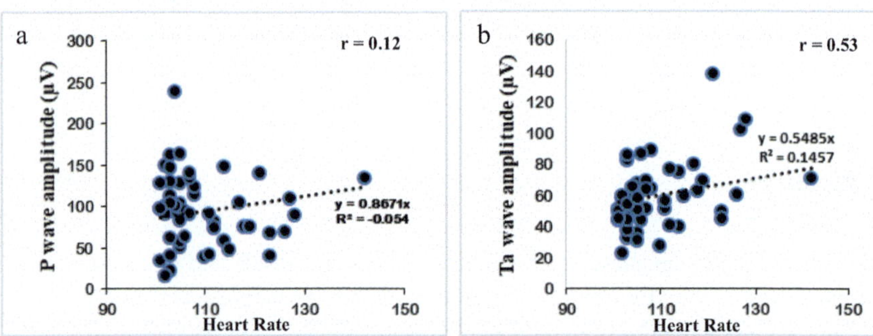

Fig. 3 Correlation between heart rate versus amplitudes **a** P wave and **b** Ta wave in ST population

Table 2 Age-wise distribution of QRS inter-segment duration in SR volunteers

Age group	21–30	31–40	41–50	51–60	61–70
QRS Inter-segment duration (ms)					
Mean ± SD	243 ± 23	247 ± 28	256 ± 34	254 ± 30	278 ± 42
Median	242	248	259	256	277
Max; Min	282; 210	294; 202	302; 200	314; 202	360; 210

Table 3 Age-wise distribution of Ta amplitude and duration in ST volunteers

Age group	21–30	31–40	41–50	51–60	61–70
Ta amplitude (μV)					
Mean ± SD	51 ± 13	61 ± 18	59 ± 20	58 ± 14	69 ± 34
Median	51	56	64.5	57	57.5
Max; Min	75;28	102;40	89;31	87;41	138;23
Ta duration (ms)					
Mean ± SD	236 ± 13	250 ± 21	235 ± 17	260 ± 19	234 ± 26
Median	234	248	233	261	230
Max; Min	254; 214	290; 222	272; 204	292; 236	286; 198

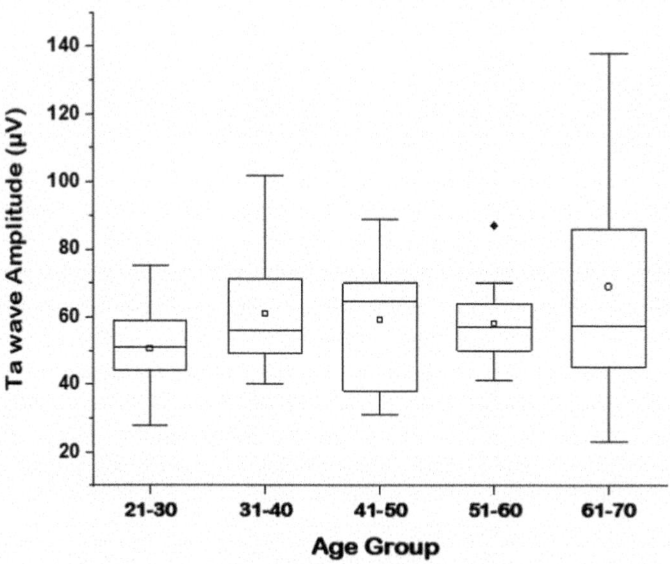

Fig. 4 Box plot of Ta wave amplitude distribution with increasing age in ST

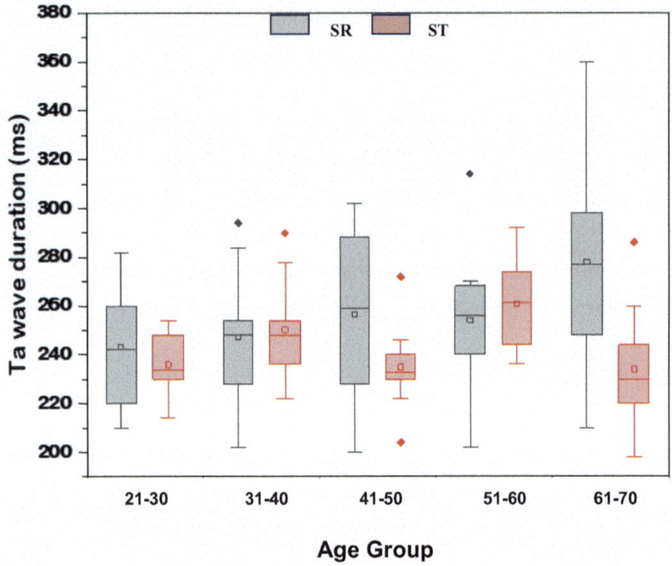

Fig. 5 Box plot of age-wise distribution comparison of Ta wave duration and equivalent QRS inter-segment duration in ST and SR volunteers, where gray represents the SR and pink represents the ST

4 Discussion

Atrial repolarization wave has scarcely been used as the indicator of atrial disorders compared to its homologous counterpart in ventricles [20, 21], i.e., T wave. This is attributable to its low amplitude, and sheltered nature due to the successive ventricular depolarization wave. Absence of Ta wave in 12-lead ECG recording have motivated various authors to study and device new methods and lead system to witness the Ta wave [22, 23]. Methods such as signal averaging [2–4], ECG of AVB patients [7, 14] and lead system such as MLL ECG [4, 5, 24] have given us far greater insights on Ta wave characteristics compared to the 12-lead ECG. However, these methods scantly provide any information of Ta wave within the QRS complex. Hence, this study attempts to identify the structure of Ta wave in QRS complex and further determine its morphological characteristics such as peak amplitude, peak location, and duration along with other atrial parameters.

It is observed in this study that the duration of Ta wave extends up to the ST segment as observed through ST ECG. Ta wave in SR recorded using standard 12-lead is not visible; however, a previous study [18] noted the presence of Ta wave until beginning of T wave using MLL ECG placement [5]. Moreover, the peak location of interpolated Ta wave is noted in QRS complex. The previous study [18] by the same authors noted that Ta wave amplitude exaggerates in higher heart rate condition which is in line with [25].

As ST showed higher Ta wave amplitude by calculating the percentage difference between the PR segment depression amplitudes in SR and ST, a similar result was performed by calculating the percentage difference between ST segment depression amplitudes in SR and ST. This method may prove to be helpful as a potent marker for the atrial disorders and arrhythmias. It is evident from this study that a stronger correlation between the heart rate and atrial features in ST condition compared to SR condition. Since, the heart rate and the temporal elements of the ECG are correlated, and the correlation was stronger in the ST condition, this suggests that an elevated heart rate may change the atrial characteristics, which can be further investigated to identify markers for common and potentially fatal atrial illnesses.

A significant result based on feature correlation plot also extends the understanding that PTI related more to Ta wave duration compared to P wave duration, which signifies that the variations of Ta wave duration affect the PTI more than the P wave duration for which vice versa also true. This interpretation helps in studying the Ta wave through the changes in PTI for any disease conditions. This study also introduces the QRS inter-segment, where QRS is sandwiched between the two segments (PR & ST), duration in SR condition, where Ta wave is obscured, however through this marker, we can interpret the variations in Ta wave durations.

Ta wave interpolation and related markers may have a clinical scope in reducing the false positive detection of myocardial ischemia and infarction. A comparative study on the understanding of the morphology of PR and ST segment depression as an effect of Ta wave or due to ischemia will be possible when Ta wave can be visible. Hence, this study introduced an interpolation method for Ta wave to study its effect on ECG.

4.1 Future Scope

Study on the Ta wave characteristics as the potential marker for early detection or prediction of atrial disorder such as AF, the most common arrhythmia throughout the world, is timely. Ta wave markers interpretation is necessary for better clinical diagnosis of AF [26]. Various studies have mentioned the occurrence of P and Ta waves changes in pre-AF patients. Reference [21] stated that the early Ta wave peak in PR segment may be considered as the AF marker. Recent studies included Ta wave dispersion [27] and Ta wave alternans [28] in AF patients as a marker. A systematic review and meta-analysis [29] on the artificial intelligence for classification and prediction of AF also stated the need to include the new predictive markers to reduce global AF burden. This study has contributed to introduce a few Ta wave-related indicators for SR and higher heart condition, which may prove to be useful to general practitioners for earlier disease interpretation.

5 Conclusion

This study implemented the spline interpolation method in ECG to interpolate the Ta wave in the QRS complex using the PR and ST segment data points. Additionally, QRS inter-segment and PT interval are also introduced as new markers. Understanding the Ta wave characteristics change as the potent marker may prove as a boon for early prediction of AF. Interpolating the Ta wave within the QRS complex will help to understand the characteristics of atrial repolarization which may help in the prognosis of various atrial arrhythmias. Further study on Ta wave features and indices may also help in the early prediction of AF. As the transition between the different types of AF occurs at a higher frequency among population; therefore, it is necessary to predict it in its benign and early stages for better strategical treatment of the patients.

Acknowledgements The authors acknowledge the support from the Ministry of Education, Government of India, to carry out this research.

Funding This work was supported by financial grants from the Science and Engineering Research Board (SERB), Government of India (CRG/2023/007815).

References

1. Langley P (2007) Analysis of the atrial repolarisation phase of the ECG in health and in atrial fibrillation. Compt in Cardiol 34:785–788
2. Giacopelli D (2015) Characteristics of the atrial repolarization phase of the ECG in paroxysmal atrial fibrillation patients and controls. Acta Cardiol 70:672–677
3. Holmqvist F (2009) Detailed ECG analysis of atrial repolarisation in humans. Ann Noninvasive Electrocardiol 14:13–18
4. Sivaraman J (2014) A study on atrial Ta wave morphology in healthy subjects: an approach using P wave signal-averaging method. J Med Imaging Health Inf 4:675–680
5. Sivaraman J (2013) A novel approach to determine atrial repolarization in electrocardiograms. J Electrocardiol 46:e1
6. Sivaraman J (2015) Unmasking of atrial repolarization waves using a simple modified limb lead system. Anatol J Cardiol 15:605–610
7. Debbas NM (1999) Human atrial repolarization: effects of sinus rate pacing and drugs on the surface electrocardiogram. J Am Coll Cardiol 33:358–365
8. Manne JRR (2018) Atrial repolarization waves (Ta) mimicking inferior wall ST segment elevation myocardial infarction in a patient with ectopic atrial rhythm. Case Rep Med 2018:1–3
9. Sapin PM (1991) Identification of false positive exercise tests with use of electrocardiographic criteria: a possible role for atrial repolarization waves. J Am Coll Cardiol 18:127–135
10. Tanabe J (2020) False-positive ST-segment elevation. Eur Heart J Case Rep 4:368–369
11. Sprague HB (1925) Clinical observations on the T wave of the auricle appearing in the human electrocardiogram. J Clin Invest 1:389–402
12. Tranchesi J (1960) Atrial repolarization its importance in clinical electrocardiography. Circulation 22:635–644
13. Childers R (2011) Atrial repolarization: its impact on electrocardiography. J Electrocardiol 44:635–640

14. Bhardwaj A (2022) Insights on atrial electrocardiogram in sinus rhythm and atrioventricular block for improved clinical diagnosis. Biomed Eng Appl Basis Commun 34:2250004
15. Sivaraman J (2012) A modified chest leads for minimization of ventricular activity in electrocardiograms. In: 2012 international conference on biomedical engineering (ICoBE). Penang, Malaysia, pp 79–82
16. Sivaraman J (2017) Effects of sinus rhythm on atrial ECG components using modified limb lead system. In: 2017 4th international conference on signal processing, computing and control (ISPCC). Solan, India, pp 527–530
17. Zheng J (2020) Optimal multi-stage arrhythmia classification approach. Sci Rep 10:2898
18. Bhardwaj A (2022) Exaggerated amplitude and peak location of Ta wave in tachycardia as an indicator for atrial disorders. In: 2022 computing in cardiology (CinC). Tampere, Finland, pp 1–4
19. Macfarlane PW (2015) Normal limits of the electrocardiogram in Indians. J Electrocardiol 48:652
20. Narayan SM (2006) T-wave alternans and the susceptibility to ventricular arrhythmias. J Am Coll Cardiol 47:269–281
21. Swerdlow CD (2008) High amplitude T-wave alternans precedes spontaneous ventricular tachycardia or fibrillation in ICD electrograms. Heart Rhythm 5:670–676
22. Holmqvist F (2009) Noninvasive evidence of shortened atrial refractoriness during sinus rhythm in patients with paroxysmal atrial fibrillation. Pacing Clin Electrophysiol 32:302–307
23. Hayashi H (1970) The experimental study of normal atrial T wave in electrocardiograms. Jpn Heart J 11:91–103
24. Arumughan J (2020) Stability analysis on the effects of heart rate variability and premature activation of atrial ECG dynamics using ARMAX model. Phys Eng Sci Med 43:1361–1370
25. Bhardwaj A (2021) Study of P wave indices in sinus rhythm and tachycardia. In: 2021 seventh international conference on bio signals, images, and instrumentation (ICBSII). Chennai, India, pp 1–5
26. Narayan SM (2016) Atrial fibrillation: can electrograms be interpreted without repolarization information? Heart Rhythm 13:962–963
27. Li Z (2001) Dispersion of atrial repolarization in patients with paroxysmal atrial fibrillation. Europace 3:285–291
28. Comtois P (2012) Atrial repolarization alternans as a path to atrial fibrillation. J Cardiovasc Electrophysiol 23:1013–1015
29. Bhardwaj A (2023) A holistic overview of artificial intelligence in detection, classification and prediction of atrial fibrillation using electrocardiogram: a systematic review and meta-analysis. Arch Computat Methods Eng 30:4063–4079

Evaluation of Acoustic Cepstral Coefficients in Differentiation of Alzheimer's Condition Using Spontaneous Speech and Hyperparameter Tuned EfficientNet Model

Meghana Manvitha Venna and Rohini Palanisamy

Abstract Continuous monitoring is imperative for the early detection of Alzheimer's Disease (AD). Speech data is a promising modality for continuous monitoring of AD, as it can be collected in real time and evaluated for subtle changes that may indicate disease progression. This work presents comparison of different cepstral encoding techniques for spontaneous speech signals that can aid in Alzheimer's disease classification using EfficientNet architecture. For this, spontaneous speech samples are obtained from publicly available dataset. These audio samples are preprocessed and spectrograms such as logmel, Mel-Frequency Cepstral Coefficients (MFCC), Gammatone Frequency Cepstral Coefficients (GFCC), and BFCC (Bark Frequency Cepstral Coefficients) are obtained. The spectrograms and their derivatives are given as input into a pre-trained EfficientNet-B0 model with a combination of fully connected networks. This model is evaluated using various performance metrics and cross-validated for five folds. The best-performing cepstral feature and its derivatives are selected for further hyperparameter tuning such as optimizer, batch size, and loss function. Results indicated that GFCC performed better among the other cepstral coefficients. Hyperparameter tuning with the Adamax optimizer resulted in a median accuracy of 68.1%. Among loss functions, Binary Cross-entropy (BCE) showed promising results with a median accuracy of 68% and minimal accuracy variation of 9%. Despite all the batch sizes providing comparable accuracy, batch size 32 showed a lower variation of 4.5%. Thus, the EfficientNet model is trained using the Adamax optimizer with a learning rate of 0.001 using BCE loss in batch sizes of 32. Five-fold cross-validation of this model gives 73% mean accuracy and precision rate of 0.73. As GFCC is capable of capturing transient sounds which are crucial for finding distinguishing patterns in AD, the EfficientNet architecture

M. M. Venna · R. Palanisamy (✉)
Department of Electronics and Communication Engineering, Indian Institute of Information Technology, Design and Manufacturing, Kancheepuram, Chennai, Tamil Nadu, India
e-mail: rohinip@iiitdm.ac.in

D.-I. Stroe et al. (eds.), *Emerging Electronics and Automation*, Lecture Notes in Electrical Engineering 1202, https://doi.org/10.1007/978-981-97-3090-2_12

with GFCC gives a better performance in differentiating Alzheimer's condition. Thus, this framework could be used for acoustic signal-based classification of Alzheimer's disease.

Keywords Speech signal · Acoustics · Alzheimer's disease · EfficientNet · Cepstral coefficients

1 Introduction

Alzheimer's dementia is a neurodegenerative disease which leads to a gradual decline in memory, language, reasoning, and social skills. According to a World Health Organization report, AD is the seventh leading cause of death globally. Common diagnosis methods for AD include neurological exams, brain imaging (MRI, CT, and PET), and cerebrospinal fluid or blood tests [1, 2].

These diagnostic techniques take place over time and are not done on a regular basis. As Alzheimer's dementia progresses swiftly, it is critical to continuously monitor and track onset and development of disease. The present diagnostic techniques are unable to meet this need. As a result, innovative diagnostic methods are needed. Despite memory loss being the primary indicator of AD, studies point out that the brain regions controlling speech and language are also among the earliest to be influenced by dementia [3]. As speech could be used in real-time continuous monitoring, it can be a valuable biomarker in early diagnosis of Alzheimer's dementia [4].

Speech data consists of two main parts: acoustic speech and linguistic speech. Acoustic speech includes sound properties of spoken language, whereas linguistic speech focuses on language properties like grammar, semantics, and meaning. Many studies investigated speech and language features for detection of AD [5–7]. Though research in linguistic features shows promising accuracy, it is constrained to a specific set of languages and cannot be extended globally. On the contrary, acoustic features are independent of the language being spoken and thus have the scope of universal applicability.

Temporal features like loudness, jitter, intensity, speech rate, duration of vocalizations, and frequency-based features like fundamental frequency, spectral harmonicity, and spectral flux are explored in speech recognition tasks, but they fail to capture frequency and temporal characteristics simultaneously which is crucial in efficient encoding of audio signals.

Spectrograms act as effective audio representations by combining both temporal and frequency domains. Cepstrals like Mel-Frequency Cepstral Coefficients (MFCC) [8, 10], Gammatone Frequency Cepstral Coefficients (GFCC) [9], and Bark Frequency Cepstral coefficients (BFCC) [10] are exploited this way in various tasks like emotion recognition, audio classification, and speaker identification [11]. However, very few works explored spectrograms for AD classification task. Meghanani et al. [8] used logmel spectrograms and MFCC with CNN-LSTM and

ResNet [12]. ResNet is a convolutional neural network which uses skip connections to overcome vanishing gradient problem. Since then, there have been advancements in image-based deep learning architectures. EfficientNet [13] is an advanced CNN technique that achieved maximum accuracies with a minimum number of parameters in image classification task. This utilizes compound scaling coefficient for uniform scaling of neural networks in image classification tasks. Thus, EfficientNet has the potential to be used for spectrogram classification in AD.

In this work, exploration of different spectrogram-based acoustic features in Alzheimer's dementia detection has been carried out by employing EfficientNet. The comprehensive approach will shed light on the most effective ways to leverage different spectrogram features within the context of EfficientNet.

2 Methods

The schematic of workflow is represented in Fig. 1.

2.1 Database

The Interspeech ADReSS Challenge dataset containing audio descriptions of the Cookie Theft picture from the Boston Diagnostic Aphasia Exam is considered for the study. The dataset contains speech samples of 87 AD and 79 non-AD subjects. The dataset is balanced for age and gender to prevent risk of bias in classification.

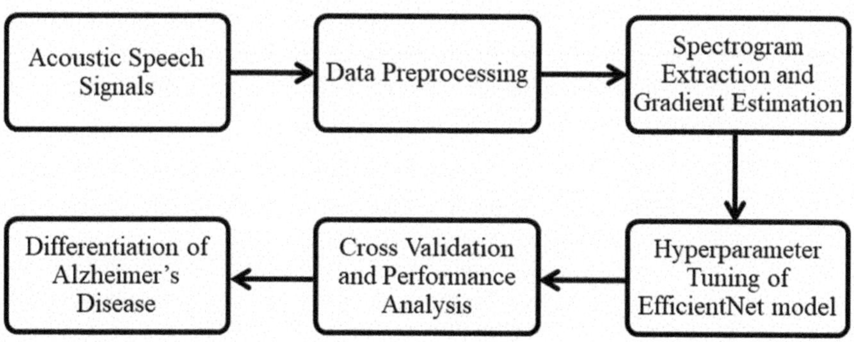

Fig.1 Schematic of workflow

2.2 Preprocessing

Stationary noise in the audio samples were removed using Noisereduce Python Library. Audacity software was used to remove non-stationary noise and to edit interviewer parts. Each sample was further divided into 20 secs audio segments. Spectrogram features like logmel, MFCC [8, 10], GFCC [9], and BFCC [10] are extracted using filter-banks which are collections of efficient human-cochlear simulation filters. First- and second-order gradient across time for each spectrogram is estimated.

2.3 EfficientNet Model

Spectrogram of each audio sample, along with its first- and second-order derivatives can be given as a three-channel RGB like input to the EfficientNet model. EfficientNet is a CNN architecture that uses an optimized scaling method to scale the network dimensions such as depth, width, and resolution uniformly. EfficientNet-B0 is based on inverted bottleneck residual blocks of MobilNetV2 [14] along with squeeze and excitation blocks. In this work, the EfficientNet-B0 model with frozen ImageNet weights is used for feature extraction. Additionally, modifications have been done to the Efficientnet-B0 model by adding the pooling layer with dropout normalization which is connected to a dense layer with SoftMax activation. There are several hyperparameters that are necessarily to be tuned in EfficientNet-B0 model for AD classification task. They include loss function, optimizer and batch size.

Binary Cross-Entropy

Binary Cross-entropy or log loss (BCE) is widely used in classification due to its effectiveness in quantifying the discrepancy between predicted probabilities and true labels of a dataset. Higher penalties are imposed for larger errors in predictions. The standard binary cross entropy loss function is given by

$$\mathrm{BCE} = \frac{1}{N}\sum_{i=0}^{N} y_i \log(f(y_i)) + (1 - y_i)\log(1 - f(y_i)) \tag{1}$$

Here, N indicates the number of training samples. y_i and $f(y_i)$ represent the true label and prediction probability of the ith training sample, respectively.

Focal Cross-Entropy

Focal cross-entropy (FL) addresses the issue of class imbalance during the training of the dataset. Focal loss reduces weightage to majority samples and conversely adds weightage to minority samples. Further, the weight of the easy-to-classify samples and tough-to-classify samples is controlled by adding a focusing parameter γ. This way model focuses more on tough-to-classify models and thus performs better overall. This is achieved by adding modulating factor $(1 - p_t)$ to cross-entropy loss with balancing factor α.

$$\mathrm{FL}(p_t) = -\alpha(1 - p_t)^{\gamma} \log(p_t) \tag{2}$$

Here, α represents the balancing factor, p_t indicates the estimated probability for category t, and γ is the focusing parameter

Squared Hinge Loss

Squared hinge loss (SL) is a rectified version of hinge loss, where it makes sure the loss is always non-negative by introducing squaring. Due to squaring, loss is more sensitive to large errors, particularly when the data point that is misclassified is far from the decision boundary. Overall, squared hinge loss penalizes classification errors more strongly when they occur further from the decision boundary.

$$\mathrm{SL} = \sum_{i=0}^{N} (\max(0,1 - (y_i * f(y_i)))^2 \tag{3}$$

Here, N indicates the number of training samples. y_i and $f(y_i)$ represent the true label and prediction probability of the ith training sample, respectively.

Various optimizers, namely, Adam, Adagrad, RMSProp, Adadelta, Adamax, Nadam, and stochastic gradient descent are employed for testing the performance of the EfficientNet model. In addition, EfficientNet model is also trained with different batch sizes of 8, 16, 32, and 64 to test the optimum size.

2.4 Performance Metrics

Dataset is split into train, test, and validation data (80–10–10). Train, test, and validation accuracies and losses are evaluated from each epoch for understanding the performance of the model. Confusion matrix is plotted to analyze classification results. Fivefold cross-validation technique is used to find average accuracy and loss.

3 Results and Discussions

Representative audio signals of normal and AD subjects are shown in Fig. 2. Distinguishing AD and normal subjects, purely based on audio representations is challenging since they mostly focus on audio amplitude variations over time. There can be cases when a normal subject talks with a feeble voice or an AD subject has a higher amplitude, which can lead to misdiagnosis. To find out the intricate patterns that

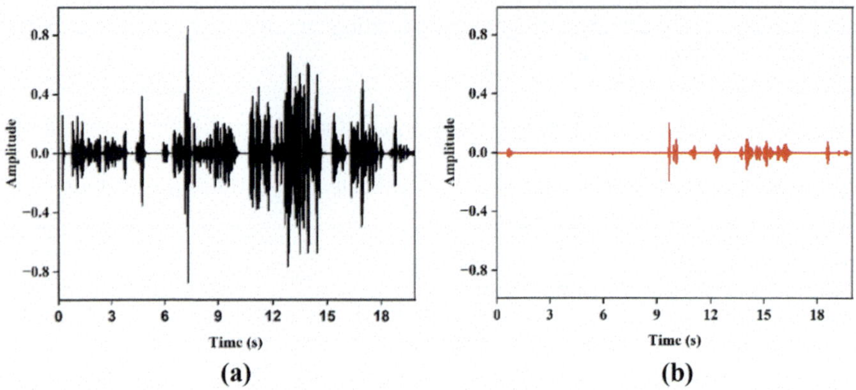

Fig. 2 Representative audio sample of **a** Normal **b** AD subject

separate AD subjects from normal subjects, time–frequency-based representations are employed.

Extracted spectrogram features namely logmel, MFCC, GFCC, and BFCC are used to depict an audio signal in its component frequencies along a time frame. Spectrograms are represented as heat maps of audio amplitudes across time and frequency domains as shown in Figs. 3 and 4 for normal and AD, respectively. Although variations in spectrograms are observed, capturing intricate patterns that identify Alzheimer's is challenging. Thus, these spectrogram images are given as input to EfficientNet-based deep learning model.

Figure 5a demonstrates the results of initial training of AD classification using EfficientNet model with different cepstral encoding techniques like logmel, MFCC, GFCC, and BFCC. From the figure, it is observed that GFCC and BFCC show lower variation and higher mean accuracy of 73% when compared with logmel and MFCC. Although both GFCC and BFCC demonstrate comparable performance, GFCC is preferred due to its capability of capturing impulsive signal classes like transient sounds which are often observed in AD subject which act as crucial biomarkers in classification.

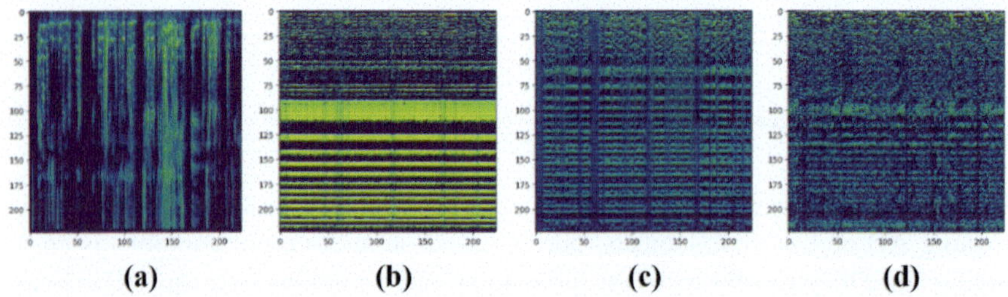

Fig. 3 Spectrogram representations of **a** logmel **b** MFCC **c** GFCC **d** BFCC for Normal subject

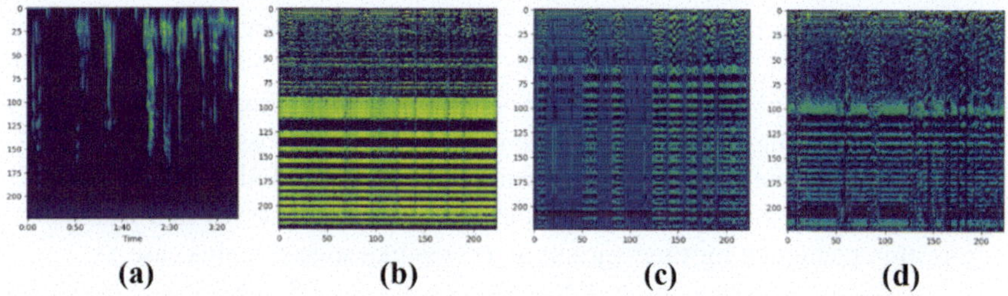

Fig. 4 Spectrogram representations **a** logmel **b** MFCC **c** GFCC **d** BFCC for AD subject

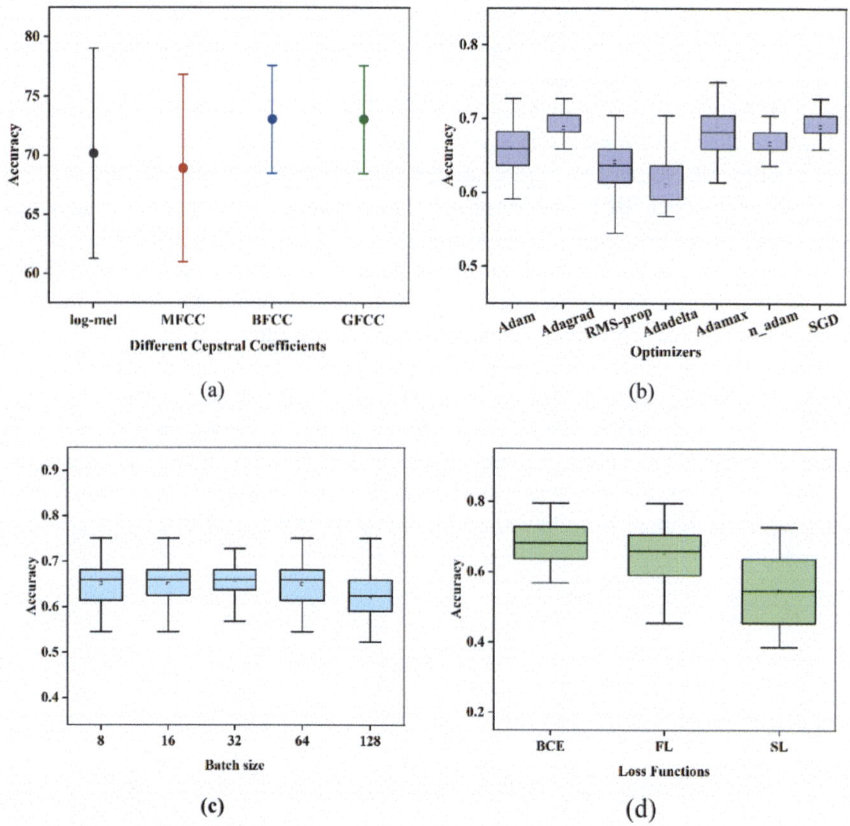

Fig. 5 **a** Cross-fold validation accuracies for different cepstral encoding techniques. **b** Distribution of validation accuracies across different optimizers. **c** Distribution of validation accuracies across different batch sizes. **d** Distribution of validation accuracies across different loss functions

The performance of the EfficientNet-B0 model for different optimization techniques is shown in Fig. 5b. From the figure, it can be seen that Adamax shows higher median accuracy of 68.1% with symmetric distribution of accuracies. This symmetry indicates that Adamax provides consistent performance across different subjects, making it a good optimizer to build robust classification models. Since Adamax adjusts the learning rate based on characteristics of data, it serves as a well-suited optimization technique for time-variant processes like speech in this case.

Distribution of accuracy for each batch size is illustrated in Fig. 5c. It is evident from the figure that all batch sizes show similar median accuracies around 66%. However, it is clear that batch size 32 stands out with significantly lower variation of 4.5% when compared with other batch sizes. Smaller batch size, such as 32, play an important role in model generalization. They maintain balance by improved generalization, still remaining large enough to save the model from expensive computations caused by frequent gradient updates.

Effect of loss function on AD classification using the EfficientNet-B0 model is shown in Fig. 5d. Notably, BCE gives better results with higher accuracy of 68% and lower variance of 9%. BCE is a probabilistic logarithmic loss whose value increases as predicted probability diverges from actual label. This way it penalizes erroneous predictions encouraging the model to make more confident predictions with higher probabilities. The probabilistic nature of BCE makes it a popular choice for medical diagnosis [15] since it predicts with confidence estimations.

Based on above observations, EfficientNet-B0 model is fixed with Adamax Optimizer with learning rate of 0.001 and trained for 60 epochs in batches of size 32 using BCE loss function and the results are as shown in Fig. 6. Table 1 illustrates results of five-fold cross validation training of the dataset. Notably, results yield a mean accuracy of 73%. From the table it can be said that the model performs consistently better which is evident from low variation is true positives and true negatives over five runs.

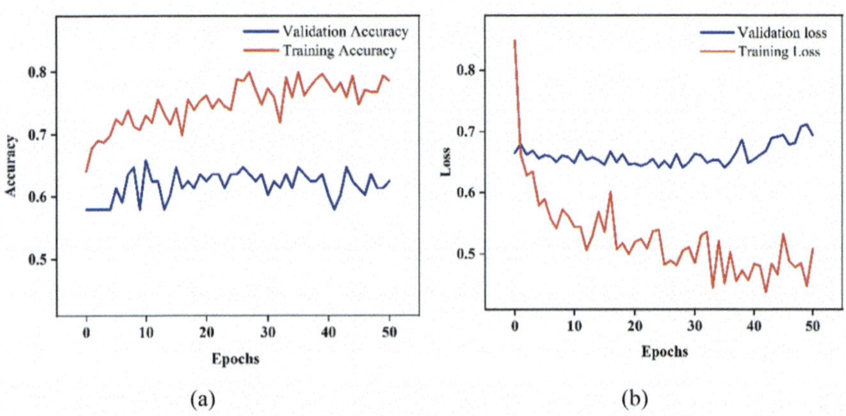

Fig. 6 Distribution of **a** Training and Validation accuracies **b** Training and Validation Losses

Table 1 Five cross-validation training results

Run	True positive	False negative	False positive	True negative	Accuracy (%)
1	11	9	5	24	71.4
2	10	10	4	25	71.4
3	9	11	5	24	67.4
4	13	7	3	26	79.6
5	10	10	2	27	75.5
Average	10.6	9.4	3.8	25.2	73.06

Meghanani et al. extracted logmel spectrograms and employed the ResNet model followed by LSTM and achieved 62.5% accuracy in AD classification. EfficientNets are introduced as an advancement over ResNet where, the base model is uniformly scaled across dimensions such as width, depth, and resolution using a set of fixed scaling coefficients. This helps in improving the receptive field and adds more channels to capture more fine-grained patterns. This way dimensions are well balanced and thus better accuracies can be achieved. Thus, compared to ResNet, significant improvement in accuracy to 73% can be observed when EfficientNet is used.

4 Conclusion

Continuous monitoring plays a crucial role in early diagnosis of Alzheimer's. However, many current diagnostic methods do not offer this capability and speech data addresses this gap in AD diagnosis. In this work, different cepstral encoding techniques for speech audio data are explored for Alzheimer's classification task. Deep learning-based EfficientNet-B0 model is trained with cepstrals like logmel, MFCC, GFCC, and BFCC. Results illustrated that the performance of GFCC is comparatively better than other cepstral encoding techniques. Best set of hyperparameters is investigated and the model is trained with a batch size of 32 using binary cross entropy as loss function with a learning rate of 0.001 using the Adamax optimizer function. Fivefold cross-validation of the model yielded a mean accuracy of 73% with 0.73 precision rate. Based on these results, it can be concluded that GFCC spectral encoding yields promising results in AD classification when transfer learning-based deep learning architectures, like EfficientNet model, are employed. The consistency in model performance makes the model suitable to be extended to real-time evaluation for AD diagnosis.

References

1. Rohini P, Sundar S, Ramakrishnan S (2019) Characterization of Alzheimer conditions in MR images using volumetric and sagittal brainstem texture features. Comput Methods Programs Biomed 173:147–155
2. Shaji S, Palanisamy R, Swaminathan R (2023) Structural irregularities in MR corpus callosal images and their association with cerebrospinal fluid biomarkers in mild cognitive impairments. Neurosci Lett, 137329
3. Fraser KC, Meltzer JA, Rudzicz F (2016) Linguistic features identify Alzheimer's disease in narrative speech. J Alzheimer's Dis 49(2):407–422
4. de la Fuente GS, Ritchie CW, Luz S (2020) Artificial intelligence, speech, and language processing approaches to monitoring Alzheimer's disease: a systematic review. J Alzheimer's Dis 1:1547–1574
5. Luz S, De La Fuente Garcia S, Albert P (2018) A method for analysis of patient speech in dialogue for dementia detection. In: Resources and processing of linguistic, para-linguistic and extra-linguistic data from people with various forms of cognitive impairment. European Language Resources Association (ELRA), Paris, France, pp 35–42
6. Mirheidari B, Blackburn D, Walker T, Venneri A, Reuber M, Christensen H (2018) Detecting signs of dementia using word vector representations. Proc Interspeech 2018:1893–1897
7. Haider F, de la Fuente GS, Luz S (2020) An assessment of paralinguistic acoustic features for detection of Alzheimer's dementia in spontaneous speech. IEEE J Sel Top Signal Process 14:272–281
8. Meghanani A, ACS A, Ramakrishnan AG (2021) An exploration of log-Mel spectrogram and MFCC features for Alzheimer's dementia recognition from spontaneous speech. In: 2021 IEEE spoken language technology workshop (SLT), Shenzhen, China, pp 670–677
9. Lella KK, Pja A (2022) Automatic diagnosis of COVID-19 disease using deep convolutional neural network with multi-feature channel from respiratory sound data: cough, voice, and breath. Alex Eng J 61(2):1319–1334
10. Kumar C, Ur Rehman F, Kumar S, Mehmood A, Shabir G (2018) Analysis of MFCC and BFCC in speaker identification system. In: 2018 international conference on computing, mathematics and engineering technologies (iCoMET), Sukkur, Pakistan, pp 1–5
11. ur Rehman F, Kumar C, Kumar S, Mehmood A, Zafar U (2017) VQ based comparative analysis of MFCC and BFCC speaker recognition system. In: 2017 international conference on information and communication technologies (ICICT), Karachi, Pakistan, pp 28–32
12. He K, Zhang X, Ren S, Sun J (2015) Deep residual learning for image recognition. In: 2016 IEEE conference on computer vision and pattern recognition (CVPR), pp 770–778
13. Tan M, Le Q (2019) EfficientNet: rethinking model scaling for convolutional neural networks. In: Proceedings of the 36th international conference on machine learning, vol 97, 6105–6114
14. Sandler M, Howard A, Zhu M, Zhmoginov A, Chen L-C (2018) MobileNetV2: inverted residuals and linear bottlenecks. In: Proceedings of the IEEE conference on computer vision and pattern recognition (CVPR), June 2018
15. Bengio Y (2012) Practical recommendations for gradient-based training of deep architectures. In: Montavon G, Orr GB, Müller KR (eds) Neural networks: tricks of the trade. Lecture notes in computer science, vol 7700. Springer, Berlin, Heidelberg

Approach for Analysis of Diabetic and Normal Electrocardiogram Signals

Anuja Jain, Anurag Verma, and Amit Kumar Verma

Abstract Diabetes or diabetes mellitus (DM) is a chronic clinical condition marked by the body's inability to regulate blood sugar (glucose) levels effectively. This condition often leads to various cardiovascular complications and can have discernible effects on electrocardiogram (ECG) signals. Diagnosis of DM typically relies on parameters such as fasting blood sugar levels, oral glucose tolerance tests, or hemoglobin A1C levels. ECG, on the other hand, is primarily employed to assess cardiac function and detect potential cardiac abnormalities that may be associated with diabetes-related complications. In this study, combination of the Short-Time Fourier Transform (STFT) and Gray-Level Co-occurrence Matrix (GLCM) techniques is explored for distinguishing between normal and diabetic ECG signals. STFT involves the transformation of 1D ECG data into a time–frequency (T–F) matrix using STFT. From each T–F matrix, four distinct GLCM features are extracted. To determine the most discriminative GLCM features for distinguishing between normal and diabetic ECG signals, we employ the Kruskal–Wallis (K–W) algorithm. We validate the effectiveness of our proposed method using an ECG database that we have compiled. The experimental findings from our research illustrate the substantial impact of this approach in identifying diabetes-related alterations in ECG signals.

Keywords ECG signals · Diabetes · Analysis

A. Jain (✉) · A. Verma
College of Pharmacy, Teerthanker Mahaveer University, Moradabad, UP 244001, India
e-mail: Jain.anuja49@gmail.com

A. Verma
e-mail: principal.pharmacy@tmu.ac.in

A. K. Verma
Mahatama Jyotiba Phule Rohilkhand University, Bareilly, UP 243006, India
e-mail: amit.verma@mjpru.ac.in

D.-I. Stroe et al. (eds.), *Emerging Electronics and Automation*, Lecture Notes in Electrical Engineering 1202, https://doi.org/10.1007/978-981-97-3090-2_13

1 Introduction

Diabetes, often referred to as diabetes mellitus, where the blood contains high levels of glucose in the human body. Insulin Resistance: In type 2 diabetes, the body's cells become less responsive to insulin. While the pancreas may produce insulin, the cells do not effectively use it to transport glucose from the bloodstream into the cells. This also leads to elevated blood sugar levels [1]. Diabetes can lead to various health complications if not properly managed, including heart disease, kidney problems, vision issues, nerve damage, and more. Management typically involves lifestyle changes (such as diet and exercise), medication (including insulin), and regular monitoring of blood sugar levels to keep them within a healthy range [2].

Non-invasive methods aimed to detect diabetes or monitor blood glucose levels without the need for needles or invasive procedures. Researchers have been exploring optical techniques such as spectroscopy and Raman spectroscopy to measure glucose levels non-invasively. These methods analyze how light interacts with body tissues to estimate blood glucose levels. Some studies have investigated the use of breath analysis to detect volatile organic compounds (VOCs) associated with diabetes. Changes in these compounds could potentially indicate elevated blood sugar levels. While not entirely non-invasive, urine testing is less invasive than blood tests. Researchers have been working on developing urine-based tests that could provide information about glucose levels and other diabetes-related biomarkers. Infrared imaging of the skin has been explored as a non-invasive method to estimate blood glucose levels. Changes in skin temperature and blood flow patterns can be associated with variations in glucose levels. There are several compelling motivations for developing non-invasive methods for the detection of diabetes: Traditional methods of monitoring blood glucose levels often require painful finger stick tests or continuous glucose monitoring devices with sensors inserted under the skin.

Non-invasive methods can reduce the discomfort and inconvenience associated with these procedures, leading to an improved quality of life for individuals with diabetes. The motivation for non-invasive methods for the detection of diabetes is rooted in the desire to improve the lives of individuals with diabetes, enhance the accuracy and convenience of glucose monitoring, and ultimately contribute to better health outcomes and reduced healthcare costs. There is a significant relationship between diabetes and heart health. Diabetes patients are at higher risk of cardiovascular disease (CVD). This connection between diabetes and heart health is well-established, and it is one of the major complications associated with diabetes. The features from facial images were extracted for diagnosis of diabetes [3]. Features extracted from heart rate (HR) signals are explored for diagnosing diabetes [4]. HRV analysis can be useful for the clinical evaluation of autonomic diabetic neuropathy [5]. Analysis of autonomic neuropathy in diabetic patients, the HRV signals have been useful as an indicator [6]. Spectral analysis of HRV has been explored for assessment of neuropathy due to diabetics using HRV signals [7]. Higher order spectral features are explored for analysis of HRV for detection of diabetes [8]. ECG signal-based features have been explored for classification of diabetic and healthy ECG signals [9].

Detecting diabetes using Electrocardiogram (ECG) signals is an emerging area of research that explores the relationship between diabetes and the electrical activity of the heart. While ECG is primarily used to diagnose and monitor heart-related conditions, there is evidence to suggest that certain ECG patterns may be associated with diabetes or prediabetes. Here's how ECG signals can potentially be used for diabetes detection. Researchers are investigating various other ECG features and parameters that may be correlated with diabetes, such as repolarization abnormalities and alterations in waveforms.

2 Methodology

In this section, we are presenting the methodology. Figure 1 depicted steps for proposed method.

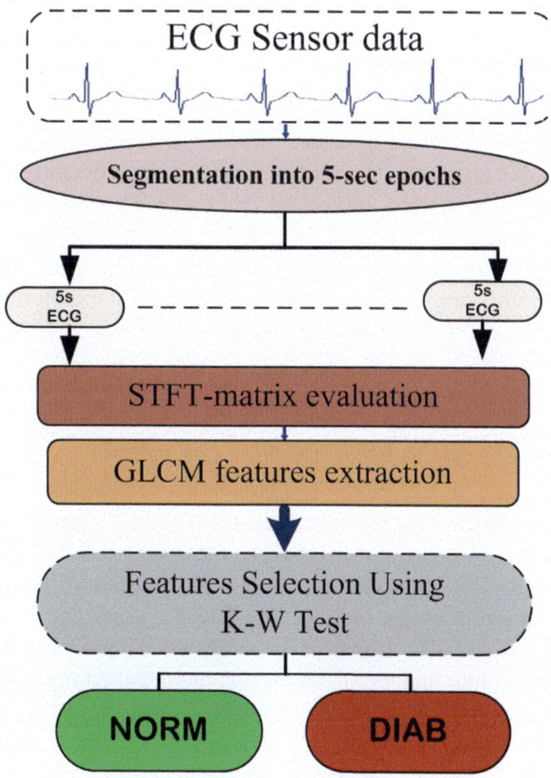

Fig. 1 Layered diagram of the proposed discrimination system

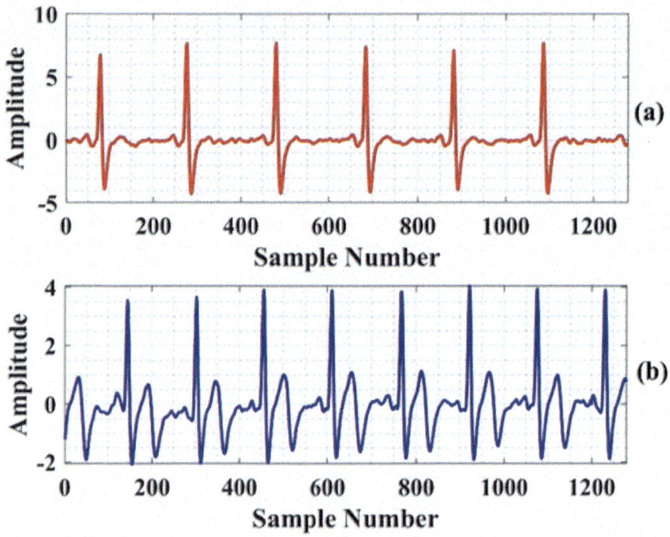

Fig. 2 Example of ECG signal of diabetic and normal subjects

2.1 ECG Signals

The ECG signals have been taken by submitting EULA form [10]. 51 normal and 35 diabetic patients with type-2 diabetes participated in this study. If blood glucose level is more than 160 mg/dL, then they are considered as diabetic ECG signals. In this study, 24,630 ECG signals out of that 9560 diabetic ECG and 15,070 healthy ECG signals are used for analysis [10]. The candidates involved in this work are not suffering from any BP and cardiovascular abnormalities. The example of ECG signals of diabetic and healthy are shown in Fig. 2.

2.2 Short-Time Fourier Transform

The Short-Time Fourier Transform (STFT) provides a way to analyze how the frequency content of a signal changes over time. It is a time–frequency representation of a signal, which means it allows you to examine the signal's frequency components at different points in time. The STFT is particularly useful in various applications, including audio signal processing, speech analysis, and image processing (for 2D signals like images, a similar technique called the STFT can be applied along both dimensions). The STFT is a mathematical transformation used to analyze the time-varying frequency content of a signal. The formula for computing the STFT of a continuous signal $x(t)$ is as follows [11]:

$$\mathrm{STFT}(t,f) = \int s(t)w(t-\tau)e^{-2jf\tau}d\tau$$

where

$\mathrm{STFT}(t, f)$ is the complex-valued result of the STFT.

$x(t)$ is ECG signals.

$w(t)$ is the Hamming window function.

τ (tau) represents a time variable used for integration.

f is the frequency variable.

In practice, the STFT is often implemented using discrete data and numerical methods. The resulting STFT values are typically represented in a 2D matrix and the matrix entries hold the complex values representing the signal's frequency content at each time and frequency point. These complex values can then be used to compute the magnitude and phase information for analysis and visualization.

2.3 Gray-Level Co-occurrence Matrix Features

The GLCM is a powerful tool used in image processing and computer vision for texture analysis. It characterizes the spatial relationships between pixel values in an image and extracts valuable information about texture patterns. GLCM features are derived from this matrix and provide valuable insights into the texture properties of an image.

GLCM Formation: GLCM is constructed by counting the occurrences of pairs of pixel values at a specified spatial relationship or distance within an image. The matrix is typically square, with each entry representing the frequency of a particular pixel pair occurring at a given offset (distance and angle) within the image.

Texture Information: GLCM features capture information related to the arrangement of pixel values in an image, which is essential for texture characterization. Texture properties such as roughness, smoothness, coarseness, and directionality can be quantified using GLCM features [12].

GLCM Features:

Contrast: Measures the local variations or differences in pixel values. Higher contrast values indicate more pronounced texture variations.

Homogeneity: Indicates how closely the GLCM elements are distributed around the diagonal. Higher homogeneity values signify more uniform textures.

Energy: Represents the sum of squared elements in the GLCM. Higher energy values correspond to more ordered textures.

Correlation: Measures the linear dependency between pixel values in the GLCM. It provides insights into the directionality of texture.

2.4 *Analysis Using Kruskal–Wallis Test*

The Kruskal–Wallis test is commonly used for determine the discriminant different classes in term of probability value, but the assumptions of parametric tests like analysis of variance (ANOVA) are not met (e.g., when the data is not normally distributed or when there are unequal variances between groups). It's a robust method for comparing multiple groups, and it's widely used in various fields, including medicine, social sciences, and environmental science.

3 Results and Discussions

The objective analysis of ECG signals as input to discriminate diabetic patients ECG and healthy ECG signals. The 86 subjects ECG dataset are utilized for this study. Subject-wise, recorded ECG data is divided into fragments of 5 s and marked as either diabetic or healthy ECG signals. The STFT is applied to obtain time frequency distribution of ECG signals, and time–frequency distribution provides more inside variation in ECG signals during diabetes. If the two-dimensional time–frequency distribution is employed to extract GLCM features, then the GLCM features use an input to discriminate diabetic and healthy ECG signals using Kruskal–Wallis test. The Krushkal-wallis test can used to features selection for discrimination diabetic and normal ECG signals [13]. From Table 1, it can be seen that the p-value is less than 0.05 in all four GLCM features. The four GLCM features can be used as discriminator indicators for analysis and discrimination diabetic and normal ECG signals.

Table 1 *P*-values of GLCM features for discrimination diabetic and normal ECG signals

Features	Diabetic class	Healthy class	*P*-value
Contrast	0.1071 ± 0.0121	0.1059 ± 0.0129	3.68×10^{-14}
Homogeneity	0.9562 ± 0.0039	0.9569 ± 0.0040	8.63×10^{-56}
Energy	0.3303 ± 0.0342	0.3365 ± 0.0373	3.83×10^{-63}
Correlation	$0.9933 \pm 5.2933 \times 10^{-4}$	$0.9934 \pm 6.6346 \times 10^{-4}$	1.977×10^{-10}

4 Conclusions

In this paper, we explore the application of the STFT to distinguish between diabetic-affected and healthy control ECG signals. STFT is a nonstationary signal processing technique widely used to convert 1D data into a T–F matrix. From each of these T–F matrices, we extract four different GLCM features. To identify the most relevant features for discriminating between normal and diabetic ECG signals, we employ a non-parametric Kruskal–Wallis (K–W) test. Our findings conclude that all four extracted features—Contrast, Homogeneity, Energy, and Correlation—are highly suitable for distinguishing between these signal types. These results suggest that a methodology based on STFT–GLCM and the K–W test could be effectively employed to develop an automated screening system for diabetes identification in clinical settings. Future research will involve the utilization of suitable machine-learning algorithms to automate the detection of diabetes through the classification of ECG signals.

References

1. Sun H, Saeedi P, Karuranga S, Pinkepank M, Ogurtsova K, Duncan BB, Stein C, Basit A, Chan JC, Mbanya JC, Pavkov ME IDF diabetes atlas
2. Global, regional and country-level diabetes prevalence estimates for 2021 and projections for 2045. Diabetes research and clinical practice. 6 Dec 2021, 109119
3. Hippisley-Cox J, Coupland C (2016) Diabetes treatments and risk of amputation, blindness, severe kidney failure, hyperglycaemia, and hypoglycaemia: open cohort study in primary care. BMJ 30(352):i1450. https://doi.org/10.1136/bmj.i1450.PMID:27029547;PMCID:PMC 4816603
4. Shu T, Zhang B, Tang YY (2018) An improved noninvasive method to detect Diabetes Mellitus using the Probabilistic Collaborative Representation based classifier. Inf Sci 467:477–488
5. Acharya UR, Faust O, Sree SV, Ghista DN, Dua S, Joseph P, Tamura T et al (2013) An integrated diabetic index using heart rate variability signal features for diagnosis of diabetes. Comput Methods Biomech Biomed Eng 16(2):222–234
6. Pagani M (2000) Heart rate variability and autonomic diabetic neuropathy. Diabetes Nutr Metab 13(6):341–346
7. Risk M, Bril V, Broad Bridge C, Cohen A (2001) Heart rate variability measurement in diabetic neuropathy: review of methods. Diabetes Technol Ther 3(1):63–76
8. Pagani M, Malfatto G, Pierini S, Casati R, Masu AM, Poli M, Malliani A (1988) Spectral analysis of heart rate variability in the assessment of autonomic diabetic neuropathy. J Auton Nerv Syst 23(2):143–153
9. Jian LW, Lim T-C (2013) Automated detection of diabetes by means of higher order spectral features obtained from heart rate signals. J Med Imaging Health Inform 3(3):440–447
10. Gupta K, Bajaj V (2022) A robust framework for automated screening of diabetic patient using ECG signals. IEEE Sens J 24(22):24222–24229
11. Kwok HK, Jones DL (2000) Improved instantaneous frequency estimation using an adaptive short-time Fourier transform. IEEE Trans Signal Process 48(10):2964–2972

12. Garg M, Dhiman G (2021) A novel content-based image retrieval approach for classification using GLCM features and texture fused LBP variants. Neural Comput Appl 33:1311–1328
13. Jain A, Verma A, Verma AK (2023) Non-invasive and automatic identification of diabetes using ECG signals. Special issue on machine intelligence and deep learning in healthcare decision making of Int J Electr Electron Res (IJEER) 11(2):418–425

Utilizing Recurrent Neural Networks for Cervical Cancer Cell Classification Using Pap Smear Images

Mithlesh Arya, Praveen Kumar Yadav, Megha Gupta, and Abha Jain

Abstract In India, cervical cancer is the second highest type of cancer in women after breast cancer. Due to few nonspecific symptoms in early stages routine screening by Pap smear of targeted population can be very useful for the early diagnosis of cervical cancer. So, in this paper, a detailed exploration of Recurrent Neural Networks (RNNs) for the classification of cervical cancer cell images. Through meticulous evaluation of learning dynamics, we offer insights into the model's learning capability, generalization, and potential interventions to optimize performance. The results demonstrate the model's exemplary accuracy is 97.14% with 100 epochs, highlighting its potential as a robust tool for biomedical image analysis.

Keywords Cervical cancer · Pap smear test · Recurrent Neural Networks

1 Introduction

Cervical cancer is the cancer of females [1]. In the world after breast cancer, cervical cancer is the second main cause of mortality in women. Human Papilloma Virus (HPV) is the main cause of cervical cancer. This virus transfer in human due to multiple sex partners. There are no pre-symptoms of the cervical cancer. But few symptoms are there like bleeding after intercourse, bleeding between the period duration, pelvic pain, and watery and smelly discharge. If these symptoms are there, females can go for screening test. So, the Pap smear test is the most popular method for the screening of the cervical cancer. In the developed countries, females go for the Pap smear test after every 3 year. In India, 60% of population live in the village. Women are not aware about the cervical cancer. Even they are also not aware about the hygiene during the periods. Still women are using cloths during the periods. Many cervical cancer camps have been organized by the Indian Government. But

M. Arya (✉) · P. K. Yadav · M. Gupta · A. Jain
Swami Keshvanand Institute of Technology, Management and Gramothan, Jaipur, Rajasthan, India
e-mail: mithlesh.arya@skit.ac.in

D.-I. Stroe et al. (eds.), *Emerging Electronics and Automation*, Lecture Notes in Electrical Engineering 1202, https://doi.org/10.1007/978-981-97-3090-2_14

women are not aware about it and the rate of mortality is increasing. Cervical cancer is classified into three classes according the TBS (the Bethesda System) [2]:

1. Normal
2. LSIL(Low-Grade Intraepithelial Lesion) or mild dysplasia.
3. HSIL(High-Grade Intraepithelial Lesion) or moderate and sever dysplasia.

With the help of Automated Cervical Cancer (ACC) method, we can detect cervical cancer using images only. Many methods have been proposed for this and most of the methods are using machine learning and segmentation methods for the detection of cervical cancer. But the main challenging task is to detect the abnormality in the presence of debris. When images are captured using slides. The images contain lots of other elements other than cervical cells like dye, red blood cell, and inflammatory cell. So the main objective is to detect the abnormality of cervical cell in the presence of debris. Figure 1 shows the debris in cervical cells.

Cervical cancer cell image classification plays a pivotal role in diagnostic applications. With the advent of deep learning techniques, there has been a surge in research focusing on automating this classification. In the context of cervical cancer cell image classification, where images can be viewed as sequences of pixel values, RNNs offer a novel approach. By treating images as sequential data, RNNs can potentially capture intricate patterns and relationships that might be missed by conventional methods. This research delves into the application of RNNs for cervical cancer cell image classification, exploring their efficacy, challenges, and potential areas for enhancement.

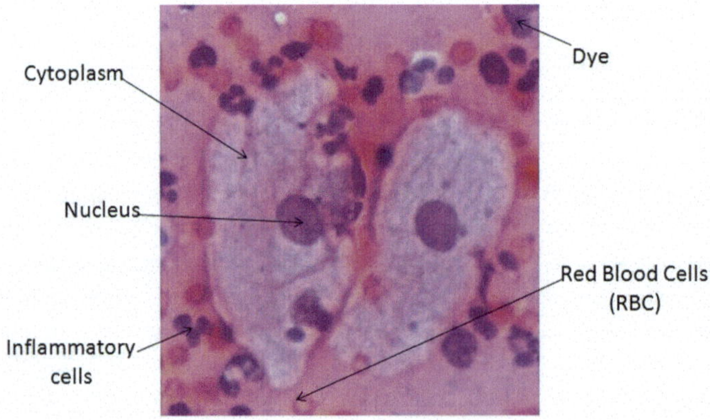

Fig. 1 Debris present in the cervical cell

2 Literature Review

In the literature, many methods have been proposed by the researchers. Basically, detection of normal and abnormal cell is done using these three steps [3].

1. In the first step, pre-processing and segmentation methods are used to find the Region of Interest (ROI). For the pre-processing of the image contrast enhancement, filters and brightness methods are used. Segmentation step used threshold, clustering, seed region growing methods.
2. In the second step after extracting ROI from image, features are extracted from the ROI. There are multiple features are extracted like texture, color, shape, and frequency.
3. In the last step, we get the table of features from the image. Now data is divided into training and testing part. Training part is used to train the model and remaining part is used to test the accuracy of the model using different classification methods like SVM, CNN, KNN, ANN, etc.

The main purpose of paper [4] was to design automated cervical cancer detection system. In this paper, two deep learning CNN architectures are proposed to detect cervical cancer using colposcopy images. The first proposed model is VGG19 (TL) and the second proposed model is CYENET (Colposcopy Ensemble Network). In this paper, dataset consists of 5679 colposcopy images and is divided into type 1, type 2, and type 3 categories. The proposed model includes three stages: 1. Data processing of dataset, 2. CNN training model, and 3. Classification of data. It includes 15 convolution layers, 12 activation layers, and 5 max-pooling layers. VGG19 is giving 73.3% and CYENET is giving 92.3% accuracy.

In paper [5], Mask Regional Convolutional Neural Network (Mask R-CNN) is used to detect and analyze the nucleus of normal and abnormal cells. Dataset they used is based on liquid-based histological slides collected from Thammasat University. Their proposed model consists of two stages. In first stage, Mask R-CNN is used to find the region of interest (ROI) and features are extracted. In stage 2, Feature map of the image is passed in Region Proposal Network (RPN) and it is used to reduce the feature map. Using this method, they are getting 91.7% accuracy.

In paper [6], machine learning method is used to classify the cells into seven classes. They used Herlev dataset for the validation of the work. First pre-processing is done on all the images using filters. After that segmentation method is used to find ROI using Active Contour model by Local Gaussian Fitting Energy (LGFE). After ROI detection, 22 geometric and texture features are extracted. In geometric features, area of cell, area of nucleus, nucleus and cytoplasm ratio, and eccentricity of cell and nucleus are extracted. For texture features, first-order and second-order statistical features are extracted. On the extracted 22 features PCA is applied and features are reduced. On the reduced features SVM with linear, Gaussian RBF, Polynomial and Quadratic kernel are used. They are clamming 95% accuracy by this method.

In paper [7], convolution neural network is used to classify the normal and abnormal cells. For the pre-processing Gaussian and bilateral filters are used on the images and these images are forwarded to convolution network with cascade structure. After that softmax layer is used to find the probability of the cells. In their proposed work using CNN they are getting only 0.562 precision.

The reviewed literature underscores the significance of deep learning techniques, particularly CNNs, in the domain of image classification and blood cell analysis [8–10]. While RNNs have been acknowledged for their prowess in tasks like speech recognition and time series forecasting, their specific application in blood cell image classification remains an area with limited exploration. This presents an opportunity for further research, especially given the unique sequential nature of flattened image data, which RNNs might be adept at handling.

3 Methodology and Result Analysis

The novelty in the work lies in the application of Recurrent Neural Networks (RNN) for cervical cancer cell classification using Pap smear images. Pap smear images inherently contain sequential information related to the progression of cell structures. Unlike traditional image classification approaches that may treat images as independent entities, the use of RNNs acknowledges and leverages the temporal dependencies present in cell sequences. This allows the model to capture nuanced changes and patterns over time, potentially enhancing the accuracy of classification.

In our pursuit to develop a robust classifier for cervical cancer cell images, we embarked on a systematic approach encompassing data pre-processing, model architecture design, and training dynamics. Figure 2 shows the flowchart of our proposed method.

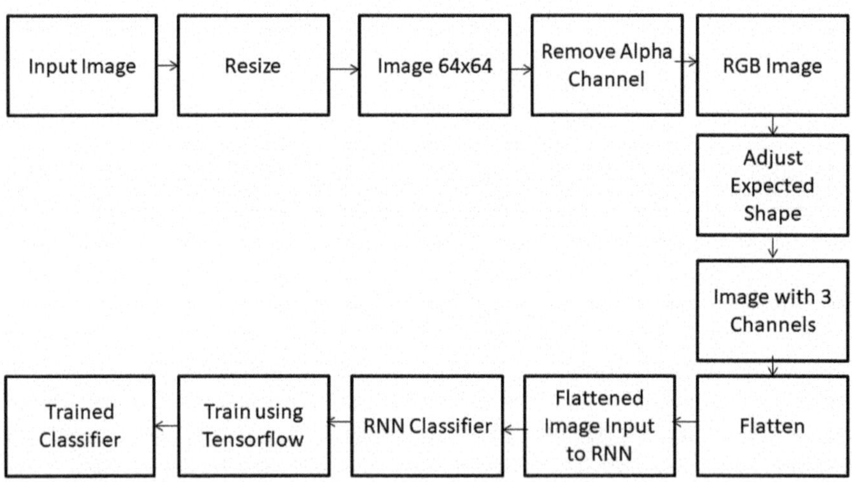

Fig. 2 Flow of proposed method

3.1 *Data Pre-processing*

To ensure that our model received consistent and standardized input, we developed a pre-processing pipeline tailored for cervical cancer cell images. The images were sourced from a specified directory, and each image underwent a series of transformations:

- RGB Conversion: Given the variability in image formats, we standardized all images to the RGB format, ensuring a consistent three-channel input.
- Resizing: Recognizing the potential computational challenges of processing high-resolution images, we resized all images to a uniform dimension of 64×64 pixels. This dimension was chosen as a balance between retaining critical image details and computational efficiency.
- Normalization: To facilitate faster and more stable convergence during training, we normalized the pixel values of the images to a range between 0 and 1. This step is crucial in mitigating potential issues related to varying scales in the input data.

3.2 *Model Architecture*

With the pre-processed data in hand, we designed a deep learning model leveraging the capabilities of Recurrent Neural Networks (RNNs):

- Input Layer: The model was architected to accept flattened image arrays of size 12288, corresponding to the reshaped dimensions of our 64×64 images.
- Simple RNN Layer: Recognizing the sequential nature of our flattened image data, we incorporated a Simple RNN layer with 64 units. This layer, equipped with a ReLU activation function, was pivotal in capturing the inherent sequential patterns in the data.
- Dense Output Layer: To produce class probabilities for our binary classification task, we employed a dense layer with two units. A softmax activation function was used to ensure the output values were in the range [0, 1] and summed to one.

3.3 *Training Dynamics*

The model was compiled using the Adam optimizer, a popular choice for deep learning tasks due to its adaptive learning rate capabilities. We employed the sparse categorical cross-entropy loss function, suitable for classification tasks with integer labels. Training was conducted over 100 epochs, with batches of 32 images. We also incorporated a validation dataset into our training regimen. This allowed us to monitor the model's performance on unseen data, providing insights into its generalization capabilities and aiding in early detection of potential over fitting (see Figs. 3, 4, 5 and 6).

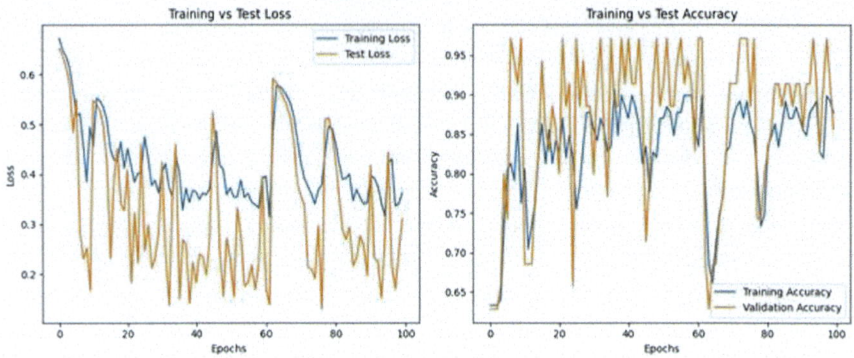

Fig. 3 Training versus test loss and test accuracy

	precision	recall	f1-score	support
0	1.00	1.00	1.00	11
1	1.00	1.00	1.00	13
accuracy			1.00	24
macro avg	1.00	1.00	1.00	24
weighted avg	1.00	1.00	1.00	24

Fig. 4 Precision, recall, F1-score, and support

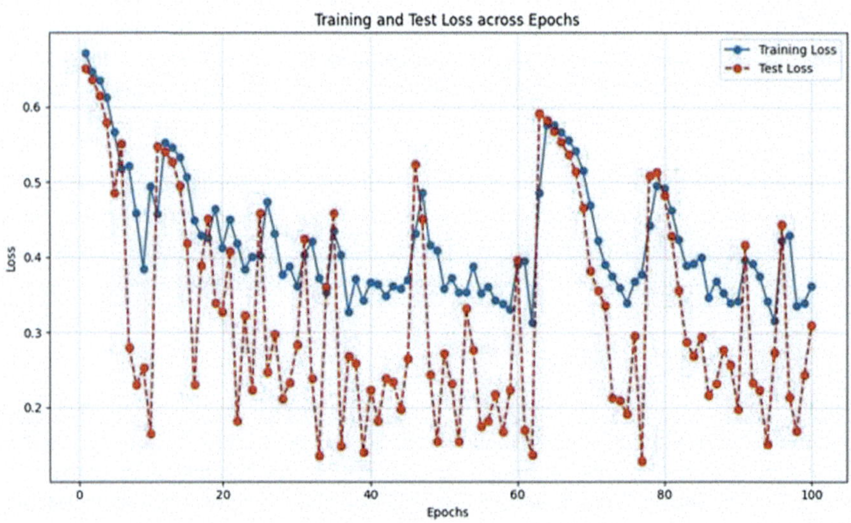

Fig. 5 Training and testing loss across epochs

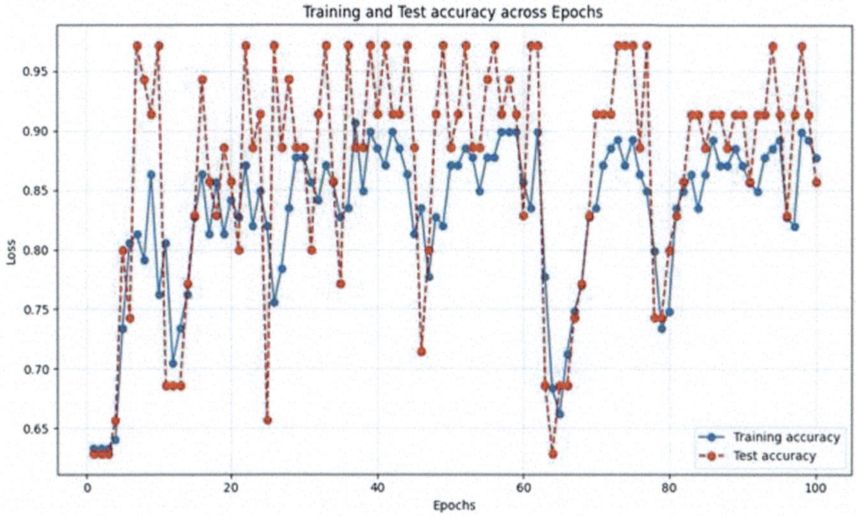

Fig. 6 Training and testing accuracy across epochs

The model's performance metrics attest to its reliability and robustness in classifying cervical cancer cell images. The high precision and recall values for both classes underscore its potential in real-world diagnostic applications. Strategies such as early stopping and regularization were considered to counteract over fitting, ensuring the model's optimal performance.

We are getting 97.14% accuracy with 100 epochs value. In paper [9], Mask R-CNN is used for the classification of cervical cells and the achieved accuracy is 91.7%. The choice of RNNs in this context is likely driven by the temporal nature of cell sequences and the inherent sequential information present in image data. RNNs, with their ability to capture dependencies over time, could prove beneficial in modeling the dynamic characteristics of cell structures and changes.

Compared to other AI schemes, the use of RNNs might offer an advantage in handling sequential patterns and relationships within the image data. For instance, traditional convolutional neural networks (CNNs) are powerful in capturing spatial features but may overlook temporal dependencies. RNNs, being inherently sequential models, can capture both spatial and temporal dependencies in the data, making them suitable for tasks where the order of information matters.

4 Conclusion

The utilization of RNNs for cervical cancer cell image classification has shown promising results. The detailed evaluation of the model's learning dynamics provided valuable insights into its strengths and areas for improvement. With its exemplary performance metrics, the model stands as a testament to the potential of deep learning techniques in advancing biomedical image analysis. However, like all research, ours is not without its limitations. While RNNs offer a novel approach to image classification, there remain avenues for further optimization and enhancement. The potential integration of attention mechanisms, exploration of hybrid architectures, and the incorporation of advanced pre-processing techniques are just a few directions that future research could venture into. In summation, our research underscores the potential of RNNs in revolutionizing cervical cancer cell image analysis. As we stand on the cusp of a new era in medical diagnostics, it is imperative to continue exploring and harnessing the capabilities of advanced computational techniques. Our study serves as a stepping stone in this direction, paving the way for more nuanced and effective diagnostic tools in the future.

References

1. www.mayoclinic.org/diseases-conditions/cervical-cancer/symptoms-causes
2. Cibas Edmund S, Ali Syed Z (2009) The Bethesda system for reporting thyroid cytopathology. Thyroid 19(11):1159–1165
3. Arya Mithlesh, Mittal Namita, Singh Girdhari (2020) Three segmentation techniques to predict the dysplasia in cervical cells in the presence of debris. Multimed Tools Appl 79:24157–24172
4. Chandran V, Sumithra MG, Karthick A, George T, Deivakani M, Elakkiya B, Subramaniam U, Manoharan S (2021) Diagnosis of cervical cancer based on ensemble deep learning network using colposcopy images. BioMed research international 2021
5. Sompawong N, Mopan J, Pooprasert P, Himakhun W, Suwannarurk K, Ngamvirojcharoen J, Vachiramon T, Tantibundhit C (2019) Automated pap smear cervical cancer screening using deep learning. In: 2019 41st annual international conference of the IEEE engineering in medicine and biology society (EMBC). IEEE, pp 7044–7048
6. Arora A, Tripathi A, Bhan A (2021) Classification of cervical cancer detection using machine learning algorithms. In: 2021 6th international conference on inventive computation technologies (ICICT). IEEE, pp 827–835
7. Shindo T, Watanabe T, Yamada K, Watanabe H (2023) Accuracy improvement of object detection in VVC coded video using YOLO-v7 features. arXiv:2304.00689
8. Chitra B, Kumar SS (2022) Early cervical cancer diagnosis using Sooty tern-optimized CNN-LSTM classifier. Int J Imaging Syst Technol 32(6):1846–1860
9. Gupta A, Parveen A, Kumar A, Yadav P (2022) Advancement in deep learning methods for diagnosis and prognosis of cervical cancer. Curr Genomics 23(4):234
10. Yan Yufang, Zhao Kui, Cao Jilong, Ma Huimin (2021) Prediction research of cervical cancer clinical events based on recurrent neural network. Procedia Comput Sci 183:221–229

Machine Learning Based Citrus Fruit Disease Prediction Using Support Vector Machine Algorithm

D. Mohanapriya, S. K. Kabilesh, R. Swaranambigai, B. Saritha, and S. V. K. Deepikapriya

Abstract The recognition of fruit diseases is rapidly becoming a significant area in computer vision. Besides reducing fruit production, plant diseases also have a profound impact on the country's economy. Therefore, predicting plant illnesses and implementing remedial actions are of utmost importance. Plant illness directly affects the crop's quality and overall productivity. To achieve swift and accurate disease prediction, an automated method is used to identify and diagnose affected leaves within a short span of time. Through the appropriate use of diagnosis and classification techniques, potential losses can be prevented. The features used for identifying plant diseases include color, strength, area of attack, and dimension. The current research primarily focuses on guava leaf disease prediction using K-means clustering techniques. In this proposed study, we aim to identify orange fruit and leaf diseases. To achieve this, K-means clustering is applied to segment the orange fruit images into distinct clusters. We leverage the gray-level co-occurrence matrix (GLCM) approach with the clustered data to extract the global features as feature vectors, and the orientation gradient approach is utilized to uncover real features. Implementing machine learning algorithms has significantly enhanced the accuracy of the system, achieving a remarkable 96% accuracy rate.

Keywords Gray Level Co-occurrence Matrix (GLCM) · Machine learning · Support Vector Machine (SVM) algorithm

D. Mohanapriya (✉) · R. Swaranambigai
Jai Shriram Engineering College, Tiruppur, Tamil Nadu 638660, India
e-mail: mohanapriya@jayshriram.edu.in

S. V. K. Deepikapriya
KGISL Institute of Technology, Coimbatore, Tamil Nadu 641035, India

S. K. Kabilesh
Jansons Institute of Technology, Coimbatore, Tamilnadu 641659, India

B. Saritha
Erode Sengunthar Engineering College, Erode, Tamilnadu 638057, India

D.-I. Stroe et al. (eds.), *Emerging Electronics and Automation*, Lecture Notes in Electrical Engineering 1202, https://doi.org/10.1007/978-981-97-3090-2_15

1 Introduction

In the field of processing plant images, the growth of both the national and global economies depends on agricultural production [1]. Fruit trees play a vital role in the economic growth of every state. Among the well-known fruit plant species, the citrus plant stands out, being rich in vitamin C and popular in the Middle East, Africa, and the Indian subcontinent [2]. Oranges, a significant source of citrus fruit, are estimated to be around 157.98 million tons by the Food and Agriculture Organization of the United Nations FAO (FAOSTAT 2019) [3]. Producing high-quality, disease-free fruits at a minimal cost is achievable through proper mechanized standards and preventive techniques [1, 4]. However, citrus farming is currently facing threats from various fruit diseases. To prevent the spread of diseases during post-harvest processing and shipment, an automatic early disease prediction system is crucial. Identifying the severity of the disease and its effects is a significant factor in determining the scope of the issue. The development of mechanized agriculture is hindered by the high cost and ineffectiveness of human illness assessment [5]. Some of the most prevalent diseases affecting citrus fruits include citrus black spot, citrus bacterial canker, citrus blight, citrus scab, and greening [6, 7]. The existing methods for diagnosing these diseases rely on manual perception by individuals or experts, making the process time-consuming and expensive [8]. However, computer vision researchers have recently developed automated algorithms that can identify fruit diseases without human intervention [7, 9]. Automating the diagnosis and classification of fruit disorders is not only essential for increasing agricultural yields but also for protecting fruits and plants from infections [10]. The process of diagnosing fruit diseases involves various steps, such as raw image preprocessing, disease segmentation using image processing methods like K-means and saliency methods, feature extraction (such as color, shape, and texture), reduction of redundant features, and classification using machine learning (ML) classifiers (such as support vector machines (SVM), K-nearest neighbor (KNN), and a few others). Image processing algorithms such as super pixel clustering, histogram of orientation gradient (PHOG), and K-mean clustering [11, 12]. The procedures that follow boost accuracy when a limited dataset is available. These methods, however, did not perform well for large image datasets, which decreased classification accuracy. However, they are only capable of performing the sequential image processing tasks of image segmentation using clustering and other techniques [5, 6], feature extraction [7, 11], and pattern recognition using SVM [8, 9] and the KNN method [13]. Conventional ML techniques are very successful at identifying and diagnosing plant diseases. The Department of Agriculture and Nutrition projects that there will be 10 billion people on the planet by 2050. As a result, farming buildings must rise by up to 75% in order to accommodate developed occupants and residents. On the other hand, the excessive use of chemicals like disinfectants and nematicides to treat infections on plants and leaves poses a potential threat to the environment [14, 15]. In contrast, deep learning algorithms for recognition utilize multiple representational layers [16]. These layers are formed by assembling straightforward, non-linear components that transform presentations at various levels into higher, slightly more abstract representations. These adaptable adjustments can be combined to learn more intricate functions. In this study, the identification and categorization of orange leaf disease were achieved using a database. Pre-processing techniques were employed to acquire images, incorporating area-based, edge-based, and model-based image processing approaches [17, 18]. Subsequently, boundary or area detection and spot detection algorithms were used for image augmentation. Furthermore, the image was segmented using the K-mean clustering method resulting in three different clusters (the final cluster, cluster 1, and cluster 2) [3, 18]. The number of indices, clusters, and distances were specified based on the maximum and minimum settings [2]. To extract global features from the clustered data, the gray-level co-occurrence matrix method was used to generate feature vectors.

2 Proposed System

ML detection methods offer an effective approach for identifying infected fruit. The process involves several key steps, starting with image collection, followed by preprocessing, segmentation, feature extraction, and finally, classification based on the predicted type of disease. Figure 1 shows the block diagram of the Citrus disease prediction system.

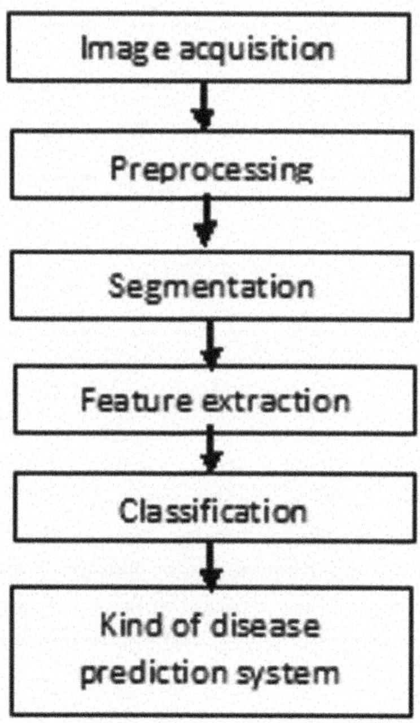

Fig. 1 Block diagram

2.1 Research Motivation

Existing research on the classification of citrus fruit diseases from images [7, 17, 18] employs convolutional neural networks (CNN) and deep learning algorithms. Jeon and Rhee [18] utilized a deep learning network to classify citrus diseases from photographic pictures. However, this neural network model experienced performance loss due to poor parameter and layer selection. To improve the identification of citrus ailments in both fruit and leaf photos, a new set of layers and parameter values are proposed in the suggested SVM technique. Comprehensive tests are performed by employing several ML model variants, and the outcomes are compared with the benchmark study. Based on the results, CNN models are highly recommended for accurately categorizing citrus diseases from fruit and leaf images. The suggested method demonstrates the capability to detect various illnesses in citrus plants, including black spot, canker, scab, greening, and melanose.

2.2 Problem Formulation

In this study, we aim to address a classification issue that involves identifying citrus plant diseases from images of both fruits and leaves. The objective is to create a model or classifier capable of accurately assigning the appropriate disease class to each citrus disease (CFi = CF1, CF2, CF3....CF4). To achieve this, we will explore and compare various ML classifiers as well as the CNN deep learning model. In contrast to existing deep learning classifiers that often rely on conventional feature representation methods, our proposed deep learning classifier will leverage a suitable number of layers and a well-organized arrangement of parameters.

2.3 Our Contributions

The planned research offers several noteworthy contributions: (i) The implementation of a cutting-edge SVM model that can effectively classify citrus diseases into five classes, namely black spot, canker, scab, greening, and melanose. (ii) The utilization of a well-structured ML technique with an adequate number of layers, which enhances the accuracy and reliability of the disease classification process. (iii) A comprehensive comparison of the efficiency of the proposed model with other existing methods, providing valuable insights into its effectiveness and superiority. The proposed approach holds the potential to advance the development of more sophisticated and practical applications for identifying plant diseases based on visual indications.

2.4 Related Work

There has been extensive research conducted over the years on the diagnosis of leaf and fruit diseases. Studies have engaged in discussions and analyses of various ML and pattern recognition techniques, all aimed at enhancing the accuracy of the diagnosis system. These innovative techniques have found applications in different types of crops, including wheat [19], rice [17], and maize [15]. Neural network techniques have also been extensively studied for the identification and categorization of diseases from images of plant leaves and fruit, as proposed by Kerkech et al. [20]. Additionally, Kabilesh et al. [21] successfully detected citrus canker and Huanglongbing using an SVM and a fluorescent imaging system. The approach correctly classified Huanglongbing and zinc deficiency with an accuracy of 95%, while correctly classifying citrus canker and scab with an accuracy of 97.8%. Padmavathi and Thangadurai discuss recursively separated weighted histogram equalization techniques to more precisely distinguish citrus disorders. The noise in the photos of citrus is removed in the second stage. The remedies provided improve the citrus photos' quality so that they may be used for more in-depth study. The K-means algorithm is offered to identify the affected region by segmentation in pre-processed orange images, as discussed by Patel et al. Using information from the training set, the SVM classifier was used to categorize the color, texture, and shape of the afflicted area. The accuracy generated by GLCM models was 67.74%. SVM, linear discriminant analysis (LDA), K-nearest neighbors (KNN), and multi-layer perceptron (MLP) methods were used to predict the citrus fruit disease by Singh et al. in 2020. Color and texture traits were gathered while sick leaf regions were segmented using the K-means clustering technique. The analysis of variance F-test is employed to determine which traits are the most important. Pathogens are identified using the above-mentioned techniques. The study comprised 236 sick citrus leaves in total. Precision was increased by combining color and textural features. The accuracy of the methods LDA, MLP, KNN, and SVM was 84.32%, 81.36%, 77.12%, and 80.93%, respectively, for color and texture features. This system is based on an understanding of the citrus disease [17]. Deep learning algorithms are preferred for feature extraction and threshold segmentation, which help to predict and classify citrus diseases in a short span of time [18]. Xing et al. represented a model for citrus disease and pest identification by using a weakly connected convolutional network. They applied different CNN models to a self-dataset for citrus. The SENet-16 model's test precision was 88.36% for the NIN-16 model, whereas it was 80.93% for the SENet-16 model. Numerous techniques have been used in literature to identify fruit illnesses. Citrus illness is identified with an automated technique by utilizing scant data, as discussed by Athmaja and Hanumanthappa [22]. For precisely diagnosing citrus diseases, they proposed a patch-based identification network with an embedded module, a cluster prototype module, and a simple neural network classifier. The accuracy of this prediction model was 95.04%, which is high compared to other techniques, according to the proposed technique. Jhuria et al. [14] presented a computer vision-based automated technique for diagnosing citrus illnesses. The five stages of the suggested technique are preprocessing, segmentation, feature extraction and reduction, fusion, and classification. The noise was removed from the images, and sick spots were found using the watershed technique. The final classification is performed using a multi-class SVM classifier. The datasets used were those from the citrus image gallery, plant village, and self-collections; the average accuracy achieved was 95.5%. Cruz et al. [13] introduced a weekly DenseNet16 CNN architecture for the identification of citrus diseases. Even though the method given can be employed on mobile phones, it was successfully used. The results showed that the weak DenseNet16 network has improved accuracy with a limited number of parameters. Additionally, they used the idea of reuse to overcome an optimization difficulty. Kong et al. [15] described an algorithmic method for categorizing images of citrus fruits. Orange and Kinnow are the two fruits they have classified. In this method, images of Orange and Kinnow are acquired and preprocessed. Those images of fruits were then segmented using edge detection and segmentation techniques. The final recognition is then performed using three different CNN models, with the best accuracy being 92.25%. An automated deep learning-based approach for classifying fruit illnesses was presented by Hamuda et al. [17]. They constructed a deep-convolutional neural network for precise classification after taking into account issues like the similarity between infections caused by various fruit diseases. The scientists took into account five illnesses, including bacterial spots, black rot, scab, rust, and powdery mildew. The experimental process indicated that the proposed strategy had enhanced accuracy. A brand-new strategy for categorizing two different forms of fruit diseases, such as apple and grape diseases, was presented by Jahan Mohan et al. [23]. They took into account difficulties including concave edges, color inconsistency, irregularity, visibility, scale, and origin. An end-to-end system has been developed based on contrast stretching, disease diagnosis, feature extraction and fusion, feature selection, and feature classification to address these issues. First, contaminated zones are identified, and an Adaptive and quartile deviation-based approach is used to generate the binary images. Fusion is done by the weighted coefficient of correlation. Then, classification is accomplished utilizing multi-class SVM, rank-based correlation, and the choice of meaningful features. The accuracy of the presented method was 97.7%, which was higher than that of more recent studies. ML-based end-to-end architectures have been used to categorize plant diseases. While traditional ML algorithms involve a step known as handcrafted feature extraction for the classification of plant diseases, deep learning automatically extracts features from raw pictures. Furthermore, the features that SVM techniques derive from binary images are not necessary for ML [20]. This approach differs from the above-mentioned methodologies in that an improved optimization algorithm is employed for optimal feature selection.

3 Proposed Methodology

This study suggests a multi-layer CNN for classifying the citrus fruit images and leaf images affected by various diseases.

3.1 Dataset Acquisition and Splitting

Datasets are needed at every stage of the image analysis research process, from algorithm training to algorithm evaluation. In this study, 200 images were collected from the plant village and citrus datasets. The plant village benchmark provides images related to various plant illnesses. Each of the three groups of tainted images featured a distinct citrus fruit or leaf ailment. The datasets looked at melanoses, black spots, cankers, scabs, and other disorders. In this study, only 2 diseases, such as canker and black spot, were detected in both orange fruit and leaves (see Table 1).

The steps in the suggested methodology for identifying citrus fruit illnesses include obtaining the image from the dataset and detecting the disease on the orange fruit plant. The image will be pre-processed by region-based, edge-based, and model-based image processing techniques. Following the stage of picture pre-processing, image segmentation will be carried out using K-means clustering. Boundary detection and spot detection algorithms are used for segmentation. The number of feature vectors and image pixels, as well as the global kind of feature sets, have all been counted using the SVM approach. To find the categories and detect the disease in citrus leaves, classification methodology was utilized.

Table 1 No of sample images taken for the system

Category	Disease	No. of images
Citrus leaf	Black spot	40
	Canker	40
	Healthy	20
	Total	100
Citrus fruit	Black spot	40
	Canker	40
	Healthy	20
	Total	100

3.2 Data for Training

Depending on the requirements of the experiment, up to 80% of the training data may be used to develop an SVM model.

The system can learn the features by training on a set of data. The dataset includes both defective and healthy fruit and leaf images as shown in the Figs. 2, 3 and 4.

Fig. 2 Dataset samples for canker

Fig. 3 Dataset samples for black spots

Fig. 4 Fresh fruit samples

4 Result and Discussions

The dataset is segregated into a 20% testing data set and an 80% training data set. The SVM model is trained on 80% of the dataset to extract the features and classification of the disease. The affected image is taken as input, and the system is evaluated.

4.1 Validation Information

The reduction of overfitting and underfitting happen based on the training phase's efficiency. The performance of the system may suffer while evaluating new data, which can also be accomplished by using data validation. As a result, to avoid efficiency errors when changing settings, a 10% validation set is used. In this regard, automated dataset validation is used, which allows for unbiased model assessment and lowers overfitting. The dataset for defective images of citrus fruit and leaf diseases is chosen and given in the results and discussion.

A natural image of a fruit is given as the first input to the system. Figures 5, 6 and 7 shows the result for black spot-diseased fruit, Canker diseased leaf and Normal fruit respectively. Once the regions of interest have been isolated, the isolated image is subjected to pre-processing followed by segmentation procedures. Subsequently, the features of the fruit are identified from the segmented regions, such as shape, color, texture, and size. After analyzing the accuracy and specifications of the input image, the system outputs relevant information along with the name of the specific disease

Fig. 5 Black spot diseased fruit

Fig. 6 Canker diseased leaf

detected in the fruit. This information includes details such as the probability of the disease being present, the severity of the disease, and any other relevant information that can aid in identifying and treating the disease.

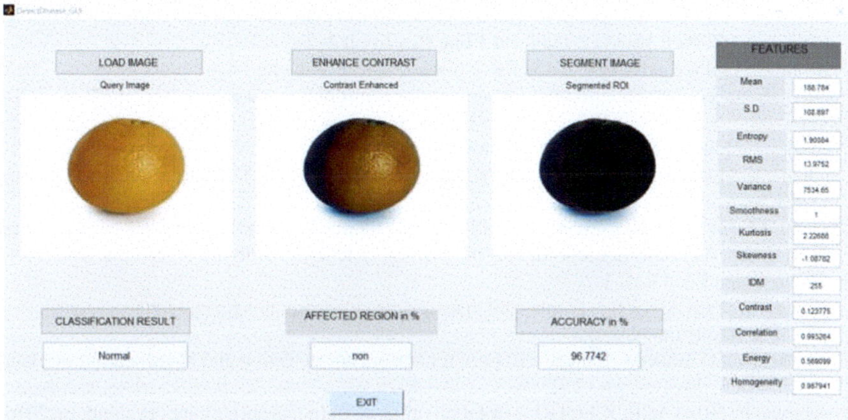

Fig. 7 Normal fruit

5 Conclusion

Agriculture is a crucial sector in India's growing economy, and plant infections pose a significant challenge to farmers. A SVM algorithm was trained on both normal citrus fruit and defected citrus fruits and leaves in order to provide the solution for this problem. The program effectively identified healthy fruits that had been affected by illnesses. To detect contaminated fruits early and avoid severe losses to agriculture, the economy, and the environment, various techniques were used, such as k-means clustering and object tracking. In order to increase the precision of disease detection, the image was segmented using the k-means clustering technique, which separated the image into smaller clusters. The classification of diseases is done by SVM, which improves the accuracy of the system to 96% and reduces the prediction time. This system helps to improve the cultivation of the orange fruit through performance parameters such as accuracy, precision, false acceptance, and rejection rate.

References

1. Sankaran S, Mishra A, Ehsani R, Davis C (2010) A review of advanced techniques for detecting plant diseases. Comput Electron Agric, 2–3
2. Ali H, Lali MI, Nawaz MZ, Sharif M, Saleem BA (2017) Symptom based automated detection of citrus diseases using color histogram and textural descriptors. Comput Electron Agric, 3–4
3. Mangla M, Mehta V, Mohanty N, Sharma N, Preetham A (2022) Statistical growth prediction analysis of rice crop with pixel-based mapping technique. Int J Artif Intell Soft Comput, 5–7
4. Jamdar AV, Patil AP (2017) Detection and classification of apple fruit diseases using K-means clustering and learning vector quantization neural network. Int J Sci Dev Res, 8–12
5. Satya Priya N, Nivetha E, Khilar R (May 2016) Efficient knowledge based system to detect disease in lemon leaf. ImpAl J Interdiscip Res, 3–6

6. Patil AR, Patil VI, Panchbhai BS (Apr 2017) Detection of plant diseases using image processing tools. Asha R. Patil Varsha I. Patil. Int J Eng Res Appl, 4–5
7. Khirade SD, Patil AB (Feb 2015) Plant disease detection using image processing. In: 2015 international conference on computing communication control and automation. IEEE, pp 768–771
8. Pandey C, Sethy PK, Behera SK, Vishwakarma J, Tande V (2022) Smart agriculture: technological advancements on agriculture—a systematical review. Deep Learn Sustain Agric, 13–15
9. Zhang M, Jin Y, Qiao H, Zheng F (2017) Product quality asymmetry and food safety: Investigating the 'one farm household, two production systems' of fruit and vegetable farmers in China. China Econ Rev, 1–6
10. Jamdar AV, Patil AP (2017) Apple fruit disease detection using image segmentation algorithm. Int J Res Trends Innov, 3–5
11. Al Bashish D, Braik M, Bani S (2016) A framework for detection and classification of plant leaf and stem diseases, pp 12–23
12. Agarwal P (2016) AGROCLOUD—Open surveillance of Indian agriculture via cloud. In: 2016 international conference on information technology (InCITe)-the next generation IT summit, pp 10–13
13. Cruz A, Ampatzidis Y, Pierro R, Materazzi A, Panattoni A, De Bellis L, Luvisi A (2019) Detection of grapevine yellows symptoms in Vitis vinifera L. with artificial intelligence. Comput Electron Agric 157:5–6
14. Jhuria M, Kumar A, Borse R (Dec 2013) Image processing for smart farming: detection of disease and fruit grading. In: 2013 IEEE second international conference on image information processing (ICIIP-2013). IEEE, pp 9–10
15. Kong L, Tan J, Huang J, Chen G, Wang S (2022) Edge-computing-driven internet of things: a survey. ACM Comput Surv, 10–17
16. Cruz P, Achir N, Sorbonne U, Nord P, Viana AC (2022) On the edge of the deployment: a survey on multi-access edge computing. ACM Comput Surv, 12–17
17. Hamuda E et al (2018) Improved image processing-based crop detection using Kalman filtering and the Hungarian algorithm. Comput Electron Agric 148:13–16
18. Jeon W-S, Rhee S-Y (Mar 2017) Plant leaf recognition using a convolution neural network. Int J Fuzzy Log Intell Syst 17(1):2–4
19. Revathi P, Hemalatha M (Jul 2012) Advance computing enrichment evaluation of cotton leaf spot disease detection using Image Edge detection. In: 2012 third international conference on computing, communication and networking technologies (ICCCNT'12). IEEE, pp 1–5
20. Kerkech M, Hafiane A, Canals R (2018) Deep leaning approach with colorimetric spaces and vegetation indices for vine diseases detection in UAV images. Comput Electron Agric 155:18–19
21. Kabilesh SK, Mohanapriya D, Suseendhar P, Indra J, Gunasekar T, Senthilvel N (2023) Research on artificial intelligence based fruit disease identification system (AI-FDIS) with the Internet of Things (IoT). J Intell Fuzzy Syst (Preprint), 1–16
22. Athmaja S, Hanumanthappa M (2016) Applications of mobile cloud computing and big data analytics in agriculture sector—a survey. In: International journal of advanced research in computer and communication engineering ICRITCSA MS Ramaiah Institute of Technology, Bangalore, vol 5, Special Issue 2, pp 1–10
23. Jahan Mohan K, Balasubramanian M, Palanivel S (2016) Detection and recognition of diseases from Paddy Plant leaf images, 1–6

Light Weighted YOLO Vehicle Tracker for Toll Gate Using Centroid Algorithm

Pankaj Pratap Singh, Shitala Prasad, Gyanjyoti Kalita, Jahnu Das, and Pallab Jyoti Mahanta

Abstract This paper focuses on vehicle detection at highway toll gates. The study involves the detection of vehicles within a custom dataset, which comprises four vehicle classes: car, truck, bus, and motorcycle. The YOLOv4 model is utilized for training and is complemented by the implementation of the Centroid tracking algorithm. In the proposed vehicle track dataset, each class comprises 1500 annotated images for training and 300 images in the testset. These images are sourced from the Google Open Images V6 dataset, and the OIDv4 toolkit is employed to scrape dataset images. The primary objective of this study is to train a vehicle detection model and evaluate the effectiveness of various tracking algorithms, with a specific focus on the centroid tracker for real-time application at the toll gates. Importantly, this model is designed to operate efficiently on conventional and cost-effective GPUs and demonstrates a commendable level of precision.

Keywords Vehicle detection · Centroid tracker · YOLOv4 · Batch size · Deep learning

1 Introduction

In recent years, the rise in the number of vehicles has led to escalating traffic issues [1], necessitating reliable monitoring through advanced analytics. This surge in vehicles, coupled with reckless driving behavior, has resulted in various problems, including

P. P. Singh (✉) · G. Kalita · J. Das · P. J. Mahanta
Department of Computer Science and Engineering, CIT Kokrajhar, Assam, India
e-mail: pankajp.singh@cit.ac.in

S. Prasad (✉)
School of Mathematics and Computer Science, IIT Goa, Goa, India
e-mail: shitala@iitgoa.ac.in

D.-I. Stroe et al. (eds.), *Emerging Electronics and Automation*, Lecture Notes in Electrical Engineering 1202, https://doi.org/10.1007/978-981-97-3090-2_16

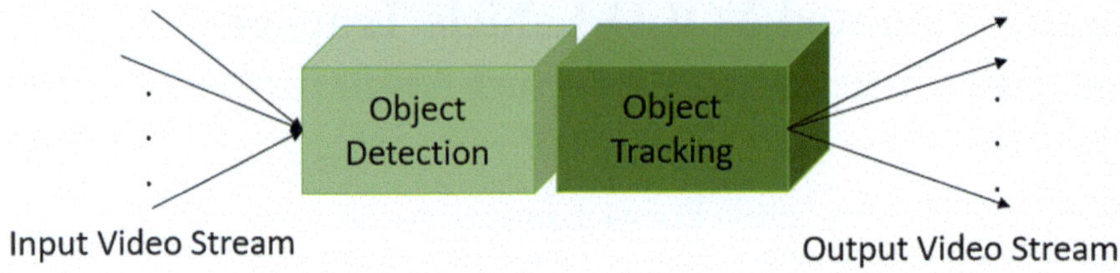

Fig. 1 A generic block diagram for detection and tracking of vehicles

increased traffic congestion, rising road accidents, hit-and-runs, and speeding violations. Toll gates, in particular, are accident-prone areas, often witnessing collisions due to vehicles failing to slow down before entering [2]. These accidents lead to casualties and damage to public property. A timely warning system could potentially save lives and prevent such incidents, alerting toll gate employees of incoming vehicles' over-speeding a few seconds before a collision. Advanced technologies like object detection and tracking have emerged, leveraging automated surveillance cameras to collect data for monitoring peak traffic hours and serving as evidence against traffic law violators.

The capability to detect and track vehicles in intersections, toll gates, and highways is crucial. This involves assigning a unique ID to each vehicle, achieved through a tracking algorithm combined with detection, see Fig. 1. Real-time object detection on standard Graphics Processing Units (GPUs) enables widespread adoption at an affordable cost. Although highly accurate neural networks like Region-based Convolutional Neural Networks (R-CNN) [3], Region-based Fully Convolutional Network (R-FCN) [4], Convolutional Neural Network (CNN) [5], and Recurrent Neural Network (RNN) [6] exist. They do not operate in real-time and require multiple GPUs for training with large mini-batch sizes. This paper adopts the YOLOv4 technique for vehicle detection and employs the Centroid tracker as the tracking algorithm. YOLOv4, a CNN-based deep model, operates in real-time with very low latency on conventional GPU and is easily trainable on custom datasets, making it an ideal choice for this application.

1.1 YOLO Model

The preference for YOLOv4 over YOLOv5 was based on several factors. Firstly, the availability of well-documented research papers on YOLOv5 was limited, while YOLOv4 had more comprehensive documentation [7]. Secondly, YOLOv5 differs significantly from YOLOv4, as it is constructed on the PyTorch framework and features an entirely distinct architecture. Figure 2 illustrates the architectural distinctions between YOLOv4 and YOLOv5. The YOLO detection model offers several advantages over classifier-based systems. It excels in multi-object detection, allowing it to identify multiple objects within a single frame. During testing, the neural network

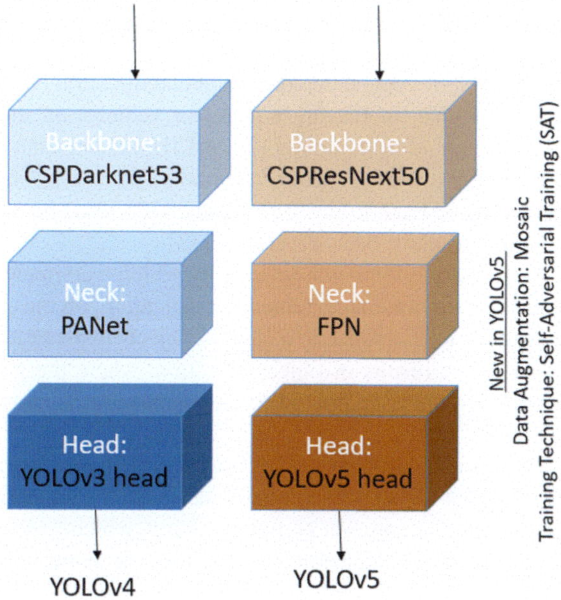

Fig. 2 A generic block diagram for detection and tracking of vehicles

examines the entire image, which informs its predictions based on the global context within the image. YOLO can make predictions using a single network evaluation, in contrast to systems like R-CNN [8], which require thousands of evaluations for a single image. This design makes YOLO exceptionally fast when compared to R-CNN [9], Fast R-CNN [9], and Faster R-CNN [10]. Moreover, YOLO's architecture supports end-to-end training and real-time processing speeds while maintaining a high mean average precision [11].

In this research paper, we have harnessed the capabilities of the YOLO detection model and integrated it with the Centroid Tracker, creating an end-to-end solution for object detection and tracking. Our focus was on addressing the challenge of vehicle detection and tracking at toll gates, especially when dealing with high density and speed. Our contributions can be summarized as follows:

- Introduction of a Subset of Vehicle Object Detection and Tracking Dataset. We have curated a specific dataset tailored to vehicle object detection and tracking.
- We have successfully combined the Centroid tracker and YOLOv4 detector to enhance vehicle detection at toll gates for an end-to-end learning.
- Through rigorous experimentation on our curated dataset, we have achieved state-of-the-art (SOTA) results, demonstrating the effectiveness of our methods.

The structure of the paper is as follows: Sect. 2 provides an overview of related works in the field of vehicle detection, Sect. 3 details our proposed system, and Sect. 4 presents the experimental results and analysis. Finally, we conclude our findings in Sect. 5, highlighting future research directions.

2 Related Works

The subject of vehicle detection and classification has garnered significant attention due to the ever-increasing number of vehicles on the road [12]. Various sensing methods, including computer vision, LIDAR, and radar, have been employed to detect and classify different types of vehicles. The origins of image processing techniques can be traced back to the 1960s and 1970s, and since then, numerous methods and technologies have been developed [13].

Object detection is the process by which a computer precisely identifies specific objects within an image. Object classification, on the other hand, seeks to recognize an object's resemblance to a class of objects in general. Object tracking is facilitated by using the results of object detection, where the position of an object from the previous frame is compared to its position in the current frame to determine if it's the same object.

This study's primary goal is to create a system capable of automatically detecting and tracking vehicles in both moving and static images and videos. Vehicle detection falls within the realm of computer vision, which focuses on how machines perceive the world rather than how humans do. The main objective of a vehicle detection model is to identify multiple vehicles in a video.

In this study, a vehicle detection and counting system is proposed. The highway road surface in the image is initially segmented into remote and proximal areas using a novel segmentation method [14]. Both areas are input into the YOLOv3 model to detect the location and type of vehicles. The directions of the vehicles are obtained using the ORB algorithm to count different vehicles and determine their driving direction. The methods proposed are validated using various highway surveillance videos in different scenes. The segmentation method is shown to offer high detection accuracy, particularly for small vehicle objects. The study also introduces object detection and tracking methods for highway surveillance scenes and establishes a high-definition vehicle object dataset. The YOLOv3 object detection algorithm is used to create an end-to-end highway vehicle detection model based on the extracted highway surface area [7].

In other scenarios, vehicle detection was performed using methods such as Fast R-CNN in various city traffic scenes [15], the introduction of the YOLOv5-NAM model with a normalization-based attention module [16], and the application of YOLOv4 for real-time vehicle detection and tracking in surveillance camera footage [17]. Additionally, a detection method based on Faster R-CNN with target tracking involving Camshift and Kalman filters was proposed [18]. To enhance vehicle classification accuracy, a novel bounding box (Bbox) was developed, and a vehicle tracking algorithm was created [19].

3 Proposed System

YOLO, short for "You Only Look Once," is a cutting-edge object detection algorithm. In contrast to earlier object detection methods like R-CNN and its improved version, Faster R-CNN, YOLO only requires a single pass of the image through its network. Traditional methods relied on scanning various regions of the image for detection. YOLO, as previously discussed, employs a distinctive neural network approach that leverages the entire image's features to simultaneously predict multiple boxes, each containing a distinct object.

3.1 Custom Dataset

In this study, we utilized a custom dataset that encompasses four distinct vehicle classes: Car, Bus, Truck, and Motorcycle. Each class was comprised of 1500 annotated images for training and 300 images for testing, see Table 1. To construct this dataset, we sourced images from the Google Open-Images V6 Dataset [20], a vast collection containing approximately 9 million annotated images with both image-level labels and object bounding boxes. The Open-Images dataset spans 600 classes, providing extensive coverage of various objects [21]. To gather images for our specific classes, we employed the OIDV4 toolkit, a Python3-based toolkit hosted on a GitHub repository maintained by the user EscVM. This toolkit streamlined the process of scraping images from the Open Images v6 dataset, given that the images were already annotated. However, it's worth noting that not all the collected images met our quality standards, and we manually curated the dataset by removing low-quality images.

3.2 A Framework Description

The system's architecture is visually represented in Fig. 3. We established this architecture by employing the YOLOv4 pre-trained weights file alongside a tailored "cfg" file, which allowed us to create an object detection class. This class plays a pivotal role within our vehicle tracking system.

In the context of object tracking, each bounding box is assigned a unique identifier by comparing the positions of their centroids between consecutive frames. When

Table 1 Customized vehicle detection dataset

Class name	#Training	#Testing
Car	1500	300
Bus	1500	300
Truck	1500	300
Motorcycle	1500	300

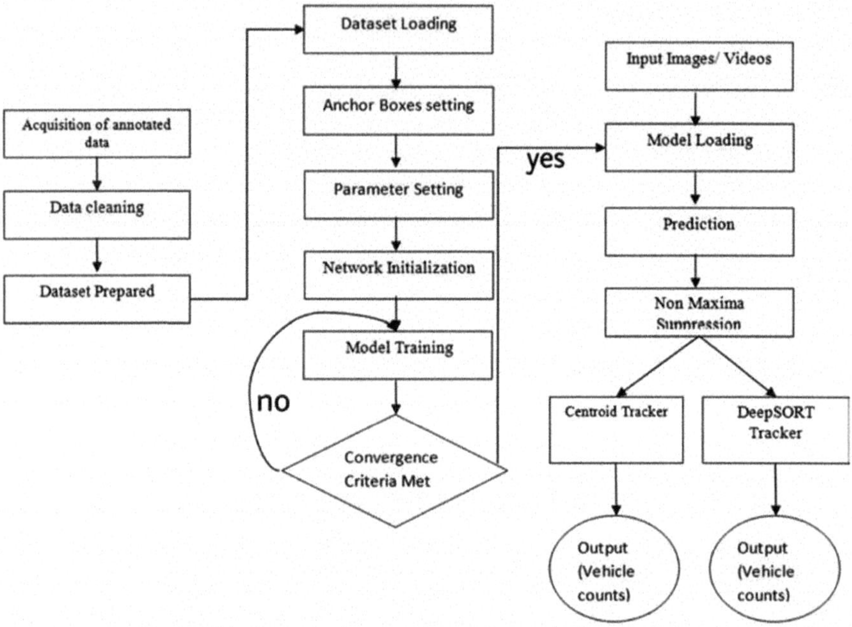

Fig. 3 A designed framework based on centroid tracer

the centroids are found to be in close proximity, the respective boxes are grouped together as part of the same object. The training process for our model, as depicted in Fig. 3, involves continuous iterations using our customized dataset until convergence is achieved. The convergence point is influenced by various hyperparameters and the underlying hardware configuration, which will be detailed in the subsequent subsection.

Following the model's training, it is loaded and subjected to testing using the designated test dataset for prediction purposes. Subsequently, the tracking algorithm is applied. In this study, we have explored and compared the performance of two tracking algorithms: DeepSORT and the Centroid tracker, as highlighted in Fig. 3.

3.3 Implementation Details

YOLOv4 training was conducted in the Google Colab cloud environment using an NVIDIA Tesla K80 GPU. In this experiment, Google Colab 2 was employed, utilizing a Python 3.1 environment. The system configuration included an AMD RyzenTM 4-core 3.6GHz CPU and 8GB RAM. The learning rate and batch size used in this paper are 0.001 and 32, respectively. The momentum was set to 0.9 for Adam optimizer.

4 Results and Analysis

After configuring the custom files, the model underwent training on the custom dataset, encompassing 8000 instances. This training process extended for over 12 h. During the initial stages, there was a notable surge in mean average precision, coupled with a significant decrease in loss. Around the 2500th iteration, the graph's progress leveled off, with mean average precision (mAP) steadily rising to 91% (refer to Fig. 4). Ultimately, the model achieved a mean average precision of 91.33% (Table 2).

The batch size, a critical hyperparameter, determines the number of samples processed before updating the internal model parameters. In cases where system resources, like a 16 GB RAM, are insufficient to accommodate the entire image

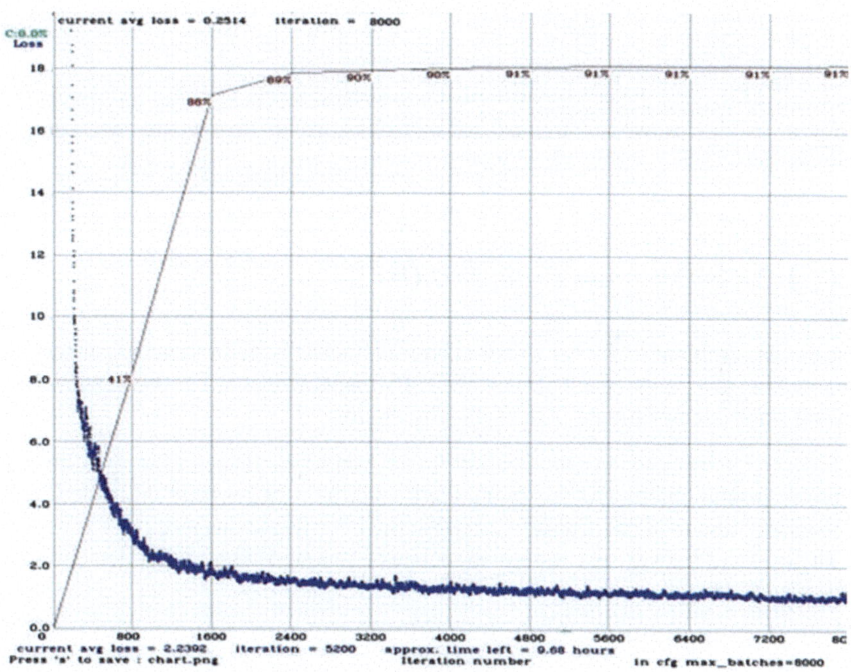

Fig. 4 Mean average precision (mAP) versus loss graph of our model

Table 2 Accuracy assessment of detected vehicle

Class Id	Class name	True positive	False positive	Average precision (%)
0	Car	297	19	93.98
1	Bus	286	31	89.63
2	Motorcycle	295	28	91.15
3	Truck	227	27	90.81

Fig. 5 Detected vehicles at toll gate showing centroid

dataset (24 GB), the dataset is partitioned into several batches or sets. Subsequently, the loss function value is computed, and weight updates are executed accordingly (Fig. 5).

4.1 Vehicle Detection Using YOLOv4

The model combines YOLOv4 with the centroid tracker to facilitate vehicle detection and tracking at a highway toll gate using CCTV camera videos. The process involves several stages demonstrated in the following figures:

- Figure 6a showcases an initialization frame, where the detector draws bounding boxes around objects. At this point, the tracker has not yet been initialized, as it requires more than one frame to establish the inter-frame distance.
- The centroid tracker successfully associates a unique ID with each detection, as depicted in Fig. 6.

Fig. 6 Detected vehicles at toll gate showing unique id using tracker

– In Fig. 6b, the centroid tracker assigns a new ID to the same vehicle, indicating continued tracking.
– Figure 8a and b demonstrate the centroid tracker effectively assigning a new ID to "car 4."

Subsequently, we proceeded to implement and evaluate alternative object detection models. Specifically, we conducted training sessions for YOLOv3-SPP and YOLOv5(L), where "L" denotes the large model variant of YOLOv5, utilizing our custom dataset. Figure 7 illustrates the comparative results of these models.

Among the models tested, YOLOv3-SPP achieved the highest average precision (AP); however, this came at the cost of increased latency. On the other hand, YOLOv5-large displayed improvements in terms of both latency and AP when compared to its predecessor. Notably, YOLOv4 exhibited a balanced performance in latency and AP.

Consequently, our choice for adoption centered around YOLOv4-CSP, featuring a leaky activation function. This model exhibited the highest AP while maintaining minimal latency, surpassing all other models in our evaluation (Table 3).

However, there were occasions when the Centroid tracker lost track of a vehicle and assigned a new ID to the same vehicle (as depicted in Fig. 6b). These situations led to inaccurate vehicle counts and subpar tracking, especially when

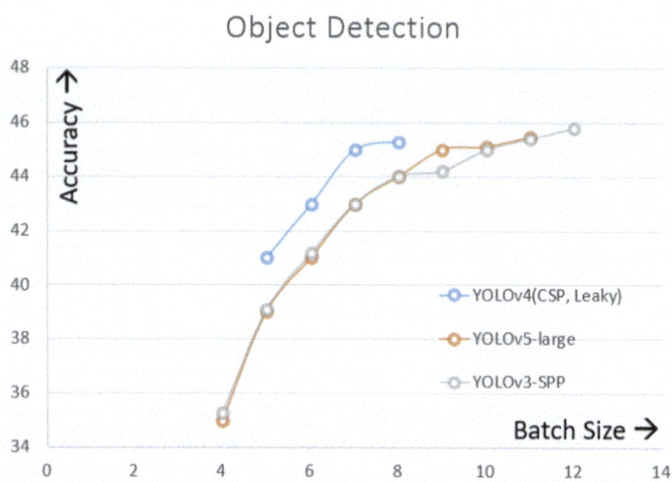

Fig. 7 Comparison graph for different YOLO models on COCO object detection

Table 3 Evaluation of quantitative parameters

Confidence threshold	Precision	Recall	F1-score	TP	FP	mAP
0.25	0.93	0.90	0.91	1106	105	0.913294

Fig. 8 Detected vehicles at toll gate showing new id for vehicle 4

Fig. 9 Detected vehicles at toll gate different approaches

Fig. 10 Detected vehicles at toll gate different approaches

objects were positioned near the frame's edges, corners, or partially within the frame.

In Fig. 9, we present additional qualitative comparisons of different trackers employed with YOLOv4. Each tracker showcases distinct detection and tracking features, as observed visually.

It's important to note the presence of a green box in Fig. 9b, utilized for vehicle counting. Vehicles that cross this designated area are considered positive counts for that specific vehicle, while any vehicles outside of this region are disregarded.

The outcomes presented in Fig. 10 exhibit diverse vehicles such as cars, trucks, buses, and others. This demonstration confirms the capability of our proposed methods in effectively detecting and quantifying these vehicles, paving the way for subsequent processing steps.

5 Conclusion

This paper focuses on utilizing YOLOv4 for vehicle detection in toll gate footage and incorporates a Centroid tracker algorithm for vehicle counting as they pass through the toll gate. After conducting comparative studies involving YOLOv4, YOLOv5, and YOLOv3, it was determined that YOLOv4 offers slightly superior mean average precision and lower latency, albeit with a trade-off involving larger model storage size. Following training on a custom dataset featuring four vehicle classes, the model achieved an impressive mean average precision of up to 91.3294%. These results highlight the promising potential of YOLOv4 in the domain of vehicle detection and tracking within toll gate areas. While the combination of the vehicle detector and Centroid tracker yielded somewhat inconsistent tracking and vehicle counting results. Further exploration is warranted to pair the trained vehicle detection model with an alternative tracking algorithm that also incorporates speed estimation.

References

1. Increasing vehicular density a major issue. https://timesofindia.indiatimes.com/readersblog/just-news/increasing-vehicular-density-a-major-issue-50196/. Accessed: 2023-03-15
2. Abraham A, Nagavarapu SC, Prasad S, Vyas P, Mathew LK (2022) In: 2022 17th international conference on control, automation, robotics and vision (ICARCV). IEEE, pp 397–404
3. Girshick R, Donahue J, Darrell T, Malik J (2015) Region-based convolutional networks for accurate object detection and segmentation. IEEE Trans Pattern Anal Mach Intell 38(1):142–158
4. Dai J, Li Y, He K, Sun J (2016) R-fcn: object detection via region-based fully convolutional networks. Adv Neural Inf Process Syst 29 (2016)
5. O'Shea K, Nash R (2015) An introduction to convolutional neural networks. arXiv preprint arXiv:1511.08458
6. Zaremba W, Sutskever I, Vinyals O (2014) Recurrent neural network regularization. arXiv preprint arXiv:1409.2329
7. Bochkovskiy A, Wang CY, Liao HYM (2020) Yolov4: optimal speed and accuracy of object detection. arXiv preprint arXiv:2004.10934
8. Girshick R, Donahue J, Darrell T, Malik J (2014) In: Proceedings of the IEEE conference on computer vision and pattern recognition, pp 580–587
9. Girshick R (2015) In: Proceedings of the IEEE international conference on computer vision, pp 1440–1448
10. Ren S, He K, Girshick R, Sun J (2015) Faster r-cnn: towards real-time object detection with region proposal networks. Adv Neural Inf Process Syst 28
11. Gautam V, Prasad S, Sinha S, Joint-yodnet: a light-weight object detector for uavs to achieve above 100fps
12. Slimani I, Zaarane A, Atouf I (2023) Traffic monitoring system for vehicle detection in day and night conditions. Transport Telecommun 24(3):256–265
13. Chen Z, Wang C, Wen C, Teng X, Chen Y, Guan H, Luo H, Cao L, Li J (2015) Vehicle detection in high-resolution aerial images via sparse representation and superpixels. IEEE Trans Geosci Remote Sens 54(1):103–116
14. Song H, Liang H, Li H, Dai Z, Yun X (2019) Vision-based vehicle detection and counting system using deep learning in highway scenes. Eur Transp Res Rev 11(1):1–16

15. Fan Q, Brown L, Smith J (2016) In: 2016 IEEE intelligent vehicles symposium (IV). IEEE, pp 124–129
16. Wang J, Dong Y, Zhao S, Zhang Z (2023) A high-precision vehicle detection and tracking method based on the attention mechanism. Sensors 23(2):724
17. Amrouche A, Bentrcia Y, Abed A, Hezil N (2022) In: 2022 7th international conference on image and signal processing and their applications (ISPA). IEEE, pp 1–5
18. Zhang Y, Wang J, Yang X (2017) In: Journal of physics: conference series, vol 887. IOP Publishing, p 012068
19. Azimjonov J, Özmen A (2021) A real-time vehicle detection and a novel vehicle tracking systems for estimating and monitoring traffic flow on highways. Adv Eng Inform 50:101393
20. Kuznetsova A, Rom H, Alldrin N, Uijlings J, Krasin I, Pont-Tuset J, Kamali S, Popov S, Malloci M, Kolesnikov A, Duerig T, Ferrari V (2020) The open images dataset v4: unified image classification, object detection, and visual relationship detection at scale. IJCV
21. Open images dataset v7 and extensions. https://storage.googleapis.com/openimages/web/index.html. Accessed: 2023 04 15

Coconut Maturity Level Identification Using Acoustic Vibration Information and Deep Learning

Apoorva N. Desai, Meenaxi M. Raikar, and Vishwanath P. Baligar

Abstract Coconut maturity identification plays a significant role in postharvest industrial processing and quality control. The manual process of identifying the coconut maturity includes acoustic vibrations, colour, size, and weight features, which require human expertise. The proposed system uses deep learning models to classify the coconut into two classes. In this paper, acoustic vibration information of the coconut, colour, and size are used to identify the coconut maturity levels. The two classes of the coconut classification considered are tender and mature. In determining the maturity of the coconut, three deep learning models used are ShuffleNet, GoogLeNet, and ResNet50. The captured dataset includes both the acoustic vibration information and image data. Both image and acoustic vibration (audio) based classification is performed. The accuracy obtained for ShuffleNet is 75.79%, GoogLeNet is 87.89%, and ResNet50 is 82% when image data is used. The acoustic-based classification is performed using Artificial Neural Network. The accuracy obtained for the audio data is 64%. The limitation of the present model is training and testing is performed on the offline stored images. The future plan is to classify the coconuts in real-time by providing the camera interface to the system.

Keywords GoogLeNet · ResNet50 · ShuffleNet · Artificial Neural Network (ANN) · Tender coconuts · Acoustic vibrations

1 Introduction

Coconut is a horticulture crop used for various purposes, grown mostly in the coastal area across the world. The scientific name for coconut is 'Cocos Nucifera'. The coconut is spherical or oval in shape. The outer part of the coconut is covered with the fibrous husk, and beneath is the hard shell, endosperm, and water content. The different types of coconut are tender coconut, semi-husked, husked, and copra. The tender coconut is mostly used as a refreshing drink. When the coconut is cut open, the

A. N. Desai · M. M. Raikar (✉) · V. P. Baligar
K. L. E. Technological University, Hubballi, 580031, India
e-mail: mmraikar@kletech.ac.in

D.-I. Stroe et al. (eds.), *Emerging Electronics and Automation*, Lecture Notes in Electrical Engineering 1202, https://doi.org/10.1007/978-981-97-3090-2_17

meat thickness depends on the maturity level of the fruit. The meat in tender coconut is soft and tasty. As the fruit maturity increases, the meat thickness increases. The tender coconut has more water content and soft meat. Mature coconuts have less water and thick meat. Based on the maturity level, it can be used as a refreshing drink or as an ingredient in salads and desserts. The over-mature coconut is used in cooking, whereas the copra is majorly used in oil extraction. The nutritional level of the coconut varies based on the maturity level, geographical location, soil, and environmental factors. Classification or sorting of the fruit aids in the export market and maintains quality measures. The traditional method of identifying the coconut maturity level is either by listening to the fingernail tapping sound or color and size. Human identification of the maturity level of coconut is time-consuming and error-prone in the mass production and shipment process. Hence there is a need to automate the process of classification, which increases the quality, economy, and productivity.

An experienced coconut grower is able to identify the maturity level of the produce by its look, sound, and mechanical properties, as shown in Fig. 1. Acoustic information data is an indicator that is used to identify the maturity of the coconut. Coconuts come in a variety of shapes and sizes and are popular all over southern India. Tall and short (dwarf) variations are the two types of trees. There are several different types of tall and dwarf plants, including west coast tall, tiptur tall, chandrakalpa and chowghat orange dwarf. Coconut can be identified by its color, shape, and size.

India is the highest producers of coconut in the year 2018, as per the data of Food and Agriculture Organization (FAO) United Nations (UN), with a production of 11,706,343 tons [21]. In 2020, 16.82 million tons of coconuts were produced. In year 2022, Indonesia is the highest coconut producer in the world. In 2021, India has produced 14.68 million tons of coconuts. Coconut cultivation has traditionally been concentrated in Kerala, Karnataka, and Tamil Nadu. Coconuts are mostly consumed in southern India. Andhra Pradesh, Maharashtra, Orissa, West Bengal, Assam, Gujarat, Pondicherry, and Goa, as well as the island administrations of Lakshadweep and Andaman & Nicobar, are all coconut producing states. The tree is thought to have originated in Asia's southeastern regions. It is also known as "Kalpavriksha," which means "heavenly tree." In the next section, related work is presented.

Fig. 1 Coconut maturity identification by the expert

Organization of the paper: The related work on different techniques for maturity identification of the coconut is presented in Sect. 2. The proposed system of applying the deep learning model to determine the maturity level of the coconut is presented in Sect. 3, which includes dataset collection, implementation and result analysis. Data is collected in the form of acoustic vibrations and image. In Sect. 4, the result analysis is presented which includes the comparative analysis of three deep learning models in term of accuracy and precision in detecting the maturity level of the coconut. In Sect. 5, the conclusion is presented detailing the accuracy obtained for each of the models.

2 Related Work

Machine vision is used for image classification, video classification, object identification, and semantic segmentation. Different applications of computer vision are face recognition and fingerprint matching, which use the dataset that contains the images, labels, and bounding boxes. The survey is performed for coconut maturity identification using different methods. In [1], the SVM is used to classify the coconuts into three categories, namely tender, mature, and copra, based on the texture and color information. Image-based identification of coconut maturity is performed in [2]. The acoustic vibration information is used in [3] to determine the maturity factor of the coconut. The three categories considered are malauhog, malakanin, and malakatad, with levels one, two, and three, respectively. The statistical parameters are considered to classify the maturity levels. In [4], the fuzzy logic concept is used in the identification of coconut maturity.

The coconut maturity identification task is carried out in [5] to improve the grading and packaging process. The physical, mechanical, and chemical characteristics are used to classify the coconut. In [6], the Arduino controller, in integration with the sound sensor, is used to detect coconut maturity. The Naïve Bayes method is used for the classification of the coconut. In [7], the audio signals are used to determine the maturity of the coconut using the Naïve Bayes, Bagging, and Sequential Minimal Optimization (SMO) techniques. The authors use the characteristics of sound to determine whether or not coconut can be used. If a coconut has sufficient amount of water and if is neither cracked nor rotted, it is considered as usable. For dataset collection, experiments are performed on granite slab surrounded by soft padding. In order to avoid any secondary sound, it is surrounded with cloth and sponge. Fast Fourier Transform (FFT) is performed on the extracted sound signal. This results in spectrum of complex values. The real part of the expression is deduced and standard deviation is performed. Other features include performing mean on the derived data. 'U' for malauhog, 'K' for malakanin, and 'T' for malakatad are the three different grades of coconut considered for classification based on sound waves [8].

The damage caused to the different types of coconut when these fall from a height is studied in [9]. The stages of the coconut are based on the age in months, 6–7 months are in M1, 7–8 months are considered as M2, and 8–9 months in category

M3 level of maturity. In [10], an improvised sound recording system is presented for the collection of the coconut acoustic vibrations using the microphone. The frequency distribution information and thickness of the coconut meat are taken for classification in [11].

The arc magnet defect is identified using acoustic information [12]. The Long Short Term Memory (LSTM) based Deep Learning (DL) model is used to identify the defect. An input audio sequence is mapped to a class or a scalar value in LSTM. The LSTM is formed of two gates: the input and forget gate. The input gate is used to pass new data flow in the network, whereas the forget gate is used to discard data that is no longer required to be retained in the network.

In [13], the sulfur content prediction is performed to identify the good quality coconut using images of the copra. The sulfur is used as a preservative for coconuts. The presence of sulfur in coconut causes breathing-related allergies in humans. The sulfur content analysis is carried out in [14] using the k-means clustering mechanism. In [15], machine and computer vision technology is used for classifying fruit and accomplished by spatial pix or spectral image processing. The freshness of the coconut is identified using the grey-stage Co-occurrence Matrix (GLCM) and K-means. Three hundred sixty-one images are present in the dataset [16]. In [17], seven distinct CNN architectures, such as VGG 16, VGG 19, Xception, Mobile Net, Inception V3, InceptionResNetV2, and ResNet50, are used to understand fundamental coconut maturity levels. In [18], the coconut stem identification is performed using image processing techniques. In [19], coconut color is analyzed using the Hue, Saturation and Value (HSV) scale. The density parameter is used to enhance the quality of the class results. The coconut dehusking device is designed to resolve the issues related to coconut [20]. In [23] deep learning models are applied to classify the network traffic data.

In [24], You Only Look Once (YOLO) deep learning model is used to count the objects in the image. YOLO DL model uses the CNN model and regression technique to identify the objects. In [25] the deep learning models are applied to identify the insect in the agricultural field for remote monitoring using the camera device. The Single Shot Detector (SSD) method is applied to detect the insects in the farm using the Internet of Things (IoT) technology. In [26], the moths are identified using the CNN model. The object tracking technique is applied to the track the moths. In [27], the combination of artificial intelligence and IoT is performed to detect the pests. The Long Short-Term Memory (LSTM) model is used for predicting the pest occurrence in the orchards. The YOLOv3 model is used to detect the pest where the live image feeds are captured by the Unmanned Ariel Vehicles (UAVs). In [28], the pest detection is automated using the deep learning and the IoT technology for precision agriculture. In [29], the insect object is detected using the deep learning model at the edge node. In [30] the classification of the surface is performed using the Robot Inertial Measurement Units (IMU) for autonomous movement of the robots.

3 Proposed Model

The proposed model is to classify the coconuts using deep learning models. The different component of the proposed system is presented in Fig. 2. The color, texture, and audio signals are captured to classify the coconut into two classes that are tender or mature. The objectives are as follows:

1. To generate the dataset for two types of coconut using color/texture and acoustic vibrations (audio signal).
2. To apply different deep learning models for the identification of coconut maturity level.
3. To evaluate the model for recall, accuracy, and precision.

The proposed system for coconut maturity identification is represented in Fig. 3. Three different models are used to classify the image data captured namely, GoogLeNet, ResNet50, and ShuffleNet. The optimizers used are Stochastic Descent Gradient Momentum (SDGM) and Adam.

3.1 Implementation

The coconut dataset is split into two sub-datasets, training having 80% of the dataset, and testing having 20% of the dataset respectively. The number of epochs specified for training the dataset is 10, with three iterations per epoch and a maximum of 30 iterations [23]. These networks are trained via transfer learning. The final layer is changed with the value set as two classes. The 'InitialLearnRate' parameter is set at 0.001. The data is validated to train the network. An epoch is a complete training cycle spanning the dataset training. The two forms of dataset collected are image and the acoustic information data.

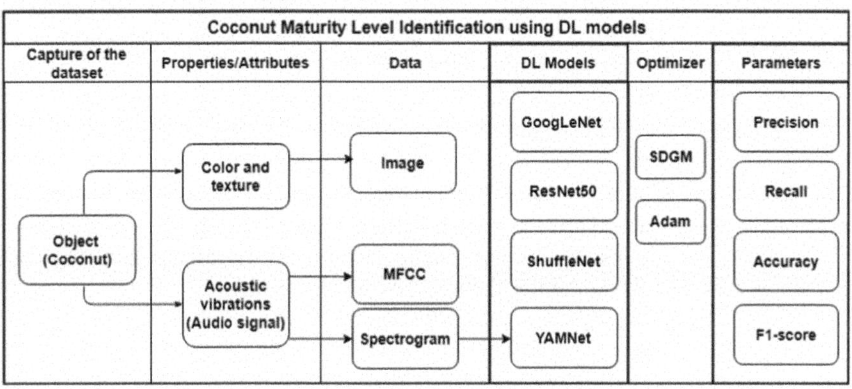

Fig. 2 Component of the coconut maturity level identification using DL models

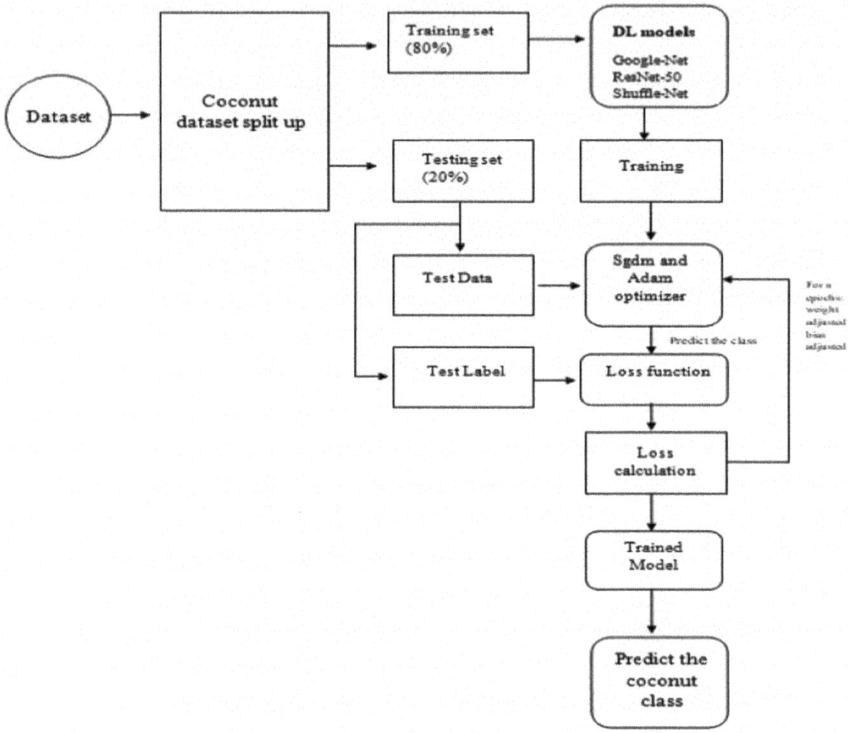

Fig. 3 Proposed system for coconut maturity identification

Acoustic vibrations (audio signal): The audio signals of the tender and copra coconut is captured using the microphone, as shown in Figs. 4 and 5. The sample audio signal is stored in the form of "*.csv" file, as shown in Table 1. The different methods used for classifying the audio signals are using spectrogram and Mel Frequency cepstral coefficients (MFCC). MFCC is also known as the most frequently considered coefficient. Mel is a standard for sound audibility for the human audio signal. An audio signal has real and imaginary parts. By performing the transforms, the imaginary part is eliminated, and only the real part is used for processing.

Spectrogram—the audio signal obtained after shaking the coconut is converted to the spectrogram. The spectrogram is a visual representation of the audio signal captured. The spectrogram is an image obtained by plotting time information against the frequency of the audio data, as shown in Figs. 6 and 7.

Transfer learning is used to identify the maturity factor of the coconut. The pre-trained network is used to identify the maturity level.

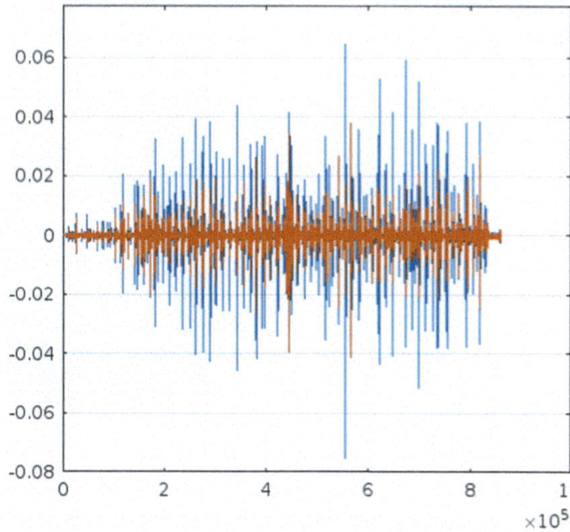

Fig. 4 Tender coconut audio signal

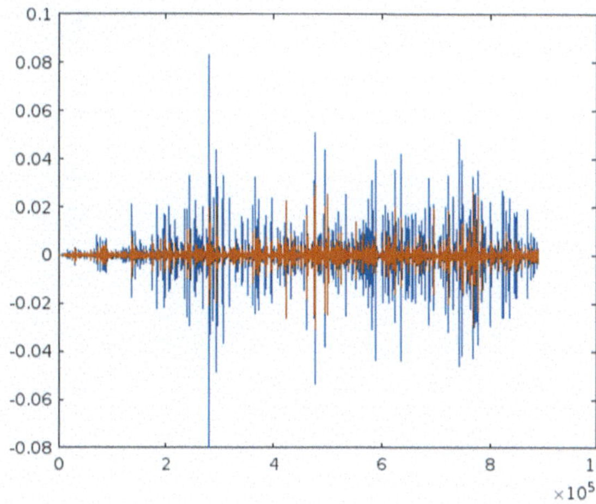

Fig. 5 Mature coconut audio signal

Table 1 Audio signal data of the coconut and its attributes stored in '*.csv' file

File	Attributes			Class
	fsID	Salience	Fold	
tendercoconut.wav	fileID	1	1	Tender
mature.wav	fileID	2	2	Mature

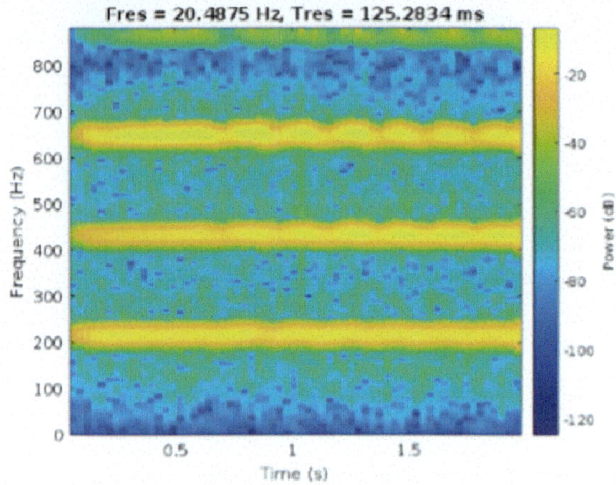

Fig. 6 Sample spectrogram of audio signal

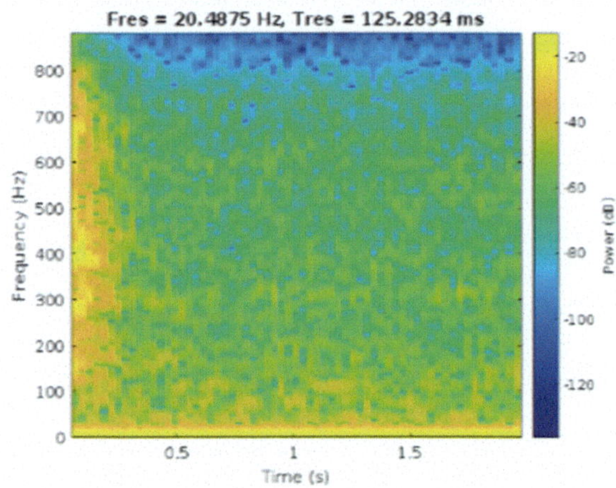

Fig. 7 Different audio signal spectrogram

3.2 Dataset

The images of coconuts are captured from the mobile digital cam 48 OIS + 15mp, with white background. 1000 images of coconuts are collected. The coconut classes are tender and mature. Each class has 500 images. The pre-processing of the coconut images is carried out to eliminate the noise. The collected images are resized to 224 X 224 X 3 pixels to generate the dataset. Coconuts with a thick brown husk are the most developed and include the least quantity of water. Because it's loaded only with water, a young green coconut should not make any sort of noise. If coconut sounds hear the liquid sloshing around, the coconut wall is probably too old (Table 2).

Table 2 Sample coconut dataset collected

Tender				
Mature				

3.3 Training and Testing

The features and target labels are included in the audio training dataset. The features are named as the audio file paths (X), and the target labels are named like the class names (Y). The sound sample and splitter retrieve features from the train and test sets. The Keras sequential API is used to create an ANN model. The categorized cross-entropy is used as the loss function, accuracy score as the metric, and Adam as the optimizer. In HDF5 format the model is created, and then trained. The model is trained using a batch size of 32 and 100 epochs. It serves as a checkpoint to determine how long it took to train the data.

The pre-processing of the "*.wav" file is performed utilizing 'Librosa' before loading it and extracting the MFCC characteristics. The image dataset collected is divided into two folders with the name tender and mature. The folder name given as "tender" and "mature" form the label for the class. In the folder tender all the images collected under category tender coconut are stored, similarly the images of category mature are stored in the folder "mature". The augmentation is applied to generate the dataset using the already collected data. The operations applied on the image are translation and scaling.

4 Result Analysis

The comparative analysis of the three DL models is presented. The plot of percentage accuracy and loss for ten epochs is presented in Table 3. The confusion matrix for classifying the mature and young coconuts using GoogLeNet, ResNet50 and ShuffleNet is presented in Table 3. The precision of 87%, 93% and 67% is obtained for GoogLeNet, ResNet50 and ShuffleNet respectively using the SDGM optimizer. The precision using the Adam optimizer for GoogLeNet, ResNet50 and ShuffleNet are 61, 74 and 64% respectively. The recall parameter obtained using the SDGM optimizer for GoogLeNet, ResNet50 and ShuffleNet is 89, 70 and 81% respectively. The recall parameter obtained using the Adam optimizer for GoogLeNet, ResNet50

Table 3 Performance comparison of DL models used in coconut identification

DL model	Training graph for the DL models	Confusion-Matrix
GoogLeNet	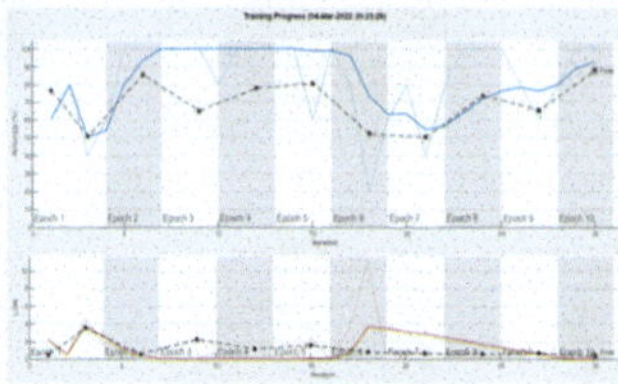	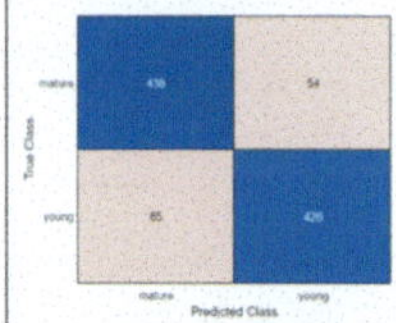
ResNet50	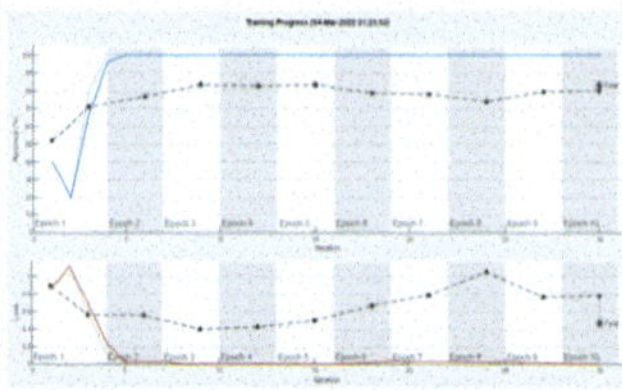	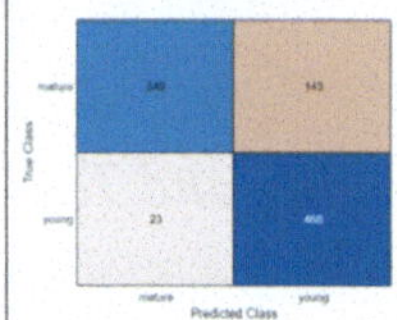
ShuffleNet	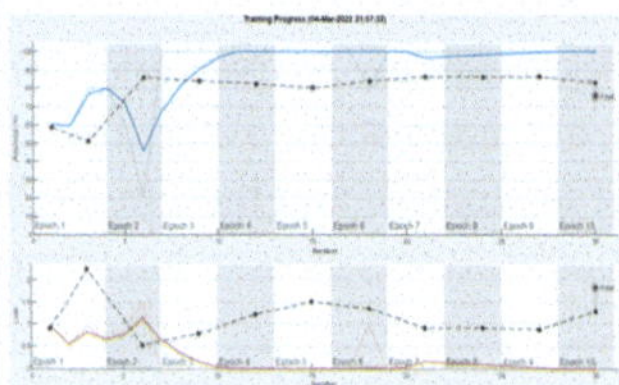	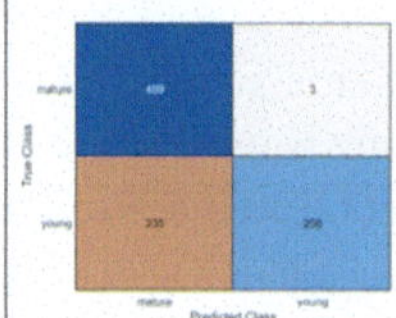

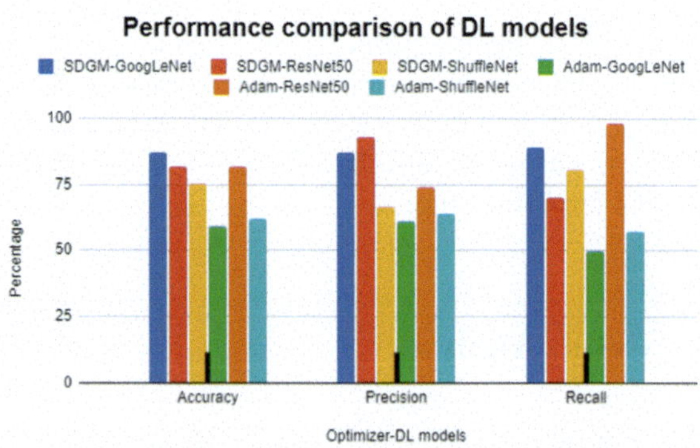

Fig. 8 Performance comparison of DL models

ShuffleNet is 50%, 98% and 57% respectively. The performance comparison of the three DL models is shown in Figs. 8, 9 and 10.

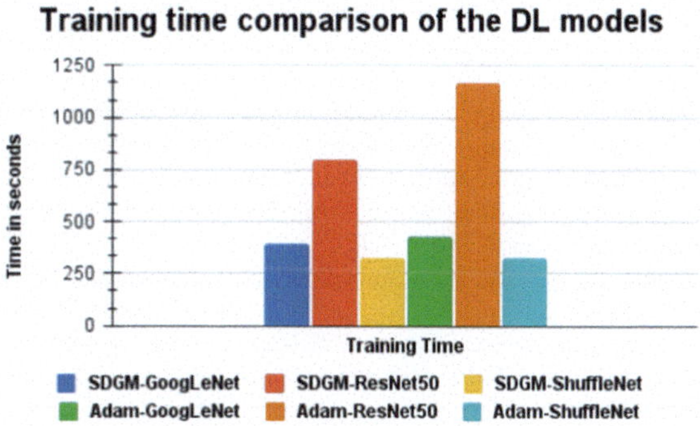

Fig. 9 Training time of the DL models

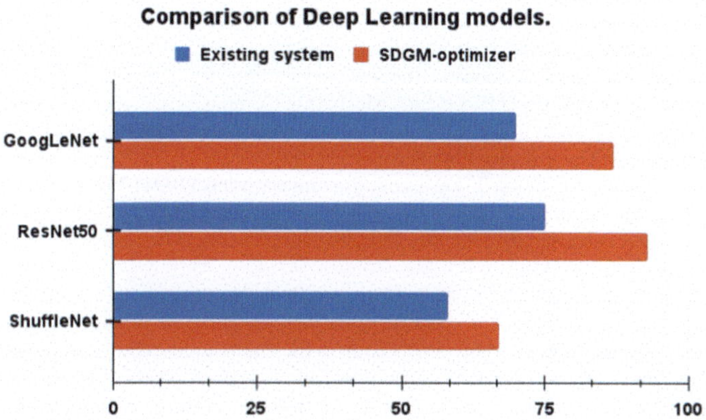

Fig. 10 Comparison of proposed system with existing system [22]

5 Conclusion

The image and audio signal information is used to classify the coconut. The expert knowledge for identification of the coconut as tender or mature is provided in the form of data (images and audio signal). The model is trained and tested for the classification of the coconuts. The three different models, GoogLeNet and ShuffleNet, and ResNet50, are used to classify the coconut dataset. The accuracy obtained from GoogLeNet is 87%, ShuffleNet is 75.79%, and Res-net is 83.11% using the SDGM optimizer, and for Adam optimizer, the accuracy for GoogLeNet is 59%, ResNet50 is 82%, and Shuffle Net is 62.26%. The audio signals of the tender and mature coconut have variations in frequency distribution. The audio signal is converted to a spectrogram and given to the DL models for classification. The MFCCs are used for classifying the audio signals. The future plan is to form a fusion of image and audio signals for the identification of the coconut maturity level.

References

1. Siddesha S, Niranjan SK (2019) Color and texture in classification of coconut. Int J Innov Technol Explor Eng (IJITEE) 8(8):1745–1750
2. Kallapur, Shrihari, MahithHegde, Adithya D Sanil, Raghavendra Pai, Sneha NS (2021) Identification of aromatic coconuts using image processing and machine learning techniques. In: Global Transitions Proceedings 2, 2: 441–447
3. Albalos, Rhey Marc Carrera et al. (2021) Acoustic impulse response of young coconut (Cocos nucifera L.) fruits in relation to maturity
4. Javel, Irister M, Argel A Bandala, Rodolfo C Salvador, RhenAnjerome R Bedruz, Elmer P Dadios, Ryan Rhay P Vicerra (2019) Coconut fruit maturity classification using fuzzy logic. In 2018 IEEE 10th International Conference on Humanoid, Nanotechnology, Information Technology, Communication and Control, Environment and Management (HNICEM), pp. 1–6. IEEE
5. Terdwongworakul, Anupun, BunditJarimopas, SongthamChaiyapong, Sher Paul Singh, Jay Singh (2010) Determination of physical, acoustical, mechanical, and chemical properties of fresh young coconut fruit for maturity separation. ASTM Int
6. Rahmawati, Diana, Haryanto Haryanto, Fahrus Sakariya (2022) The design of coconut maturity prediction device with acoustic frequency detection using naive bayes method based microcontroller. JEEMECS (J Electr Eng, Mechatron Comput Sci), 2.1(2019): 15–20. Web. 30
7. Thomas, Prashanth, Anita HB (2017) A novel automated method for coconut grading based on audioception. J Theor & Appl Inf Technol 95(10)
8. Gatchalian, Miflora M, Sonia Y De Leon, Toshimasa Yano (1994) Measurement of young coconut (Cocosnucifera, L.) maturity by sound waves. J Food Eng 23, 3: 253–276
9. Pascua AM (2017) Impact damage threshold of young coconut (Cocosnucifera L.). Int J Adv Agric Sci Technol 4(11):1–9
10. Fadchar, Nemilyn A, Jennifer C Dela Cruz (2020) A Non-Destructive approach of young coconut maturity detection using acoustic vibration and neural network. In: 2020 16th IEEE International Colloquium on Signal Processing & Its Applications (CSPA), pp 136–137. IEEE
11. Fernandez, Edmon O, Jonalyn E Escosio, Romeo L JordaJr, Michelle Tamase, John Carlo V Puno, Louiejim Hernandez, August C Thio-ac, Mark B Cruz, Camille V Lumogdang, Eddieson Real. Arduino-based sound acquisition system for coconut maturity classification using fast fourier transform algorithm
12. Li Q, Huang Q et al (2021) Combined convolutional and LSTM recurrent neural networks for internal defect detection of arc magnets under strong noises and variable object types. IEEE Access 9:71446–71460
13. Sagayaraj A et al (2021) Prediction of sulfur content in copra using machine learning algorithm. Appl Artif Intell 35(15):2228–2245
14. Sagayaraj, Stephen A, Ramya G, Dhanaraj N (2018) Analysis of sulphur content in copra. Int J Image Video Process, (ICTACT) 9, 02
15. Naik S, Patel B (2017) Machine vision based fruit classification and grading-a review. Int J Comput Appl 170(9):22–34
16. Anushya A (2020) Freshness detection of coconut via K-Means. Int J Innovat Technol Explor Eng
17. Subramanian, Parvathi, Tamil SelviSankar (2022) Coconut maturity recognition using convolutional neural network. In Comput Vis Mach Learn Agric, 2, pp 107–120. Springer, Singapore
18. Megalingam, Rajesh Kannan, Sakthiprasad K Manoharan, Rajesh G Reddy, Gone Sriteja, Ashwin Kashyap (2017) Color and contour based identification of stem of coconut bunch. In IOP Conference Series: Materials Science and Engineering, 225(1), p 012205. IOP Publishing
19. Noppon L, Nipon P (2019) Adapting a stereo camera system to determine coconut ages. In: 2019 16th International Conference on Electrical Engineering/Electronics, Computer, Telecommunications and Information Technology (ECTI-CON), 2019, pp 802–805.

20. Istas Fahrurrazi MFA, bin I. F., bin Mohd Nasir, M. A. A., bin Kamal, A. A., & binti Mohammad Noor, H. (2022) Coconut Dehusking machine. Multidiscip Appl Res Innov 3(1):565–576
21. Coconut production in India, https://en.wikipedia.org/wiki/Coconut_production_in_India, Accessed on 18th Nov 2022.
22. Meenaxi M Raikar, Meena SM, Chaitra Kuchanur, Shantala Girraddi, Pratiksha Benagi (2020) Classification and grading of okra-ladies finger using deep learning. Procedia Comput Sci, 171, Pages 2380–2389, ISSN 1877–0509
23. Raikar MM (2021) PhD Forum: Data traffic classification using deep learning models. In: 2021 IEEE 22nd International Symposium on a World of Wireless, Mobile and Multimedia Networks (WoWMoM), Pisa, Italy, pp 219-220, https://doi.org/10.1109/WoWMoM51794.2021.00035
24. Mamdouh, Nariman, Khattab, Ahmed (2021) YOLO-Based deep learning framework for olive fruit fly detection and counting. IEEE Access. PP 1–1. https://doi.org/10.1109/ACCESS.2021.3088075
25. Ramalingam, Balakrishnan Mohan, Rajesh Elara Pookkuttath, Sathian GÃ³mez, Braulio Borusu, Charan Teng, Tey Krishnasamy Tamilselvam, Yokhesh (2020) Remote insects trap monitoring system using deep learning framework and IoT. SensS (Basel, Switz). 20. https://doi.org/10.3390/s20185280
26. Bjerge, Kim, Sepstrup, Martin, Nielsen, Jakob, Helsing, Flemming, HÃ¸ye, Toke (2020) A light trap and computer vision system to detect and classify live moths (Lepidoptera) using tracking and deep learning. https://doi.org/10.1101/2020.03.18.996447
27. Chen C-J, Huang Y-Y, Li Y-S, Chang C-Y, Huang Y-M (2020) An AIoT based smart agricultural system for pests detection. IEEE Access. 8:1–1. https://doi.org/10.1109/ACCESS.2020.3024891
28. Albanese, Andrea Nardello, Matteo Brunelli, Davide (2021) Automated pest detection with DNN on the edge for precision agriculture. https://doi.org/10.36227/techrxiv.15087729.v1
29. Le, An Dinh et al. (2021) AlertTrap: A study on object detection in remote insects trap monitoring system using on-the-edge deep learning platform. ArXiv abs/2112.13341
30. Simranjit Singh, Mohit Sajwan, Gurbhej Singh, Anil Kumar Dixit, Amrinder Mehta (2023) Efficient surface detection for assisting Collaborative Robots. Robot Auton Syst, 161, 104339, ISSN 0921-8890, https://doi.org/10.1016/j.robot.2022.104339

Comparative Study of Pruned CNN Models for Tomato and Potato Disease Classification

Akshay Dheeraj, Satish Chand, and Aditya Raj

Abstract Plant disease affects crop productivity and yield. Several machine learning and deep learning architectures have been used for plant disease classification for the last decade. Tomato and Potatoes are being grown in many parts of the world. This research study presents various pruned deep CNN models for potato and tomato disease classification where different CNN architectures were pruned, their core structure was retained, and parameter size was reduced to approximate 1.5 M. Rather than focusing solely on accuracy, this study also considers reproduction and deployment. Images of tomato and potato plants covering 13 diseases, including healthy, were taken from the PlantVillage dataset. Six deep CNN models were considered, pruned and concatenated, out of which EfficientNetB0 have a superior performance score of 99.90%, with a parameter size of 1.5 M and 0.826G computing operations (FLOPs), outperforming the remaining five pruned CNN models. The developed lightweight model can be easily deployed in mobile and portable devices with limited resources and computational power, to assist farmers in identifying potato and tomato diseases.

Keywords Plant disease · Pruned · Lightweight · Transfer learning · Deep learning

A. Dheeraj (✉)
ICAR-Indian Agricultural Statistics Research Institute, New Delhi, India
e-mail: akshay.dheeraj@icar.gov.in

A. Dheeraj · S. Chand
School of Computer and Systems Sciences, Jawaharlal Nehru University, New Delhi, India

A. Raj
National Informatics Centre, Ministry of Electronics and Information Technology, New Delhi, India

D.-I. Stroe et al. (eds.), *Emerging Electronics and Automation*, Lecture Notes in Electrical Engineering 1202, https://doi.org/10.1007/978-981-97-3090-2_18

1 Introduction

Agriculture significantly contributes to the economy of various countries, including India. Timely identification of plant disease is crucial for enhancing crop yield and productivity and feeding the increasing human population healthy food. Some plant diseases do not have visible symptoms and, thus, are difficult to be identified by the naked eye. Most plant diseases have visible symptoms that can be identified by plant pathologists through visual inspections. This is a very time-consuming process, infeasible in larger fields and prone to errors if pathologists do not have good skills [1]. Thus, an intelligent and accurate system for plant disease diagnosis is needed to increase crop yield and quality. Such systems integrated with mobile and portable devices can assist farmers who do not possess pathological skills. Advancements in artificial intelligence and digital cameras have enabled the development of intelligent systems for plant disease diagnosis. Various machine learning and deep learning models are being used to recognize plant diseases and different agricultural applications [2–5]. Machine learning based approaches use handcrafted features to identify leaf diseases. The research study proposed by Pantazi et al. [6] used local binary patterns and SVM for feature extraction and classification for healthy and diseased leaves. In another research, the authors developed a disease classification system for pepper, potato, and tomato crops. Fisher vector as several order differentiation of Gaussian distribution was employed to extract the discriminative features, and SVM was used for classification and yielded 94.35% accuracy [7]. These machine learning methods are faster but generally have poor results when the dataset becomes large. Thus, the deep learning-based models can be effectively employed for detection of plant diseases. These models automatically extract the discriminative features from the leaf images. Mohanty et al. [8] used several deep convolutional neural network (DCNN) models for plant disease diagnosis on PlantVillage data. GoogleNet outperformed the various CNN models with a 99.35% accuracy score. Ullah et al. [9] proposed a hybrid CNN model by fusing EfficientNetB3 and MobileNet for tomato disease identification. The model achieved a classification accuracy of 99.92% on a dataset consisting of 32,535 images of 11 tomato disease classes. Ahmed et al. [10] presented a tomato disease classification model using MobileNetV2 architecture and achieved 99.30% accuracy. The authors used run time augmentation techniques to avoid data leakage and class imbalance problem. Thakur et al. [11] developed a plant disease recognition system by combining VGG and Inception network. The proposed model with parameter amount of 7M, achieved 99.16% accuracy on PlantVillage dataset. Another research study proposed by Sutaji et al. [12] employed ensemble of Xception and MobileNet model. The model concatenated the feature extracted by these models and achieved 99.10% classification accuracy on PlantVillage dataset. Although the model yielded good results but is computationally complex and expensive. Chakraborty et al. [13] presented a VGG16 model for potato disease identification system and yielded 97.89% accuracy using potato disease dataset taken from PlantVillage. Chen et al. [14] presented an improved MobileNetV2 model for potato disease recognition by embedding the spatial and channel-wise attention mechanism to the network, followed by octave convolution block. The model achieved an average accuracy score of 97.73% on potato disease dataset taken from PlantVillage. Moussafir et al. [15] proposed an average weighted ensemble of two pre trained deep learning model (ResNet50 and EfficientNetB0) for tomato disease identification. The proposed approach yielded 98.10% accuracy on tomato disease dataset from PlantVillage. Another research study proposed deeper lightweight model for multi class plant disease identification. Authors utilized collective blocks to extract deep features, which results in better features fusion and propagation. This model has 6.4M parameters and an average classification score of 95.49% over four datasets [16]. Research study proposed in [17] employed an ensemble of three modified lightweight MobileNet models to detect and classify plant diseases. The authors used depthwise separable convolutions and squeeze and excitation block in MobileNet architecture to further reduce the network complexity, channel and spatial attention were employed in the architecture to extract the important features. The proposed approach achieved an accuracy score of 99.61% on the PlantVillage dataset.

The proposed research study has the following contributions:

1. We have employed six pre trained DCNN models where we pruned the model layers to reduce the network complexity and computational cost.
2. These pruned models were then replicated and then these two models (original and replicated) are concatenated and trained differently to generate the diverse set of discriminative features.
3. Out of six DCNN models employed, Pruned EfficientNetB0 has an accuracy of 99.90% for tomato and potato disease classification.

Following the Introduction, the proposed methodology is covered in Sect. 2. Section 3 describes the experiment results. Finally, the conclusion and future scope of the research study are covered in Sect. 4.

2 Proposed Methodology

This section covers the description of the dataset and methodology used in this research study. This also covers the architecture employed in this work.

2.1 Dataset

Images used in this research study have been taken from the publicly available PlantVillage dataset [18]. PlantVillage dataset has 39 classes of disease and healthy leaf images of various crops. Here, we have considered potato and tomato disease classes along with the healthy leaf images, covering thirteen classes. A total of

Table 1 Description of dataset used in present study

S.No.	Disease type	Class name	Images	# Training images	# Testing images	# Validation images
1	Potato Early blight	PEB	1000	800	100	100
2	Potato healthy	PH	152	122	15	15
3	Potato late blight	PLB	1000	800	100	100
4	Tomato Bacterial spot	TBS	2127	1702	213	212
5	Tomato Early blight	TEB	1000	800	100	100
6	Tomato healthy	TH	1591	1273	159	159
7	Tomato Late blight	TLB	1909	1527	191	191
8	Tomato Leaf Mold	TLM	952	761	95	96
9	Tomato mosaic virus	TMV	373	298	38	37
10	Tomato Septoria leaf spot	TSLS	1771	1417	177	177
11	Tomato Spider mites Two spotted spider mite	TSMTSSM	1676	1341	167	168
12	Tomato Target Spot	TTS	1404	1123	141	140
13	Tomato Yellow Leaf Curl Virus	TYLCV	5357	4286	535	536
Total			20,312	16,250	2031	2031

20,312 images covering thirteen disease and healthy leaf classes of two crops have been considered in the presented study. The distribution of data is shown in Table 1.

2.2 Data Preprocessing

All the images were resized to 224 × 224, from the original image size of 256 × 256 before they are fed as input to CNN model. To all the models employed in this work, images of dimension 224 × 224 are used.

2.3 Deep Neural Network

CNN have been extensively used for plant disease identification [19, 20]. Here, we have selected seven CNN architecture namely EfficientNetB0, DenseNet121, Xception, InceptionResNetV2, InceptionV3, ResNet50V2, MobileNetV2 initiated with pretrained ImageNet weight. Figure 1 shows the general architecture of various CNN models employed in the research work. We have used pruning technique and removed the majority of the layers of these architectures since these models have been

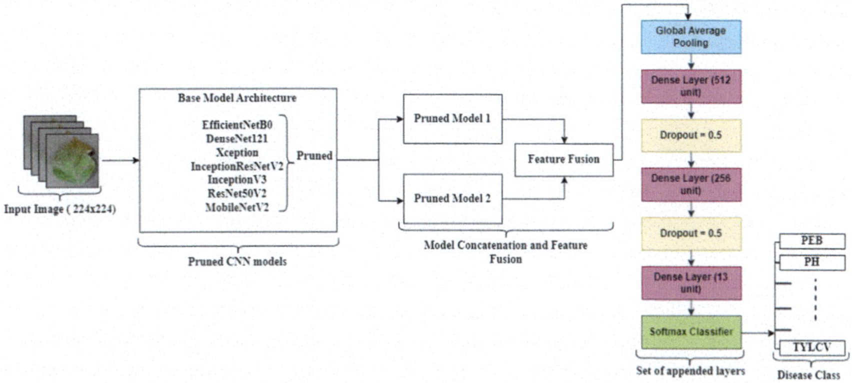

Fig. 1 Architecture of deep neural network

specifically employed for a large ImageNet dataset. By model pruning, the network complexity is reduced, which in turn reduced the number of parameters required to be trained. Then these pruned models (Pruned model 1) were replicated and named as Pruned model 2. Model concatenation technique was employed to concatenate pruned model 1 and pruned model 2. By feature fusion technique, features extracted by pruned model 1 and pruned model 2, are concatenated to generate the robust set of discriminative features. Pruned model 1 and Pruned model 2 are trained using varying approach to generate the distinct feature set. In pruned model 1, its layers are frozen while in pruned model 2, all layers are kept in an unfrozen state and have been trained on ImageNet and PlantVillage dataset to generate the diverse feature set [21].

After the model concatenation (pruned model 1 and pruned model 2), an additional set of layers is added to the network. These additional layers consist of Global average pooling or GAP [22], 512-unit dense layer [23] with ReLU as an activation function, followed by dropout with 0.5, 256-unit dense layer, dropout with 0.5 connected with a thirteen unit dense layer with SoftMax classifier [24] which gives the probabilistic score for each predicted class. GAP helps in preserving spatial hierarchy of features and improves the interpretability of deep neural networks. ReLU brings the non-linearity and induces the sparsity in the network, along with efficient propagation of the gradient in the deeper network. These additional layers help in mitigating overfitting problem and enhancing the performance of the model. The general architecture of the models is shown in Fig. 1.

2.4 Hyperparameters Finetuning

In this research study, transfer learning approach has been employed. On the top of all CNN architectures used here, an additional set of layers has been added. Main

Table 2 Settings of Hyperparameters

Hyperparameter	Value
Epoch	50
Optimize	Adam
Loss function	Categorical cross entropy
Dropout	0.5
Learning rate	0.001
Batch size	16

decisions about finetuning are described in Table 2. Maximum number of epochs were defined to be 50 and Batch size was 16. Adam and categorical cross entropy were employed as optimizer and loss function. Learning rate is set to be 0.0001 and dropout rate is set to be 0.5 to reduce the overfitting. All the models were trained with the same set of hyperparameters.

2.5 *Performance Metrics*

All the models were tested with 2031 unseen images and different performance metrics, derived from confusion matrix, were employed to assess the performance of the models. Precision, recall, F1 score, and accuracy were used. Accuracy refers to the overall classification performance of the system and is calculated as:

$$Accuracy = \frac{TP + TN}{TP + TN + FP + FN} \tag{1}$$

Precision is defined as the ratio of the correct number of predictions in the diseased class to total number of predictions made in the diseased class by the model. It is calculated as:

$$Precision = \frac{TP}{TP + FP} \tag{2}$$

Recall is used to evaluate the exactness of correct prediction made by comparing with all the actual number of observations in the class.

$$Recall = \frac{TP}{TP + FN} \tag{3}$$

F1 Score is a single value that takes both precision and recall into consideration.

$$F1Score = \frac{2 * Precision * Recall}{Precision + Recall} \tag{4}$$

Along with these performance metrics, floating point operations (FLOPs) have also been calculated using keras-flops package of python [25]. FLOPs is a measure of total number of operations needed by a model for a classification task. If FLOPs are more, more computing resources and power are needed. Therefore, it is an important measure along with parameter amount to decide for a model to be a lightweight.

3 Experiment Results

This section discusses the experiment results of all the CNN models employed in the research study along with the comparative study with the previously done research.

3.1 Experiment Setup

Six popular CNN models have been experimented on Potato and Tomato image dataset taken from the PlantVillage dataset. Transfer learning approach has been employed to train all six models with the same set of hyperparameters. The proposed work has been implemented with NVIDIA DGX GPU servers equipped with 512 GB RAM and Tesla V100 GPUs, with each GPU equipped with 32 GB capacity. Python library Keras with Tensorflow in backend was used.

3.2 Dataset Preparation

Dataset preparation consists of many steps. In the first step, the dataset was partitioned into three distinct sets, such as training, validation and testing dataset in the ratio 80%, 10% and 10% respectively. In the second step, all the images were resized to a dimension of 224×224. In the third step, all the images were normalized by dividing the intensity value of each pixel by 255 to reduce the computation complexity.

3.3 Model Performance Comparison

The performance of the pruned EfficientNetB0 has been compared with pruned architecture of DenseNet121, Xception, InceptionResNetV2, ResNet50V2, and MobileNetV2 and is demonstrated in Table 3. All the models were pruned in such a way that the network parameters count is approximately 1.5M and model performance doesn't degrade much. Among all the pruned models experimented on potato and tomato dataset, the pruned EfficientNetB0 has the highest classification accuracy of 99.90%. This model also has the highest score for precision, recall and

F1 score with 0.999. Among all the models experimented, pruned EfficientNetB0 takes the least number of FLOPs with 0.826G, along with the parameter count of 1,564,139. DenseNet121 takes the maximum number of FLOPs with 5.62G. Pruned MobileNetV2 takes the least number of computing operations with 0.822 G operations because it uses NAS technology and has a classification accuracy of 99.66% which is less than that of pruned EfficientNetB0. Xception is the worst performing model among all with value of 95.47%, 0.9607, 0.9547 and 0.9550 for accuracy, precision, recall and F1 score respectively. Total number of floating point operations in Pruned Xception is 0.826G which is equal that of Pruned EfficientNetB0.

Figure 2 demonstrates the accuracy and loss curve of the Pruned EfficientNetB0 model. As evidenced by the figure, both training and validation accuracy show a similar pattern in their values as both values increase with decrease in the respective loss values.

Table 3 Comparison of all pruned CNN models experimented

Model name	Accuracy	Precision	Recall	F1	Parameters	Training time	FLOPS (G)
DenseNet121	99.16	0.9919	0.9916	0.9916	1,364,365	33.45	5.62
Xception	95.47	0.9607	0.9547	0.9550	1,273,101	76.25	0.826
InceptionResNetV2	98.77	0.9882	0.9877	0.9877	1,353,997	31.09	3.6
ResNet50V2	95.62	0.9630	0.9562	0.9564	1,348,621	27.09	3.44
MobileNetV2	99.66	0.9966	0.9966	0.9966	1,436,045	56.16	0.822
EfficientNet B0	99.90	0.9990	0.9990	0.9990	1,564,139	68.46	0.826

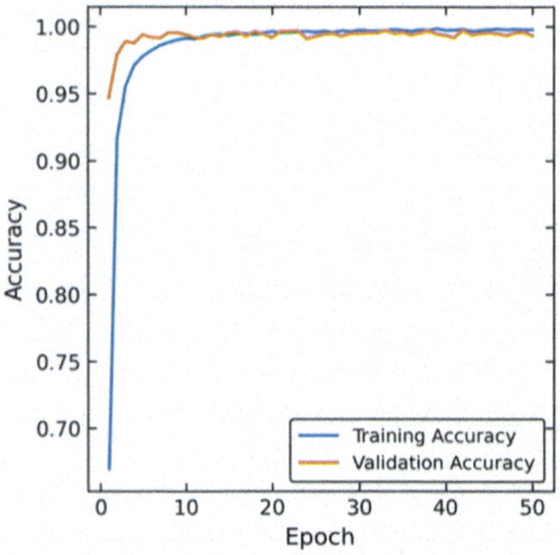

(a)

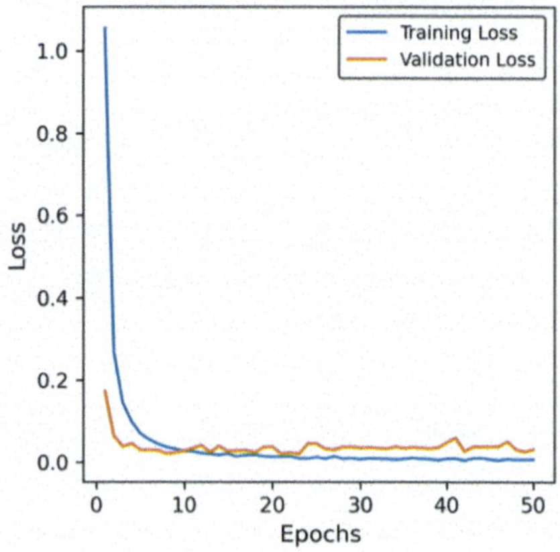

(b)

Fig. 2 A Accuracy curve **b** Loss curve of Pruned EfficientNetB0 model

Figure 3 shows the confusion matrix for Pruned EfficientNetB0 model. The model incorrectly classifies two images in TEB and TSMTSSM with 99% and 99.40% accuracy, respectively. For remaining eleven classes, the model yields 100% accuracy. ROC and Precision- Recall curve for comprehensive evaluation has been shown in Fig. 4. AUROC value of less than 0.5 indicates that the model cannot distinguish a particular case. Here, the proposed model has a consistent AUC value of 1.00 for all the classes. AUPR curve particularly focus on unbalanced data and incorrect diagnosis cases. AUPR has a value of 1.00 for all the classes.

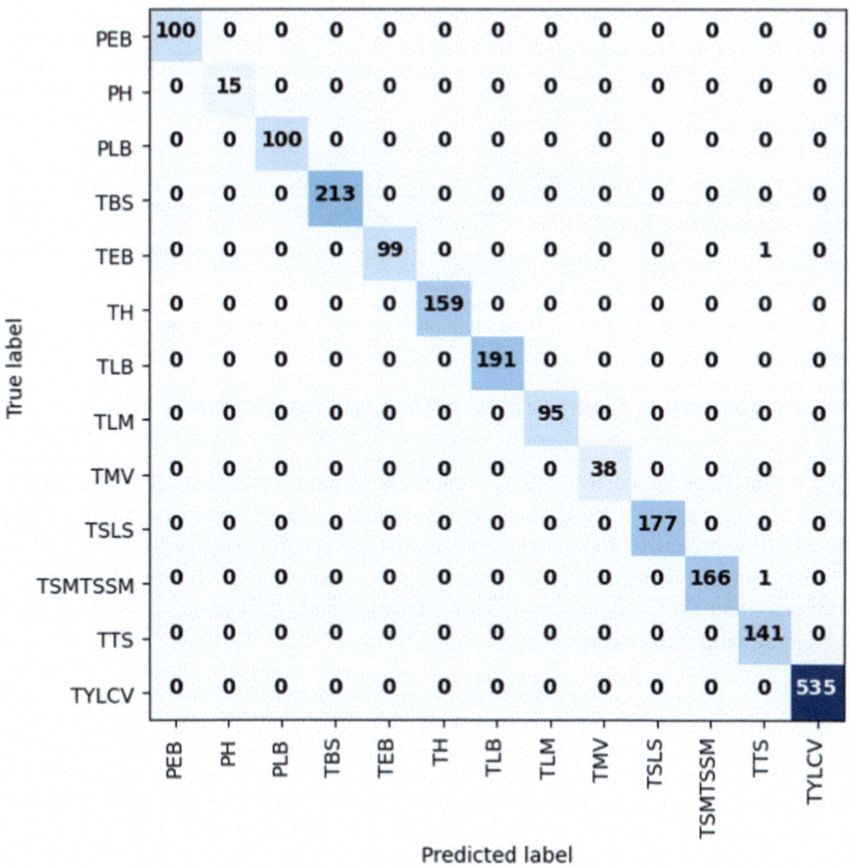

Fig. 3 Confusion matrix of Pruned EfficientNetB0 model

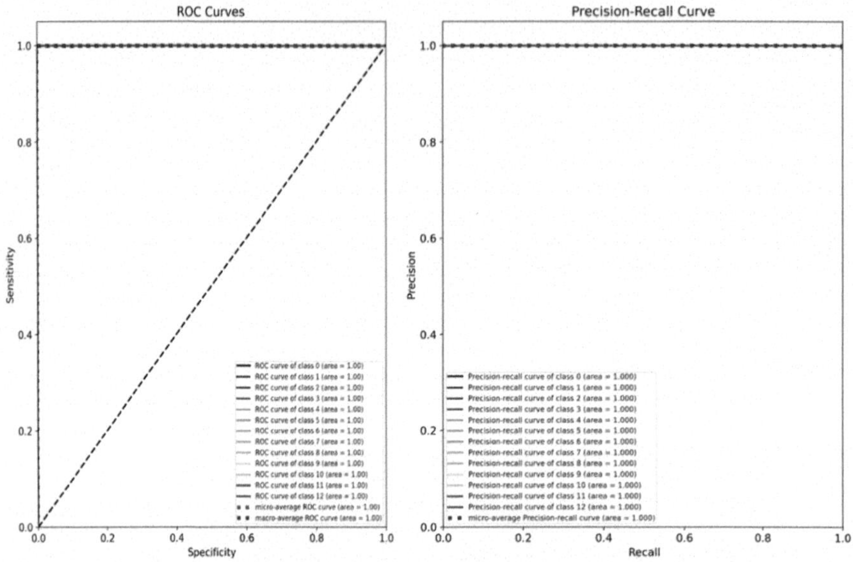

(a) (b)

Fig. 4 A ROC Curve **b** Precision- Recall curve of Pruned EfficientNetB0 model

3.4 Performance Comparison with Existing Research Work

Table 4 illustrates the performance comparison of the proposed study with existing small and compact CNN model developed for plant diseases diagnosis. Our proposed Pruned EfficientNetB0 model outperforms existing research work with computationally less complex architecture specifications. Our model takes 76% and 94% lesser parameter than the model proposed by Sharma et al. [25] and Sutaji et al. [12] respectively and outperforms those.

Table 4 Model performance comparison with existing work

Study	Dataset	Model	Accuracy (%)	# Parameters
Dheeraj and Chand [5]	Pepper, tomato, potato PlantVillage	EfficienNetB0	99.79	5.3 M
Thakur et al. [11]	PlantVillage	VGG-ICNN	99.16	6 M
Sutaji et al. [12]	PlantVillage	Ensemble of Xception and MobileNetV2	99.10	26, 567, 310 26 M
Sharma et al. [16]	grapes, cucumber, tomato, Citrus,	Custom model named DLCM-Net	99.50, 92.34, 96.56, 93.56	6.4 M
Bhujel et al. [26]	Tomato PlantVillage	Attention embedded ResNet50 model	99.69	1.5 M
Proposed work	PlantVillage Potato and Tomato data	Pruned EfficientNetB0	99.90	1.5 M

4 Conclusion

This study proposed various pruned models and experimented them on the Potato and Tomato PlantVillage dataset. These pruned models were replicated, concatenated and then these two models (original and replicated) were trained using varying techniques to generate a diverse set of discriminative features. A set of additional layers were appended to these pruned models to reduce the overfitting and enhance the network performance. Out of the six models experimented in this study, Pruned EfficienetNetB0 has the best performance with 99.90% accuracy for detecting disease associated with potato and tomato. The Pruned EfficientNetB0 model takes 1.5 M parameters and 0.826G FLOPs for forward and back propagation to process an image, making it a lightweight model. Our model also has the superior performance than the existing lightweight model developed for plant disease recognition. Thus, the lightweight pruned EfficientNetB0 model can be easily deployed to mobile and portable devices with limited computation power and assist the farmers in identifying the potato and tomato disease. In the future, the proposed work may be extended to cover more diseases and crops.

References

1. Sladojevic S, Arsenovic M, Anderla A, Culibrk D, Stefanovic D (2016) Deep neural networks-based recognition of plant diseases by leaf image classification. Comput Intell Neurosci 2016:11. https://doi.org/10.1155/2016/3289801
2. Jackulin C, Murugavalli S (2022) A comprehensive review of the detection of plant disease using machine learning and deep learning approaches. Measurement: Sensors, p 100441

3. Thakur PS, Khanna P, Sheorey T, Ojha A (2022) Trends in vision-based machine learning techniques for plant disease identification: A systematic review. Expert Syst Appl, p 118117
4. Dheeraj A, Chand S (2023) Using deep learning models for crop and weed classification at early stage. In: Shakya S, Du KL, Ntalianis K (eds) Sentim Anal Deep Learning Adv Intell Syst Comput, 1432. Springer, Singapore. https://doi.org/10.1007/978-981-19-5443-6_69
5. Dheeraj A, Chand S (2023) Deep learning model for automated image based plant disease classification. In: Sharma H, Saha AK, Prasad M (eds) In: Proceedings of International Conference on Intelligent Vision and Computing (ICIVC 2022). ICIVC 2022. Proceedings in Adaptation, Learning and Optimization, 17. Springer, Cham. https://doi.org/10.1007/978-3-031-31164-2_3
6. Pantazi XE, Moshou D, Tamouridou AA (2019) Automated leaf disease detection in different crop species through image features analysis and One Class Classifiers. Comput Electron Agric 156:96–104
7. Kurmi Y, Gangwar S, Agrawal D, Kumar S, Srivastava HS (2021) Leaf image analysis-based crop diseases classification. SIViP 15(3):589–597
8. Mohanty SP, Hughes DP, Salathé M (2016) Using deep learning for image-based plant disease detection. Front Plant Sci 7:1419
9. Ullah Z, Alsubaie N, Jamjoom M, Alajmani SH, Saleem F (2023) EffiMob-Net: A Deep Learning-Based hybrid model for detection and identification of tomato diseases using leaf images. Agric 13(3):737
10. Ahmed S, Hasan MB, Ahmed T, Sony MRK, Kabir MH (2022) Less is more: lighter and faster deep neural architecture for tomato leaf disease classification. IEEE Access 10:68868–68884
11. Thakur PS, Sheorey T, Ojha A (2023) VGG-ICNN: A Lightweight CNN model for crop disease identification. Multimed Tools Appl 82(1):497–520
12. Sutaji D, Yıldız O (2022) LEMOXINET: Lite ensemble MobileNetV2 and Xception models to predict plant disease. Eco Inform 70:101698
13. Chakraborty KK, Mukherjee R, Chakroborty C, Bora K (2022) Automated recognition of optical image based potato leaf blight diseases using deep learning. Physiol Mol Plant Pathol 117:101781
14. Chen W, Chen J, Zeb A, Yang S, Zhang D (2022) Mobile convolution neural network for the recognition of potato leaf disease images. Multimed Tools Appl 81(15):20797–20816
15. Moussafir M, Chaibi H, Saadane R, Chehri A, Rharras AE, Jeon G (2022) Design of efficient techniques for tomato leaf disease detection using genetic algorithm-based and deep neural networks. Plant Soil 479(1–2):251–266
16. Sharma V, Tripathi AK, Mittal H (2023) DLMC-Net: Deeper lightweight multi-class classification model for plant leaf disease detection. Eco Inform 75:102025
17. Chen J, Zeb A, Nanehkaran YA, Zhang D (2022) Stacking ensemble model of deep learning for plant disease recognition. J Ambient Intell HumIzed Comput, pp 1–14
18. Hughes D, Salathé M (2015) An open access repository of images on plant health to enable the development of mobile disease diagnostics. *arXiv preprint* arXiv:1511.08060
19. Ferentinos KP (2018) Deep learning models for plant disease detection and diagnosis. Comput Electron Agric 145:311–318
20. Too EC, Yujian L, Njuki S, Yingchun L (2019) A comparative study of fine-tuning deep learning models for plant disease identification. Comput Electron Agric 161:272–279
21. Montalbo FJP (2021) Diagnosing Covid-19 chest x-rays with a lightweight truncated DenseNet with partial layer freezing and feature fusion. Biomed Signal Process Control 68:102583
22. Kamal KC, Yin Z, Wu M, Wu Z (2019) Depthwise separable convolution architectures for plant disease classification. Comput Electron Agric 165:104948
23. Dahl GE, Sainath TN, Hinton GE (2013) Improving deep neural networks for LVCSR using rectified linear units and dropout. In 2013 IEEE international conference on acoustics, speech and signal processing (pp 8609–8613). IEEE
24. Fu L, Li S, Sun Y, Mu Y, Hu T, Gong H (2022) Lightweight-CNN for apple leaf disease identification. Front Plant Sci, p 1508

25. Tokusumi (2020) Keras-flops Calculator
26. Bhujel A, Kim NE, Arulmozhi E, Basak JK, Kim HT (2022) A lightweight Attention-based convolutional neural networks for tomato leaf disease classification. Agric 12(2):228

An Approach Towards Abstractive Summary Generation and Knowledge Graph Representation

Prottay Kumar Adhikary, Pankaj Dadure, Riyanka Manna, and Partha Pakray

Abstract With the proliferation of blogs, news stories, and reports, the extraction of useful information from this extensive collection of textual documents is a tedious task. Automatic text summarization provides an effective solution for condensing these documents into concise summaries while preserving essential information and meaning. Numerous noteworthy summarization models have been proposed to address various challenges, including saliency, fluency, human readability, and the generation of high-quality summaries. In this study, we introduce the Text-To-Text Transfer Transformer (T5) model for the task of abstractive summarization with knowledge representation. The experimental results demonstrate that the T5 model produces summaries that are more conceptual, comprehensible, and abstractive. To assess the quality of the generated summaries, we consider the ROUGE and BLEU scores.

Keywords Text summarization · Abstractive summary · News documents · Transformer · Knowledge graph

1 Introduction

Text summarization [1], a prominent application of Natural Language Processing (NLP), has gained significant attention in recent decades. With the rise of the digital world and its impact on the publishing industry, finding time to thoroughly read articles, documents, or books has become increasingly challenging due to time constraints. Moreover, the transition of traditional print publications to digital formats has led to an overwhelming number of web-based publications, making it difficult to

P. K. Adhikary (✉) · P. Pakray
NIT Silchar, Silchar, Assam, India
e-mail: prottay71@gmail.com

P. Dadure
UPES Dehradun, Misraspatti, Uttarakhand, India

R. Manna
Amrita Vishwa Vidyapeetham, Amaravati, Andhra Pradesh, India

D.-I. Stroe et al. (eds.), *Emerging Electronics and Automation*, Lecture Notes in Electrical Engineering 1202, https://doi.org/10.1007/978-981-97-3090-2_19

keep track of all the content available. In such scenarios, text summarization plays a crucial role in simplifying the complexity of texts. In general, text summarization involves the creation of a concise, coherent, and efficient summary that captures the main ideas of a lengthy text article [2]. There are two main categories of text summarization: Extractive Summarization and Abstractive Summarization [3]. Extractive summarization entails identifying key phrases within the original text and combining them to produce a condensed version of the document. On the other hand, abstractive summarization relies on advanced natural language techniques to paraphrase and condense parts of a document. While abstractive summarization algorithms are capable of generating new sentences and phrases to capture the essence of the source text, it is crucial to ensure grammatical accuracy when employing deep learning methods for abstraction.

In the late 1950s, the scientific community started showing interest in automatic summarization research [4]. This interest was sparked by the need for automating the production of summaries for technical documents. However, this interest declined for a few years until artificial intelligence brought up the topic again [5]. For a long time, it was believed that summarization required a deep understanding of the input text and the explicit interpretation of its semantics to identify the essential content. Summarizing text became a compelling application for testing the understanding capabilities of artificial systems. However, due to the task's complexity, interest in this method of text summarization rapidly declined, leading to a focus on the research of text understanding itself.

In this paper, we explore the use of the Text-To-Text Transfer Transformer model for abstractive summarization. The experimental results demonstrate that the Text-to-Text Transfer approach generates summaries that are more conceptual, understandable, and abstractive. The paper is organized as follows: Sect. 2 provides an overview of prior work related to text summarization. Section 3 presents a detailed account of the dataset used in the study. Section 4 describes the system architecture in depth. Section 5 outlines the experimental setup and presents the results. Finally, Sect. 6 concludes with a summary and suggests directions for future research and development.

2 Related Work

Since the 1950s, research on text summarization has focused primarily on extractive summarization, which involves analyzing the structure of words in the document [6]. However, with the advent of Recurrent Neural Networks (RNNs) and their success in image recognition, machine translation, and speech recognition, researchers began exploring their application to text summarization. Baotian et al. [7] introduced the Large-Scale Chinese Short-Text Summarization (LCSTS) dataset, which consists of nearly 2 million Chinese texts along with their summaries. The dataset also includes relevant information for 10,666 texts and their corresponding source summaries. The attentional encoder–decoder RNN model, initially successful in machine translation, has also shown promising results in text summarization. Ramesh et al. [8] applied this model to text summarization tasks using two different datasets (DUC Corpus and CNN/Daily Mail Corpus) and achieved state-of-the-art performance. Additionally, a new dataset was proposed for multisentence summarization.

Generative adversarial networks (GANs) have emerged as a recognized deep learning model for text generation. Linqing et al. [9] employed a generator as an agent of reinforcement learning, taking the text as input and generating abstractive summaries. They also developed a discriminator to distinguish between the generated summaries and the ground truth summaries. The guiding generation model combines the capabilities of abstractive and extractive summarization methods [10]. It first employs the extractive summarization method to extract keywords from the text. Then, a Key Information Guide Network (KIGN) encodes these keywords into key information, which guides the summary generation process.

Unlike previous studies that utilized a single encoder, the proposed method employs a dual encoder, consisting of primary and secondary encoders [11]. The primary encoder performs conventional coarse encoding, while the secondary encoder models the relevance of words and provides finer encoding based on the input raw text and previously generated output summarization. The two-level encodings are merged and fed into the decoder to generate a more diverse summary, reducing the repetition of long sequences. The multi-head attention summarization (MHAS) model [12] learns important information in distinct representation subspaces using a multi-head attention mechanism. To avoid duplicate repetition of terms in the generated summary, the MHAS model considers the previously predicted words while generating new words. It also incorporates a self-attention layer to effectively capture the underlying structure of the article and maintain the original data. Furthermore, multi-head attention distribution is integrated into the pointer network to enhance the model's performance.

The sequence-to-sequence framework incorporates a keyword-aware encoder (KAE) [13], which enriches the RNN-based encoder by combining keyword information with word representations. The KAE model distills important information using keywords and has shown improved performance compared to recent state-of-the-art methods in experiments on datasets with various languages and lengths. The COSUM approach [14] combines clustering and optimization techniques to identify topics in the text. It uses k-means algorithms for clustering and an optimization model to select the main sentences within the defined clusters. The optimization process aims to achieve a harmonious summary by covering and diversifying the chosen phrases. This model also ensures readability by monitoring the length of selected sentences in the candidate summary.

In extraction-based summarization, the identification of important sentences is crucial. The fuzzy logic-based text summarization approach [15] improves the quality of the summary by incorporating latent semantic features in the sentence and establishing semantic connections between concepts extracted from the original text. The classification–summarization approach [16] categorizes tweets into different classes and performs summarization. It follows a two-stage extractive-abstractive summarization framework. Initially, it selects a number of important tweets from a

Table 1 Dataset description

Data	Length of data
Training	2,86,817
Validation	13,368
Testing	11,490

wide range of information and constructs a bigram-based text graph. Then, an optimization technique is used to select the most significant tweets and paths based on various parameters, such as information and content coverage. This approach not only addresses customized summarization but also caters to critical time-sensitive information needs.

3 Dataset Description

The proposed model was tested, trained, and validated using the CNN/Daily Mail dataset from Hugging Face. Each article in the dataset contains an average of 28 sentences. Specifically, the highlighted sentences from the news articles were used to create the summarization dataset. These highlighted sentences were concatenated to form the article summaries. The training set consisted of 286,817 documents, the validation (Table 1).

4 System Architecture

The proposed approach consists of two crucial modules: abstractive summary generation and knowledge graph representation. The summary generation module is responsible for generating the summary, while the knowledge graph representation module represents the key concepts of the document using nodes and edges. In this representation, nodes represent the subjects and objects, while edges represent the verbs. The workflow of the proposed approach is illustrated in Fig. 1, where each module operates sequentially.

4.1 Summary Generation

The text summarization approach designed in this study harnesses the capabilities of the Text-to-Text Transfer Transformer (T5) architecture [17]. T5 is an encoder–decoder model specifically designed for text-to-text tasks. It has been pre-trained on

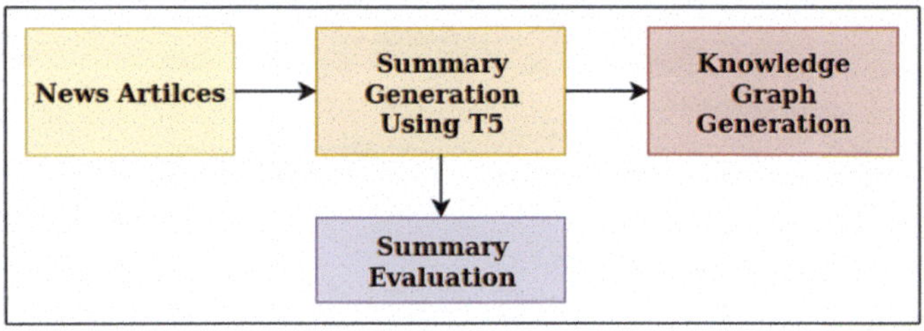

Fig. 1 Workflow of the proposed approach

a combination of unsupervised and supervised tasks. In this architecture, each task is formulated as inputting text and training the model to generate the desired output text.

Training: T5 follows an encoder–decoder paradigm that transforms all NLP tasks into a text-to-text sequence format. During training, T5 learns from a teacher by receiving input sequences and corresponding goal sequences. For the input sequence, T5 utilizes input IDs to represent the text. The target sequence, on the other hand, is preprocessed with a start-sequence token and fed into the decoder using decoder input IDs. The target sequence is then appended with an end-of-sequence (EOS) token, which corresponds to the labels in the teacher-forcing approach. The start-sequence token is represented by the PAD token. In the context of the text summarization task, the input and output sequences follow a standard sequence-to-sequence mapping. The input sequence consists of the lengthy context of the news document, while the output sequence represents the concise summary of the news document. During training, only two fields are required: input IDs (encoding of the input sequence) and labels (encoding of the target sequence). Based on these labels, the model generates the decoder input IDs, which are used for further processing.

4.2 *Knowledge Graph Representation*

The main objective of this module is to visualize the information presented in the news articles using a knowledge graph. To create the graph, the spaCy Python library is employed, specifically utilizing the largest English corpus model (en_core_web_lg). This allows for the extraction of key details from the news articles. Once the details are extracted, the NetworkX package is utilized to construct the network of nodes and edges. In this context, nodes represent the subjects and objects of the sentences in the news articles, while edges represent the verbs within the same sentence. Finally, the matplotlib.pyplot library is used to visualize the generated graph. Figure 2 provides a visual representation of the resulting knowledge graph.

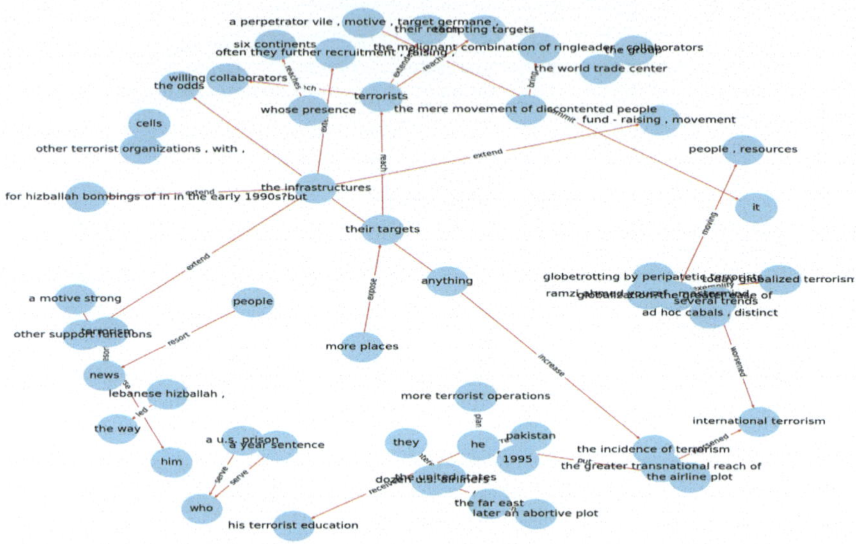

Fig. 2 Knowledge graph

Table 2 Experimental results value

Metric	Sequence-to-sequence RNN [8]	Proposed approach
ROUGE-1	35.46	40.79
ROUGE-2	13.30	18.55
ROUGE-L	32.62	34.8
BLEU	NA	43.5

5 Experimental Results

The proposed approach has been evaluated using the CNN/Daily Mail dataset, which consists of news articles along with their corresponding summaries. The effectiveness of the approach is measured using metrics such as ROUGE-1, ROUGE-2, ROUGE-L, and BLEU. These metrics are calculated for each test document, and then the average scores are computed across the entire test set. The ROUGE scores provide a measure of semantic comparison between the source text and the predicted summary. The proposed approach demonstrates significant improvements in the task of abstractive text summarization. The experimental results are compared with the state-of-the-art sequence-to-sequence RNN approach [8], where the proposed approach outperforms in terms of the achieved results. Table 2 presents a comparative analysis of the obtained results. In Fig. 3, we provide a visual representation of the comparison between the summary generated by the proposed system and the source summary. The evaluation scores range from 1 to 3, where a score of 3 indicates that the approach

Source Summary	Predicted Summary	Evaluation Score
Spanish researchers say climate change impacted human migration until 1 4 million years ago it was too cold to inhabit southeast spain but then the climate warmed to 13 c lrb 55 f rrb and became more humid this enabled hominins our distant ancestors to move to new regions	Climate change impacted human migration until 1 4 million years ago it was too cold to inhabit southeast spain but then the climate warmed to 13 c lrb 55 f rrb our distant ancestors to move to new regions	3
Jayden wingler of arizona was interviewed by fox news earlier this week about a theme park accident which left him with severe burns on his legs while on camera the youngster eloquently recalled what happened with arm actions and wide eyed facial expressions to match his emotions	Jayden wingler of phoenix arizona was interviewed by fox news earlier this week about a theme park accident which left him with severe burns on his legs while on camera the youngster eloquently recalled what happened with arm actions and wide eyed facial expressions to match his emotions	2
Pair a floral skirt with a bold pink sleeveless shirt and matching shoes accessorise a mondrian print dress with red accents brighten up monochrome stripes with a vibrant orange waistcoat	Clutch bag 145 russellandbromley co uk aurora crystal earrings 54 99 zara com top 59 95 joules com stripy trousers 35 oasis stores com sandals 38 office co uk ray ban sunglasses 135 amazon co uk bag 30	1

Fig. 3 Comparative Analysis of the Source and Generated Summary

generates a semantically correct summary without losing any important information, a score of 2 indicates partial similarity in the summary, and a score of 1 indicates that the generated summary is not similar to the source summary.

- The obtained summaries demonstrate that the proposed T5-based summarization approach is capable of understanding the coherent semantic meaning of large contexts.
- The results of the proposed approach have been compared to the existing sequence-to-sequence RNN approach, and the proposed approach has achieved superior results.
- The designed system can establish relationships between sentences in a paragraph and generate meaningful summaries based on specific text requirements.
- It should be noted that the documents in the CNN/DM dataset are primarily written in formal English. Consequently, the model may encounter challenges in generating summaries for documents written in informal English.
- The utilization of a knowledge graph provides a visual representation of textual information and highlights key terms present in the documents.

6 Conclusions and Future Scope

In this work, we have introduced the Text-To-Text Transfer Transformer model for the task of abstractive summarization, and the results have been highly promising, surpassing state-of-the-art performance on the CNN/Daily Mail dataset. Additionally, we have incorporated a knowledge graph to visually represent the key concepts found in the documents. The experimental results have demonstrated that the text-to-text transfer model generates summaries that are more conceptual, comprehensible, and abstractive. To assess the quality of the generated summaries, we have considered the ROUGE score, which yielded the highest values for ROUGE-1, ROUGE-2, ROUGE-L, and BLEU (40.79, 18.55, 34.8, and 43.5, respectively). In future work, we plan to further enhance the model's accuracy by training it on new datasets from diverse domains. Additionally, we aim to improve the visual representation of the knowledge graph by constructing nodes based on importance and frequency. This enhancement would facilitate the display of long paragraphs and provide more informative insights.

References

1. Fattah MA, Ren F (2008) Automatic text summarization. World Acad Sci, Engin Technol 37(2):192
2. Day M-Y, Chen C-Y (2018) Artificial intelligence for automatic text summarization. In: 2018 IEEE international conference on information reuse and integration (IRI). IEEE, pp 478–484
3. Allahyari M, Pouriyeh S, Assefi M, Safaei S, Trippe ED, Gutierrez JB, Kochut K (2017) Text summarization techniques: a brief survey. arXiv preprint arXiv:1707.02268
4. Luhn HP (1958) The automatic creation of literature abstracts. IBM J Res Develop 2(2):159–165
5. DeJong G (1982) An overview of the FRUMP system in strategies for natural language processing: edited by Lehnert, WG and Ringle MH, pp 149–176
6. Nenkova A, McKeown K, others, Automatic summarization: foundations and trends in information retrieval. Now Publishers, Inc, pp 103–233
7. Hu B, Chen Q, Zhu F (2015) Lcsts: a large scale chinese short text summarization dataset. arXiv preprint arXiv:1506.05865
8. Nallapati R, Zhou B, Gulcehre C, Xiang B et al (2016) Abstractive text summarization using sequence-to-sequence rnns and beyond. arXiv preprint arXiv:1602.06023
9. Liu L, Lu Y, Yang M, Qu Q, Zhu J, Li H (2018) Generative adversarial net- work for abstractive text summarization. In: Thirty-second AAAI conference on artificial intelligence
10. Li, C, Xu, W, Li, S, Gao, S.: Guiding generation for abstractive text summa- rization based on key information guide network. In: Proceedings of the 2018 Conference of the North American Chapter of the Association for Computational Linguistics: Human Language Technologies, Volume 2 (Short Papers), pp. 55-60 (2018)
11. Yao K, Zhang L, Du D, Luo T, Tao L, Wu Y (2018) Dual encoding for abstractive text summarization. IEEE Trans Cybernet 50(3):985–996
12. Li J, Zhang C, Chen X, Cao Y, Liao P, Zhang P (2019) Abstractive text sum- marization with multi-head attention. In: 2019 international joint conference on neural networks (ijcnn). IEEE, pp 1–8

13. Hu T, Liang J, Ye W, Zhang S (2021) Keyword-aware encoder for abstractive text summarization. In: International conference on database systems for advanced applications. Springer, pp 37–52
14. Alguliyev RM, Aliguliyev RM, Isazade NR, Abdi A, Idris N (2019) Cosum: text summarization based on clustering and optimization. Expert Syst 36(1):12340
15. Babar S, Patil PD (2015) Improving performance of text summarization. Procedia Comput Sci 46:354–363
16. Rudra K, Goyal P, Ganguly N, Imran M, Mitra P (2019) Summarizing situational tweets in crisis scenarios: an extractive-abstractive approach. IEEE Trans Comput Soc Syst 6(5):981–993
17. Raffel C, Shazeer N, Roberts A, Lee K, Narang S, Matena M, Zhou Y, Li W, Liu PJ (2019) Exploring the limits of transfer learning with a unified text-to-text transformer. arXiv preprint arXiv:1910.10683

Smart Soil Nutrient Prediction System

M. Saravanabalaji, V. Athappan, S. Ranganathan, E. Muthuramlingam, K. Akshaya, and V. Celcia

Abstract Agriculture is a major sector which is vital for the survival of modern man. Agriculture must embrace a digital, connectivity-fuelled transformation in order to overcome increasing demand and several disruptive forces. The major reason for the downfall in the agriculture practice is due to less crop yield. Also with the rapid development of human civilization it is important that food production should increase. As a result, modern agricultural development is critical because traditional agricultural methods cannot increase crop yields. Overuse of fertilizers and pesticides harms the crop and is the primary cause of lower crop yield and unhealthy agricultural practices. Soil analysis is an important process to determine the available plant nutrients in the soil. Plants absorb the major nutrients through soil. This project aims to control overuse of chemical fertilizers by suggesting the amount of fertilizers to be used for a specific crop and create a prediction engine for most suitable crop for a particular soil.

Keywords Soil nutrition · ML algorithm · Npk value · Crop growth · Cultivation · Fertilizer

1 Introduction

Analyzing the soil properties, its composition and climatic data plays a vital role to maximize the crop yield. There are several existing ways for the prediction of the soil properties which includes the following: [1]. The determination of soil nutrients using optical methods in which soil nutrients are predicted with and without the involvement of probe, [2] Soil quality analysis and an efficient irrigation system in which an agro sensor is used to measure the macro nutrients available in the soil

M. Saravanabalaji (✉) · V. Athappan · S. Ranganathan · E. Muthuramlingam · K. Akshaya · V. Celcia
Kumaraguru College of Technology, Coimbatore, India
e-mail: Saravanabalaji.m.eie@kct.ac.in

V. Athappan
e-mail: Athappan.v.eie@kct.ac.in

D.-I. Stroe et al. (eds.), *Emerging Electronics and Automation*, Lecture Notes in Electrical Engineering 1202, https://doi.org/10.1007/978-981-97-3090-2_20

and the method of photo-diode is used to determine the various levels, [3] Machine learning strategy for soil nutrient prediction using the spectroscopic method in which a dataset is collected from Kaggle to create a model, trained using certain methods and use a collection of dataset to predict the model, [4] Soil analysis and crop prediction in which the soil properties are analyzed and all the datasets collected are stored in IoT cloud, [5] Use of data mining in crop yield prediction in which using the WEKA tool, different categorization algorithms were tested for accuracy and error rate. On the dataset, they used four different categorization methods, including J48, LWL, LAD Tree, and JBK. They examined the accuracy rate and concluded that JBK had the highest accuracy of the four [6]. Prediction of Soil Quality Using Machine Learning Techniques in which soil types are classified using two algorithms and out of that one as an estimated figure. Important factors such as land algorithm has a better accuracy rate compared to the other, [7] Here, the effective rainfall is calculated through the use of a series of intricate mathematical calculations that have been divided into easier steps. Evaporation transpiration rates, humidity, groundwater characteristics, and temperature have all been considered **to** determine the effective amount of rainfall received in a designated area in a more straightforward manner [8]. By assessing the soil fertility and rainfall in the area that the user has entered as an input and building a prediction engine for the best suitable crop for a certain soil, one can accurately estimate the crop yield to the user [9]. Finding the suitable relationship percentage among the properties of soil using supervised learning, back propagation neural network. Back Propagation Networks (BPN) are trained using the growth characteristics of reference crops [10]. Using data from past years' crops to analyze and predict crop yield. The study also describes different methods that are being used to audit crop yield. Additionally, it contrasts several algorithms and includes their advantages and disadvantages. The proposed method yields solution in a more digital way rather than the traditional methods and provides real-time data's for more precise measurement. Comparatively our method uses JXCT soil NPK sensor which is of low cost and provides high precision data. The main benefit of this sensor is it provides extremely quick measurement and highly precise data than conventional detection techniques.

2 Proposed Idea

The proposed method uses a JXCT Soil NPK Sensor to detect the soil nitrogen, phosphorous & Potassium in a soil. The JXCT Soil NPK sensor is a portable, high-precision, low-cost sensor that operates with Modbus RS485. This sensor's benefit over more conventional detection techniques is that it provides extremely quick measurement and highly precise data. The random forest classifier algorithm, Logistic Regression, Support Vector Machines (SVM) and Naïve Bayes algorithm are used to test the accuracy. Farmers with basic information about the weather, soil type, and the previous crop that was planted can learn how much nitrogen, phosphorus, and potassium needs to be added to the soil to make it fertile for planting the following crop as well as how much fertilizer needs to be added. This study focus is to assist the farmers in cultivating the soil, producing crops and raising livestock Calculating NPK value of soil and helps to give the right amount of nutrients to the plants. The field data collected from the deployed sensors and the weather forecast data are considered for efficient germination. To create a database that will suggest the fertilizers for the specific crop to be used by calculating the soil fertility. To limit the use of chemical fertilizers and maintain the soil wealth.

3 Methodology

There are multiple methods of measuring the soil nutrient content like using optical sensors or using spectrometer. These methods are not convenient. So, here a JXCT soil NPK sensor is used to detect the nitrogen, phosphorous and potassium in the soil. The JXCT soil NPK sensor is cost effective, quick responsive, high precision and portable sensor that works with Modbus RS485. The advantage of this sensor over a traditional detection method is that it gives very fast measurement and data are highly accurate. The rotational crop method, the analysis of the soil nutrient levels after crop cultivation, the type of the soil, the climatic conditions are all taken into consideration for the creation of the dataset. Various soil samples from different regions are tested and a dataset is created. This dataset is used to create a prediction engine. GitHub is used to host the dataset. A portal was developed with the assistance of Streamlet, an online app developer. The farmers with basic information like the type of the soil, climatic condition and the crop that was harvested before, will enable them to determine whether the soil is fertile or not and it also helps to obtain the amount of NPK needed to be injected in the soil for the next cultivation of crop. This is done using ML techniques. The random forest classifier algorithm, Logistic Regression, Support Vector Machines (SVM) and Naïve Bayes algorithm are used to test the accuracy. These ML models can be used for various soil samples. Thus, the combination of NPK sensor, ML algorithm and Real time data analytics from Thing Speak enhanced by a user-friendly portal focused to simplify the efforts taken by a user to measure the fertility of a soil hence this setup enables a user to overcome the demerits of the traditional methods.

3.1 ML Algorithms

Computer models known as machine learning algorithms enable computers to recognize patterns, forecast, or reach conclusions based on data without the use of explicit programming. These algorithms serve as the backbone of contemporary artificial intelligence and are employed in several applications. The following algorithms are used here for the Prediction [10].

1. Logistic Regression
2. Random Forest
3. Naïve Bayes
4. SVM.

Logistic Regression

It is one of the ML techniques that comes under supervised learning technique, in which we take a set of independent variables to find the accuracy of dependency in the unknown data set. The working equation of logistic regression given below:

$$\log\left[\frac{y}{1-y}\right] = b_0 + b_1x_1 + b_2x_2 + b_3x_3 + \cdots b_nx_n$$

Random Forest

Random forest is one of the supervised ML learning techniques. It is used for cases including classification and regression problem. It basically contains the number of decision tree algorithms grouped as cluster, and takes the average to improve the dataset accuracy.

Naive Bayes

Bayes theorem states that the probability of an event can be predicted with the help of conditions of the system from prior knowledge, In Machine Learning, Naïve Bayes theorem used in supervised learning technique for classification purposed for the high dimensional training dataset. The representation of Bayes theorem is given by:

$$\mathrm{P(A|B)} = \frac{P(B|A)P(A)}{P(B)}$$

Support Vector Machines

It is one of the ML supervised learning techniques. It creates a line or boundary to classify and separate data of different locations in the graph, with the help of segregation of data with a line, where we can easily find the new data point in the correct category, SVM chooses the boundary using a hyperplane.

3.2 *Portal Used*

Streamlit Portal

The whole dataset is fed into different machine learning models, where these models are then simplified for user-friendly purpose. The app used for the hosting of Machine learning models is Streamlit. Python interactive web application development can be done using the open-source framework Streamlit. Without needing to write HTML, CSS, or JavaScript, Streamlit enables you to quickly construct web applications.

The portal is into 3 different parts

Use Specific NPF Value.

Crop to be planned and harvested crop.

Model to predict the accuracy.

When the user enters the NPK value of a particular region, the crop to be planted and the harvested crop with the model of choice. The algorithm first accepts the user specific NPK value and feeds it to the algorithm which has already been trained for that particular site and when a specific set of values are given, the model predicts whether the spot is fertile or not and how much fertile that spot is. The backend of the site is programmed with python code and hosted by GitHub with ML algorithms and dataset taken. Finally, the accuracy of the particular algorithm is shown as output for the user.

4 Software Implementation

The Fig. 1 gives the process flow of the NPK sensor. First, we have to initiate and declare the sensor pins as input and then map them with respective functions and get the hexadecimal value and convert them into a decimal value. Finally, NPK values are displayed as output.

5 Results and Discussion

The result from the NPK sensor are in cloud display as shown in the above Fig. 2 The value of nitrogen, potassium and phosphorous are displayed in the form of graph which is plotted between date vs the respected value from the sensor.

The NPK sensor is calibrated and a sample black soil is taken for the test. The ML Supervision techniques has been implemented in soil fertility prediction. A sample data set has been given to the machine from which various algorithms were run.

The results of the algorithms with the given data set is categorized into fertile and non-fertile. The accuracy levels of the Logistic Regression, Random Forest, Naïve Bayes and SVM algorithms are as follows.

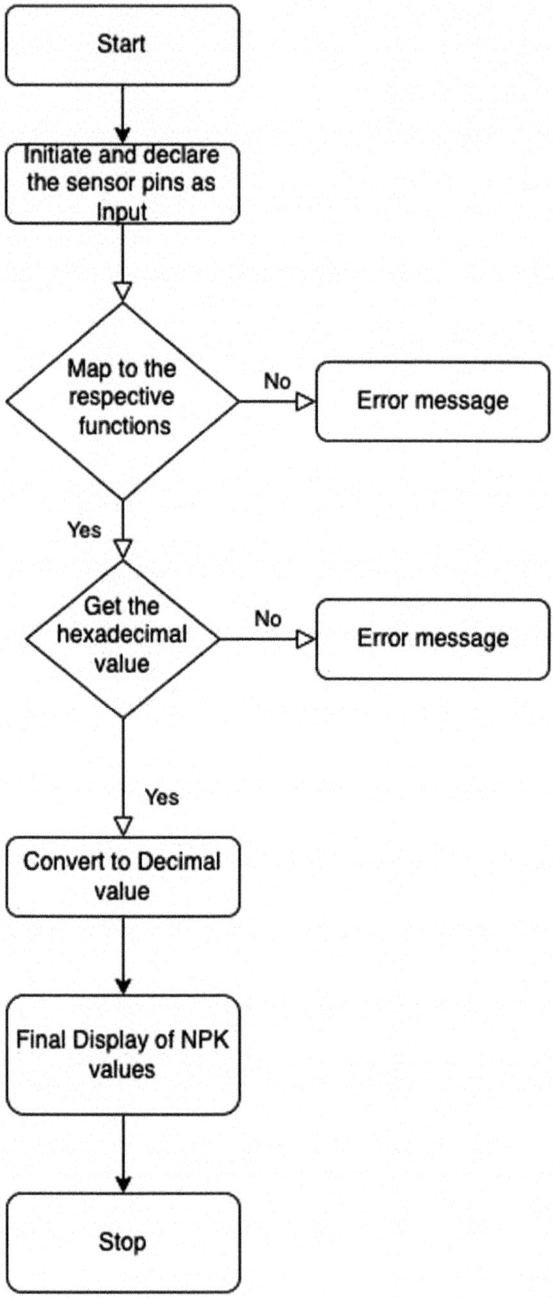

Fig. 1 Flowchart of Software Implementation

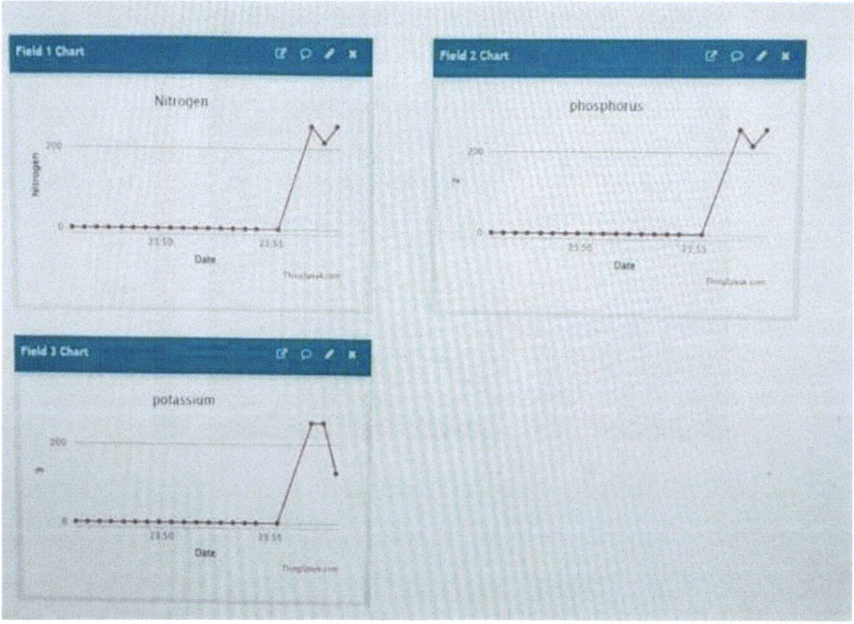

Fig. 2 Thing speak cloud display

The NPK sensor is calibrated and a sample black soil is taken for the test. The ML Supervision techniques has been implemented in soil fertility prediction. A sample data set has been given to the machine from which various algorithms were run. The results of the algorithms with the given data set is categorized into fertile and non-fertile. The accuracy levels of the Logistic Regression, Random Forest, Naïve Bayes and SVM algorithms are as tabulated in Table 1.

The co-relation matrix for the various components present in the soil is mapped with the components present in the sample data set, the co-relation matrix expresses the relationship between each component in the soil shown in Figs. 3 and 4.

The Fig. 5 shows the front end of the portal from which the desired information can be obtained When the user enters the NPK value of a particular region, the crop to be planted and the harvested crop with the model of choice.

Table 1 ML Prediction result

Algorithms	Accuracy (%)
Logistic regression	90
Random forest	93
Naïve Bayes	90
Support vector machines	56

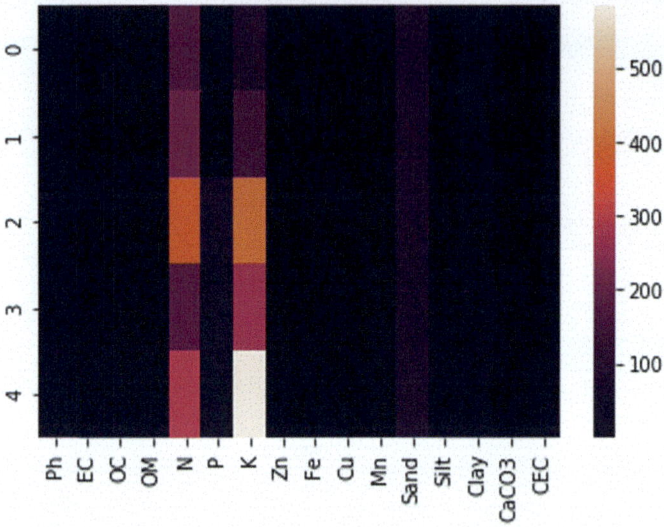

Fig. 3 Confusion matrix

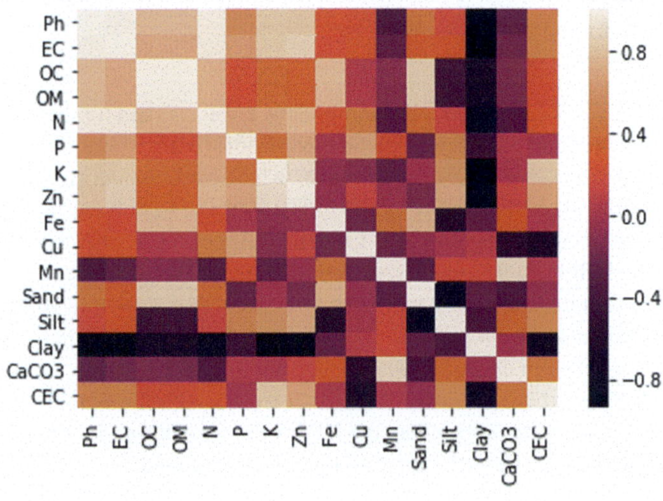

Fig. 4 Correlation matrix

The algorithm first accepts the user specific NPK value and feeds it to the algorithm which has already been trained for that particular site and when a specific set of values are given, the model predicts whether the spot is fertile or not.

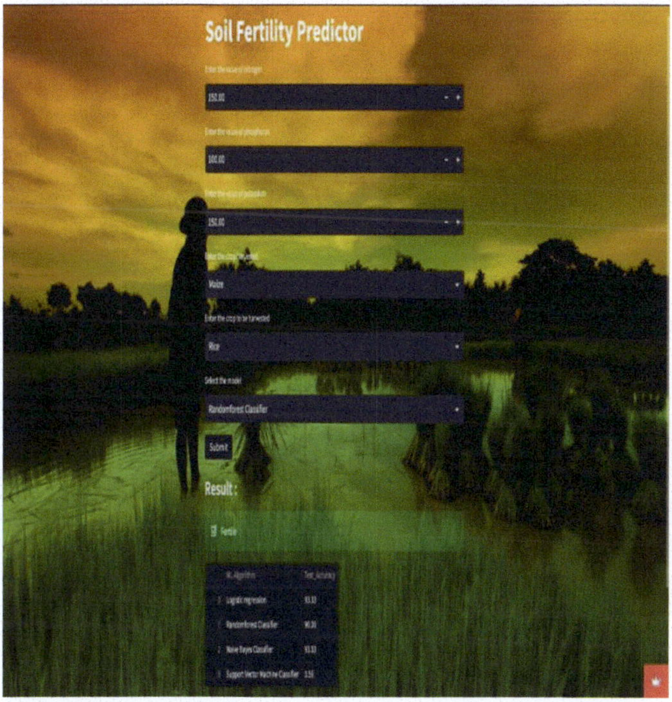

Fig. 5 Front end of the portal

6 Conclusion

In this project, we were successful in making a prototype which detects the soil nutrients present in the soil sample. Soil samples were taken from different fields and tested for multiple times. The results were constant at each trial. This was achieved due to the accuracy level of the Soil NPK sensor used. The soil nutrient's value was displayed on PC screen with the help of a serial monitor. The rotational crop method, the analysis of the soil nutrient levels after crop cultivation, the type of soil, the climatic conditions are all taken into consideration for the creation of the dataset. This dataset will be used to train the algorithms like Random Forest algorithm, Logistic Regression, Support Vector Machines (SVM) and Naïve Bayes algorithm are all trained to test the accuracy of the output.

To develop a feasible compact fertility prediction equipment that helps underlying farmers to produce healthy efficient crops with maximum production rate and to reduce the cost margin in production of fertilizers. Farmers with basic information like the type of soil, the value of the NPK at a particular point, climatic conditions and the crop that was harvested before, will enable them to know whether the soil is fertile or not. Our portal accepts values of macronutrients and feeds them to the ML algorithms to predict an approximate fertility value. Making farming smart in nutritional aspects enhances the future of farming and also makes technology user-friendly. The future for macronutrient calculation will be even image processed if sufficient data is taken. From our proposed methodology even the values of NPK can be stored for calculation purposes to make a highly predictable model than this, the training algorithms used here can be made self-sufficient by adding more crop data. In the forthcoming years, when the NPK values of specific crops are standardized by concerned authorities, the ML models can be used for a wide variety of soil samples.

References

1. Kulkarni A, Revathi Potdar, Alok Verma (2021) Determination of soil nutrients (NPK) using optical methods. In: International Conference on Plant Physiology and Growth; Plant Interations and Disease Control; 3(2)
2. Fenila Naomi J, Theepavishal RA, Madhuaravinth KS, Tharuneshwar V (2021) Soil quality analysis and an efficient irrigation system using agro-sensors. In: International Conference on Plant Nutrition; 8(5)
3. Janez Trontelj ml, Olga Chambers (2020) Machine learning strategy for soil nutrients prediction using spectroscopic method. In: IEEE International Conference on Signal Processing, Computing and Control, 7(3)
4. Shubham Prabhu, Prem Revandekar, Swami Shirdhankar, Sandip Paygude (2019) Soil analysis and crop prediction. Int J Sci Technol Res, 7(4)
5. Shruti Mishra, Priyanka Paygude, Snehal Chaudhary, SonaliIdate (2018) Use of data mining in crop yield prediction. In: International Conference on Inventive Systems and Control (ICISC), 2(3)
6. Venkat Narayana Rao T, Gaddam Rishitha Reddy (2019) Prediction of soil quality using machine learning techniques. Int J Sci Technol Res, 8(11)
7. Abishek B, Akash Eswar (2017) Prediction of effective rainfall and crop water needs using data mining techniques. In: IEEE International Conference on Technological Innovations in ICT for Agriculture and Rural Development (TIAR)
8. Sadia Afrin, Abu Talha Khan (2017) Analysis of soil properties and climatic data to predict crop yields. In: IEEE International Conference on Inventive Systems and Control
9. Shivnath Ghosh, Santanu Koley (2014) Machine learning for soil fertility and plant nutrient management using back propagation neural networks. IJRITCC, 2:2
10. Arun Kumar, Naveenkumar, Vishal Vats (2012) Efficient crop yield prediction using machine learning algorithms. Int Res J Eng Technol (IJRET), 5

Design of Intelligent Solar Cooling System with IoT Monitoring

C. Kushagra, S. Kirtimaan, P. Rushit, K. Jane Joshita, J. Divya Navamani, and A. Lavanya

Abstract Currently, IoT rules many unmanned applications to improve supervision and productivity. The proposed work concentrates on the need for a cooling system for solar Photovoltaic (PV) panels to enhance its efficiency. An increase in temperature will reduce the efficiency of the solar panel. The cooling system should be monitored, automated, and controlled without human intervention. This is achieved by developing a web App by which the cooling system can be controlled automatically. DHT11–Temperature and Humidity Sensor is used to monitor the temperature of the panel. If the module temperature increases more than 35 °C to 60 °C, the sensor sends the signal to the actuator to cool the PV panel, which will increase the output power. This technique is proposed for solar rooftop home applications where the grey water can be treated and used for cooling the solar panel rather than using fresh, clean water. Finally, a web-interface is done to cool the solar panel with water pumping through the tubes below the panel.

Keywords Solar · IoT · Cooling · Web interface · Temperature

C. Kushagra · S. Kirtimaan · P. Rushit
Department of Computer Science Engineering, SRM Institute of Science and Technology, Kattankulathur, Tamil Nadu, India

K. Jane Joshita (✉)
Department of Electronics and Instrumentation Engineering, SRM Institute of Science and Technology, Kattankulathur, Tamil Nadu, India
e-mail: Janejoshita@gmail.com

J. Divya Navamani · A. Lavanya
Department of Electrical and Electronics Engineering, SRM Institute of Science and Technology, Kattankulathur, Tamil Nadu, India
e-mail: divyateddy1@gmail.com

D.-I. Stroe et al. (eds.), *Emerging Electronics and Automation*, Lecture Notes in Electrical Engineering 1202, https://doi.org/10.1007/978-981-97-3090-2_21

1 Introduction

Many solar-powered systems need IoT integration to increase their efficiency and productivity. Figure 1 presents the various application where the intelligent solar power system is required. A Smart IoT-based system is proposed for aquaponic systems to enhance productivity. They used various sensors such as Groove-pH, and SI1145 [1]. The low efficiency in the egg hatcheries is due to human beings monitoring and controlling the processes of incubation. They had put forward the idea of using a Solar panel and a temperature humidity sensor DHT22 sensor for increasing the hatchability rates [2]. The IoT integration has made it possible to monitor and control the pump in real time, improving water consumption efficiency and making it possible to conduct practical farming activities. In order to transfer data regarding agriculture and irrigation to a Blynk IoT cloud, smart irrigation systems use a variety of microcontrollers, sensors like the DHT11, actuators, and wireless communication methods [3]. With the assistance of the HC-SR04 sensor (SONAR), the model that was created is programmed with the Arduino IDE to control the operations of the Grasscutter. The article covers an Internet of Things (IoT)-based grass cutter that can be controlled and accessed from a distance utilizing a Node MCU microcontroller, a DC motor, a rechargeable battery, a solar panel, and the Blynk (IoT) platform [4]. The sun's position and the quantity of dust deposited have an enormous effect on efficiency. Automatic panel adjustment and dust particle cleanup can improve performance. The current-to-volt-age converter in the MCU (Raspberry Pi) may function when paired with the LDR and temperature sensors in this setup [5]. Article [6] explores a process to look at and enhance a solar water heating systems (SWH) energy efficiency by utilizing the data collected from IoT with the aid of solar irradiance sensors and temperature sensors.

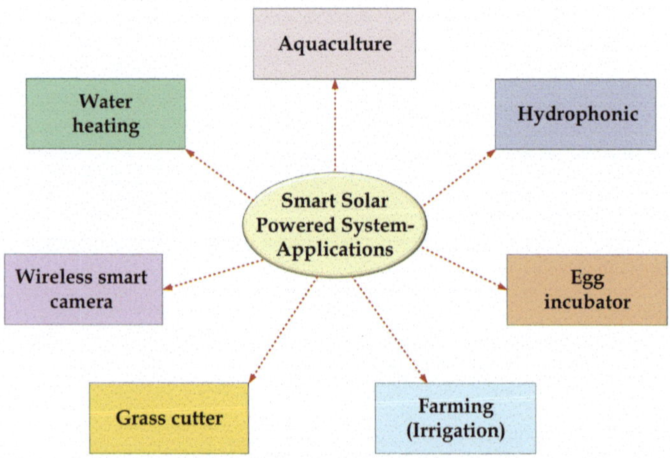

Fig. 1 Various smart solar-powered applications

A dual-axis solar tracker (DAST) is required to rotate the solar panel about its horizontal and vertical axes in order to maximize energy output. This technology will guarantee effective sun tracking and maximum solar panel energy output by using Two Light-Dependent Resistor (LDR) sensors which are monitored through the use of the internet [7]. A system is designed in such a way that remotely monitors the irrigation system that connects to a solar cell and has built-in WIFI to supply the necessary running power adapter. The float switches, soil moisture sensor, humidity and temperature sensor, and flow rate sensor are all linked to the controller and are interfaced with via the controller's analog and digital inputs. The bilge pump and diaphragm pump that control irrigation is connected to the relay board [8]. SlugCam is a wireless smart camera network that runs on solar power and may be utilized for a variety of outdoor purposes. The onboard camera of Slug Cam tends to turn off, but when motion is detected, a passive infrared (PIR) sensor switches it on [9].

1.1 Related Works

The IoT gadget is intended to measure the temperature and communicate the data to the cloud, where the MQTT lens application will gather the data from the sensors. This is so that plant fields and their employees can learn about the temperature early on to develop future plans. The temperature-based omnidirectional solar-tracking IoT gadget is presented in this study [10]. This study creates an Internet of Things (IoT) based dust detecting system to analyze the accumulation of dust and solar panel efficiency. Reasonable IoT systems were created to collect the data. The device used a cheap optical dust sensor and Wi-Fi connectivity microcontroller The device made use of a My IoT Dashboard, Cayenne devices are used to keep an eye on online data collection [11]. Choosing a location for PV plants, designing PV systems, and making financial decisions can all be confidently done using the data gathered and solar maps, and PV project analysis. This research, is majorly looked into automating the solar map's capture, processing, and capture generation utilizing the recorded solar data. Furthermore, to create, implement, and test a comparable Internet of Things module for the solar data is captured and provided in a hardware-compatible format [12].

The output of a solar panel may be impacted by the sun's position and dust. As a result, this paper developed a smart system based on the Internet of Things (IoT) that can spin the control panel to allow tracking of traits, cleaning, and watching the results. The IoT system as a whole is used by Obtaining an Arduino Uno, Wi-Fi module, and smartphone data contained in the application [13]. This gives analysis utilizing tools including a PV analyzer, Solar Power Meter, and various incorrect conditions to compare the outcomes using a thermal camera. The graphs of PV at normal, partial temperatures, efficiency, fill factor, and power are attained. Using a PV analyzer, shade and dust accumulation circumstances results from the PV analyzer and thermal pictures are contrasted in no-load, load, partial shading, and accumulation of dust conditions. The acquired information is routinely uploaded. The uploaded data can be utilized further utilizing NodeMcu on the cloud for keeping an eye on the installed plant [14]. The Development of a system is to observe numerous parameters of Pico solar systems. Once connected to a voltage and current sensor, the load is through the use of Thing Speak, the Internet of Things (IoT) cloud server. The sensor sends signals to the MCU ESP32, following which the value is uploaded to the cloud by the microcontroller server and shown it on the LCD screen as well. The device successfully kept track of the solar parameters in real-time [15].

The suggested work uses light-intensity diodes to track solar light automatically, directing the panel as necessary and if the panel has any dust or debris, an automated cleaning system is employed for cleaning. This method will result in higher power output. For detection, suitable sensors are utilized. The controller is connected to the arrangement and the dust. The action will be taken based on the controller's output and carried out in the created prototype [16]. Boltuino (Bolt IoT + Arduino) device is utilized as a controller circuit and a 20-W solar panel as the PV module in this work. Temperature, voltage, current, and other sensors for the collecting of solar parameters, and humidity sensors are employed. The Arduino Uno gathers voltage and current data. The sensor sends data from the temperature and humidity sensors through the serial transmission to the BOLT subsequent to receipt. It shows all of the sun statistics in BOLT. It creates a graph that stores all of the parameter information. The warning letter will be delivered to being able to access a registered mobile phone using SMS and observe the data coming from the PV module online to operate the system if any undesirable circumstances arise [17]. With a primary focus on using, this proposed article describes the implementation of a Smart Solar panel cleaning system that belongs to IoT (internet of Things) technology. This makes dust monitoring possible aptitude, sophisticated analysis, and system management that encourages boosting the solar PV panel's overall efficiency [18].

2 Methodology

The IoT is a wonderful platform for the creation of an intelligent monitoring system that is affordable. IoT enables the quality monitoring and control of PV systems in remote locations. This system also helps to record and analyze the collected data and real-time online system monitoring. Figure 2 presents the suggested methodology of this work. The proposed method is designed for small-scale standalone PV systems and has scope for a scaled, more extensive PV system. The developed system uses greywater management to cool the PV system to improve the performance of the PV panel. Suitable sensors are implanted in the system to monitor various parameters and control the system based on the prevailing weather conditions. The future scope is to extend the method for fault diagnosing, control and maintenance to increase the lifetime of the PV system. The salient features of this work are.

- Minimal power consumption
- Simple to implement
- Interactive website and easy updating

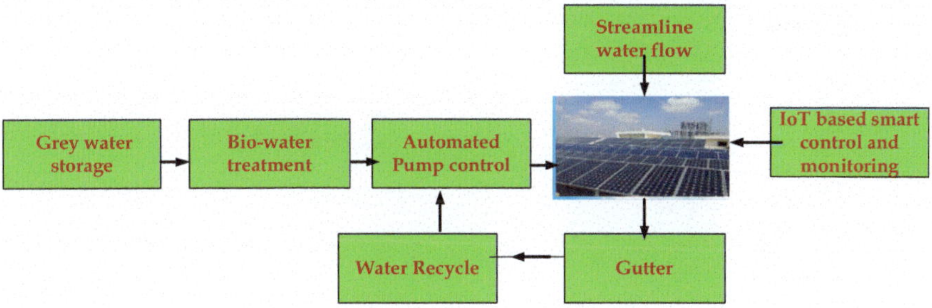

Fig. 2 Suggested methodology for this work

- On Dashboard
- Cost effective.

3 Impact of Temperature

The temperature in PV modules creates a greater impact on the solar output power. Usually, the module temperature will be greater than the ambient temperature. Depending on the solar radiation and wind speed, the module temperature will be 20 °C to 30 °C greater than the ambient temperature. The reason for the increase in the module temperature is due to the presence of glass cover which traps the infrared radiation and retains it. A generalized expression to calculate the module temperature of the PV panel is given as

$$T_{module} = T_{ambient} + KP_{in} \tag{1}$$

T_{module} is the PV module temperature in °C, $T_{ambient}$ is the ambient temperature in °C, Pin is the intensity of the solar radiation in W/m^2 and K is the constant, which depends on the wind speed. This value increases with lower wind speed. Figure 3 depicts the Global Horizontal Irradiance (GHI) for the month of July and October. It is observed that the due to this irradiance, there is a possibility in increase in high temperature. So, the need of cooling system is essential in this location (12.8230° N, 80.0444° E).

4 Components and Devices

The following components are used to conduct the experiment and the setup is shown in Fig. 4.

- Voltage Regulator

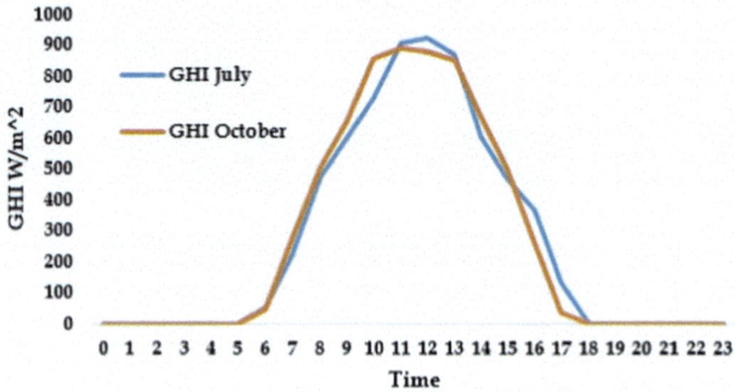

Fig. 3 Global horizontal irradiance in the tested location (12.8230° N, 80.0444° E)

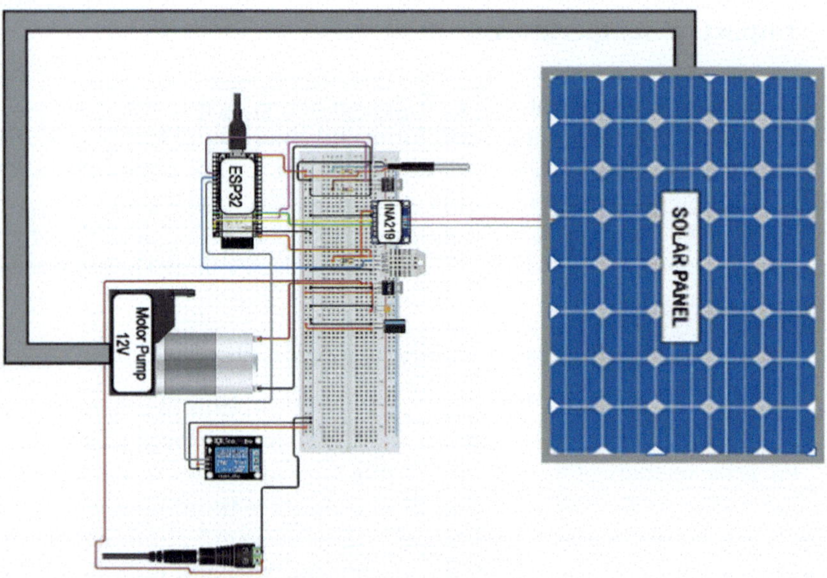

Fig. 4 Schematic circuit of the experimental setup

- DC motor water Pump: 6 V to 9 V
- DS18B20 Temperature Sensor Probe
- DHT22 Ambient Temperature Sensor
- INA219: Voltage, Current and Power Monitor.

The DS18B20 Temperature Probe is used to take the actual reading of the PV Cell. The DHT22 Temperature Sensor is used for the ambient temperature sensor reading.

It is necessary to use the INA219 Power Monitor because the max Voltage limit of taking analog reading using the NodeMcu microcontroller board is 1 V. The INA219 Monitor will facilitate the reading of Voltage, Current as well as power digitally.

5 Experimental Results

The flow mechanism of the PV panel is initiated by the treatment of grey water combined with coolants for using it in the cooling management system. The DHT sensors onboard will continuously observe both the ambient environment temperature and the Panel surface temperature and once the threshold is crossing, the monitoring and sensing unit will send the cooldown request to the autonomous cooling system to decrease the panel temperature to an optimal level in order to get the highest possible power output. The Cooling system will then start pumping coolants through the heatsink installed at the back of the PV panel. The water flow will dissipate the surface heat and bring the temperature down by 10–20 °C (depending on the ambient temperature). Once the coolant water is pumped through the heatsink, it is then collected in the coolant storage unit for reuse. The control system for the pump using IoT setup implemented is shown in Fig. 5.

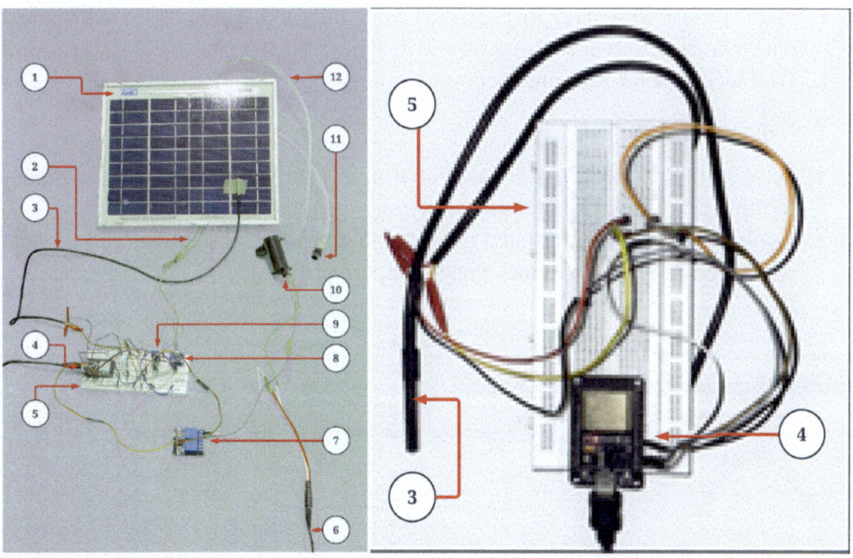

Fig. 5 Implemented prototype for this study 1. Solar Panel, 2. Solar Panel's Output connectors, 3. DS18B20 Temperature Sensor Probe, 4. NodeMCU Microcontroller, 5. Breadboard, 6. Supplementary DC power for water pump, 7. Relay, 8. INA219: Voltage, Current and Power Monitor, 9. DHT 22 Ambient Temperature Sensor, 10. DC motor water Pump, 11. Water Outlet, 12. Water Inlet

6 Results and Discussion

The live monitoring system has been created so that users can get the readings in real time on any device. At the heart of the system is the ESP32 microcontroller which is connected to all the components to command the cooling system automatically once the temperature threshold limit is crossed according to the DS18B20 sensor installed on the surface of the PV panel. The microcontroller also receives the live Voltage, Current and Power readings from the PV panel through the INA219 Power Monitor sensor. These readings are then uploaded to the cloud server where they are logged and analyzed. The live monitoring web interface gets the live data from the microcontroller as well as the data log in the cloud server. The web page developed for customizing this system performance is shown in Fig. 6a. The effect on PV panel surface temperature after implementing the cooling system is shown in Fig. 6b. The main features of this work are.

1. Data Logging:
   - Sensor Integration: Implemented DS18B20 and DHT11 sensors for recording panel surface temperature and ambient environmental temperature, respectively.
   - Electricity Measurement: Utilized INA219 power sensor integrated with a load to accurately measure electricity output generated by the solar panel.
   - Data Storage: Stored readings on the ESP32 microcontroller and synchronized data to a cloud server for logging and analysis. Also, accessible through a web GUI interface for real-time access.
2. Automated System:
   - Temperature Control: Automated cooling system activated based on predetermined thresholds to optimize electricity generation.
   - Cooling Mechanism: Initiated cooling via nodemcu, utilizing a relay system with a motor pump to reduce panel temperature.

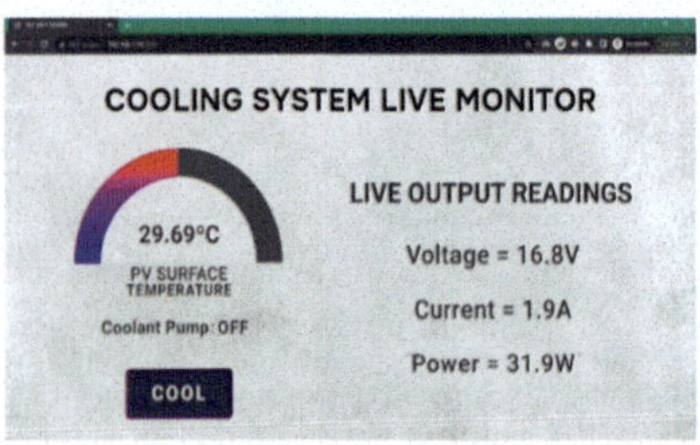

(a)

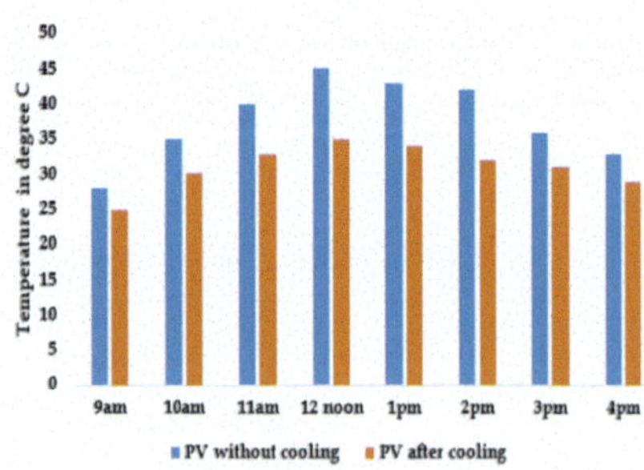

(b)

Fig. 6 **a** Live monitoring system -Web interface **b** Effect of cooling PV panel

- Manual Override: Incorporated a manual cooldown option for urgent temperature adjustments or emergency situations.

3. File System:
   - Challenges Addressed: Addressed issues related to the absence of a file system for appropriate data and file storage.
   - Solution Introduction: Implemented SPIFFS to manage and store necessary data files required for the functioning of the live monitoring system on the NodeMCU.

4. Web GUI:
   - Live Monitoring Interface: Developed a web GUI for real-time monitoring, displaying crucial data such as temperature, current, and power output through visualizations and graphs.
   - User Control: Provided user control options, including manual override functionalities, within the interface.

5. Tech Stack:
   - GUI Development: Utilized HTML, CSS, and Javascript to create a user-friendly interface showcasing real-time solar panel data.
   - Hosting Platform: Employed NGROCK for hosting due to its versatile accessibility across devices and locations, ensuring a cost-efficient and seamless user experience.
   - ESP-32 NodeMCU Functionality: Implemented the ESP-32 NodeMCU as an autonomous server, handling data transmission to the NGROCK platform and website display, consolidating server operations within the microcontroller.

6. Access Anywhere:
   - Device Compatibility: Ensured real-time data accessibility across multiple devices, offering interactive charts and graphs for optimized user engagement.
   - Cross-Platform Usability: Prioritized cross-platform usability and responsiveness, enabling seamless interaction across diverse web-enabled devices.

7 Conclusion and Future Scope

Efficient cooling mechanisms for solar panels are pivotal for optimizing their power output and ensuring their long-term functionality. The salient features of the system are.

- Efficient Data Logging
- Automation for Optimal Performance
- System Enhancement for Accessibility
- Cost-Effectiveness and Future Viability

- Improved Energy Efficiency and Environmental Impact
- User Control and Future Collaborations.

The relationship between elevated panel temperatures and diminishing efficiency underscores the significance of effective cooling strategies. Integrating a dashboard-based monitoring system within the Smart Solar Panel setup has yielded promising outcomes. This element is a crucial tool for rapid fault detection, allowing for swift and accurate identification of potential issues. Moreover, its data analytics capabilities offer insights into the system's performance trends and facilitate predictions about its future operational lifecycle. By utilizing historical and real-time data, the dashboard identifies vulnerabilities and enables proactive measures to ensure consistent power generation and system stability. The experimental results mentioned in the paper claim that the decrement in operating temperature is around 2–5 °C will increase the power output of the PV panel. Thus, this methodology will be more effective and beneficial for larger PV systems.

References

1. Taji Khaoula, Rachida Ait Abdelouahid, Ibtissame Ezzahoui, Abdelaziz Marzak (2021) Architecture design of monitoring and controlling of IoT-based aquaponics system powered by solar energy, Procedia Comput Sci, 191, Pages 493–498, ISSN 1877–0509
2. Author F (2021) Article title. J 2(5):99–110
3. Forson Peprah, Samuel Gyamfi, Mark Amo-Boateng, Eric Buadi, Michael Obeng (2022) Design and construction of smart solar-powered egg incubator based on GSM/IoT Scientific African, 17, e01326, ISSN 2468–2276
4. Roshahliza M Ramli, Waheb A Jabbar (2022) Design and implementation of solar-powered with IoT-Enabled portable irrigation system, Internet Things Cyber-Phys Syst, 2, Pages 212–225, ISSN 2667–3452
5. Babangida Ibrahim, Siva Brahmaiah V, Pradeep Sharma (202) Design of smart autonomous remote monitored solar powered lawnmower robot. In: Materials Today: Proceedings, 28, Part 4, Pages 2338–2344, ISSN 2214–7853
6. Priyadharsini K, Dinesh Kumar JR, Ganesh Babu C, Srikanth A, Sounddar V, Senthamilselvan M (2021) Elegant method to improve the efficiency of remotely located solar panels using IoT. In: Materials Today: Proceedings, 45, Part 9, Pages 8094–8104, ISSN 2214–7853
7. Wen-Tai Li, Wayes Tushar, Chau Yuen, Benny Kai Kiat Ng, Stewart Tai, Kwee Tiang Chew (2020) Energy efficiency improvement of solar water heating systems—An IoT based commissioning methodology. Energy Build, 224,110231, ISSN 0378–7788
8. Muthukumar P, Manikandan S, Muniraj R, Jarin T, Ann Sebi (2023) Energy efficient dual axis solar tracking system using IOT, Measurement: Sensors, 28, 100825, ISSN 2665–9174
9. Al-Ali AR, Ahmad Al Nabulsi, Shayok Mukhopadhyay, Mohammad Shihab Awal, Sheehan Fernandes, Khalil Ailabouni (2019) IoT-solar energy powered smart farm irrigation system. J Electron Sci Technol, 17(4) ,100017, ISSN 1674–862X
10. Kevin Abas, Katia Obraczka, Leland Miller (2018) Solar-powered, wireless smart camera network: An IoT solution for outdoor video monitoring, computer communications, 118, Pages 217–233, ISSN 0140–3664
11. Yusof ZM, Billah MM, Kadir K, Ali AMM (2019) A Temperature-based omnidirectional solar tracking system for IoT application. In: 2019 IEEE International Conference on Smart Instrumentation, Measurement and Application (ICSIMA), Kuala Lumpur, Malaysia, pp. 1–4

12. Murshiduzzaman, Kadir JA, Ismarrubie ZN, Yussof H, Hasan WZW (2021) Development of IoT based dust density and solar panel efficiency monitoring system. In: 2021 International Conference on Electrical, Computer and Energy Technologies (ICECET), Cape Town, South Africa, pp. 1–6
13. Madub P, Appavoo P, Chiniah A (2019) Evaluation of IoT for automatisation of solar map pproduction: case of mauritius and outer Islands. In: 2019 IEEE Global Conference on Internet of Things (GCIoT), Dubai, United Arab Emirates, pp. 1–4
14. Shah MMA, Parvez MS, Ahmed A, Hazari MR (2021) IoT based power monitoring of solar panel incorporating tracking system. In: 2021 International Conference on Automation, Control and Mechatronics for Industry 4.0 (ACMI), Rajshahi, Bangladesh, pp. 1–4
15. Phoolwani UK, Sharma T, Singh A, Gawre SK (2020) IoT based solar panel analysis using thermal imaging. In: 2020 IEEE International Students' Conference on Electrical, Electronics and Computer Science (SCEECS), Bhopal, India, pp. 1–5
16. Tanimun Hasan S, Sultana Shompa M, Abdur Rahman M, Abu Rasel M, Rahim Hossain Apu M, Arifur Rahman M (2022) IoT based solar power monitoring & data logger system. In: 2022 IEEE International Women in Engineering (WIE) Conference on Electrical and Computer Engineering (WIECON-ECE), Naya Raipur, India, pp 182–187
17. Vidhya H, Akshaya U, Keerthana MG, Dhivyanandhini T (2022) IoT based solar technology monitoring and cleaning system. In: 2022 International Conference on Automation, Computing and Renewable Systems (ICACRS), Pudukkottai, India, pp. 1–5(2022)
18. Mondal A, Ali MJ, Dutta P (2022) IoT enabled smart solar panel monitoring system based on boltuino platform. In: 2022 IEEE International IOT, Electronics and Mechatronics Conference (IEMTRONICS). Toronto, ON, Canada 2022:1–7
19. Rao, Neha, Santosh Kumar S (2018) Smart IoT based solar panel cleaning system. In: 2018 3rd IEEE International Conference on Recent Trends in Electronics, Information & Communication Technology (RTEICT), Bangalore, India, 2018, pp 2357–2361, (2018)

Analysis of Energy Consumption Using Support Vector Regression

V. Pavan Kumar, G. Likhith Reddy, H. Mahadev, V. Harish Kumar, and K. Krishna Prasad

Abstract Energy consumption analysis is essential for maximizing energy use and improving energy efficiency across a range of applications. In this study, we suggest a unique method for analyzing energy use based on real-time data from an energy meter. The information gathered comprises the voltage, current, power, and overall energy consumption's root mean square (RMS) values. We use the Support Vector Regression (SVR) model, a potent machine learning method known for its ability to handle non-linear connections and high-dimensional data, to reliably forecast future energy usage. A popular interactive computing tool called Jupyter Notebook has the SVR model incorporated, making experimentation and analysis simple. Our proposed methodology proves its effectiveness in anticipating future energy use through testing and evaluation. The SVR-based prediction model's findings offer useful insights into trends in energy consumption, empowering users to make deft choices that would optimize energy use, cut down on waste, and enhance overall energy sustainability.

Keywords Voltage · Current · Power · Energy consumption · Support Vector Regression (SVR)

V. Pavan Kumar (✉) · G. Likhith Reddy · H. Mahadev · V. Harish Kumar · K. Krishna Prasad
Computer Science and Engineering, B V Raju Institute of Technology, Narsapur, Telangana, India
e-mail: drpavankumarv@gmail.com

G. Likhith Reddy
e-mail: 21211A0583@bvrit.ac.in

H. Mahadev
e-mail: 21211A05A8@bvrit.ac.in

V. Harish Kumar
e-mail: 21211A05A7@bvrit.ac.in

K. Krishna Prasad
e-mail: 21211A05B7@bvrit.ac.in

D.-I. Stroe et al. (eds.), *Emerging Electronics and Automation*, Lecture Notes in Electrical Engineering 1202, https://doi.org/10.1007/978-981-97-3090-2_22

1 Introduction

Energy consumption analysis is now at the center of research and policy making because of the rising demand for energy resources, environmental concerns, and the requirement for sustainable energy practices. To optimize energy use, reduce waste, and ensure effective energy distribution across a variety of sectors, including residential, commercial, and industrial settings, it is crucial to understand energy consumption patterns [1].

Smart energy meters have made it possible to gather real-time data on energy consumption in recent years, thanks to technological improvements. The root mean square (RMS) values of voltage and current, power consumed, and total energy consumed are all detailed information that is provided by these instruments on a regular basis [2]. This abundance of data offers a fantastic chance to use machine learning algorithms to predict future energy consumption.

The analysis of energy use using real-time data from energy meters is the focus of this research. This study's main goal is to suggest a novel strategy for reliably forecasting future energy usage using the Support Vector Regression (SVR) model. SVR is a powerful machine learning method that can handle non-linear connections and complex data patterns, making it suitable for predicting energy usage [3]. The SVR model is built and tested using Jupyter Notebook's capabilities. This makes it possible to experiment with and analyze the obtained energy data with ease, enabling effective model adjustment and performance assessment.

The goal of the project is to provide an accurate and dependable method for estimating energy use, which will help improve energy management procedures. The findings of our study could provide useful information on patterns of energy use, assisting stakeholders in making decisions that will maximize energy use and increase overall energy sustainability.

The paper organized as Literature Survey is explained in Sect. 2. Next, Sects. 1.1 and 1.2 explain about problem statement and objectives respectively. Existing work is explained in Sect. 1.3. Proposed work is demonstrated in Sect. 3. Implementation modules are explained in Sect. 4. Results are discussed in Sect. 4.3 and followed by conclusions.

1.1 Problem Statement

- The literature survey is based on the technology, which continuously monitors energy consumption and sends the measured data to a cloud server. But we can only know about real-time consumption.
- Therefore, the smart energy meter can be combined with machine learning to predict future energy consumption.

1.2 Objectives

1. To provide accurate measurement of energy consumption and eliminate the need of manual reading.
2. By understanding how much energy consumed, users can reduce their consumption.
3. By using the data, we can prepare analytics of energy consumption and predict some valuable insights.
4. The user can determine his expected future costs by using his estimated energy use.

1.3 Existing Work

The current energy meter regulates energy use by connecting and controlling energy appliances and equipment in residential buildings, commercial buildings, and industrial facilities. It gathers information from a variety of sources, including sensors, energy meters, and other linked devices. This data is gathered in real-time and contains details on energy generation, storage, and consumption. It keeps track of the state of many factors, including voltage reading, current reading, temperature, etc., and has the capability of remotely adjusting those parameters and energy usage [4].

2 Literature Survey

In the article [5], the potential uses and difficulties of incorporating IoT technology into the electricity system are discussed. The five levels that make up the conceptual framework for the system are the sensor level, the communication level, the data analytics level, the decision-making level, and the actuation level. The advantages of IoT-enabled energy grids, such as increased efficiency, dependability, and cost savings, are highlighted by the author. Additionally, they examine a number of application scenarios, including distributed energy resource management, demand response, and smart meters. There are obstacles to integrating IoT technology into the electricity grid, too, including cybersecurity risks, data privacy issues, and the requirement for interoperability between various IoT platforms and devices. Collaboration among stakeholders in the energy, telecommunications, and IT industries will be necessary to address these difficulties.

The article [6] provides a complete assessment of how Internet of Things (IoT)-enabled smart grids have been implemented. It provides a general introduction to IoT and smart grid technologies and how they interact. In order to develop IoT-enabled smart grids, it uses a layered strategy that takes into account the physical level, network level, platform level, and application level. The physical layer is made up of many sensors and tools that are used to gather information on energy generation, storage, and consumption. The platform layer, which consists of cloud-based systems and analytics tools for processing and analyzing the data, receives the collected data from the network layer and processes it. The smart grid application layer is made up of a variety of applications, including grid optimization, demand response, and energy management and it also covers the difficulties and possibilities posed by the deployment of IoT-enabled smart grids. These difficulties include interoperability problems, security issues, and data privacy problems. It suggests a number of methods to deal with these difficulties, including the application of blockchain technology, data encryption, and standardization initiatives.

The paper [7] describes the layout of a ground-breaking smart house management system that employs edge computing, cloud computing, and smart grid technology in order to lower energy use and increase overall home energy system efficiency. It outlines the problems with conventional home energy management systems, such as the dearth of real-time data, the absence of flexible control options, and their poor performance. To deal with these issues, it suggests a new approach that combines edge computing, cloud computing, and smart grid technology. Three primary parts make up the proposed system: edge devices, smart meters, and a cloud-based control system. Smart metres are used to track energy consumption at the current time and provide data to edge devices. Based on customer preferences and current data, the edge devices, which are installed in the home, utilize machine learning algorithms to optimize energy consumption. The system's general management is handled by the cloud-based control system, which also offers consumers a simple interface for managing their energy use. The advantages of the suggested approach are also covered, including a better user experience, enhanced energy efficiency, and lower energy expenditures. They draw the conclusion that the suggested approach has the potential to fundamentally alter how we manage energy in our homes and advance the cause of a more sustainable future.

The article [3] describes regression approaches and machine learning algorithms to answer the urgent demand for precise and effective power consumption control. The importance of real-time monitoring and forecasting in enhancing energy consumption techniques is acknowledged by the writers. The study's authors begin by outlining the significance of power consumption monitoring and forecasting in the context of energy management. They highlight the drawbacks of traditional methods and draw attention to the potential benefits of cutting-edge data-driven solutions in addressing the complexity of power consumption patterns. Investigated machine learning algorithms include SVR, decision trees, neural networks, linear regression, and ARIMA for time series analysis. The unique qualities of each algorithm and their applicability in various situations are discussed in depth by the writers. They go over how these methods help with precise power usage forecasting and in-context monitoring in diverse settings.

The article [8] demonstrates a new IoT-based smart energy meter for smart grids. The authors stress the value of smart meters in assisting smart grids' effective energy management and consumption. The proposed smart energy meter uses IoT technology to provide continuous monitoring and control of energy consumption, as well as communication with other devices and systems in the smart grid. The design of the proposed smart energy meter which includes sensors, a microcontroller, communication modules, and cloud-based software. The article also discusses the benefits of the proposed smart energy meter, such as increased energy efficiency, improved demand response, and reduced operational costs for energy providers. It talks about the value of data security and privacy in smart grids as well as the steps taken to preserve the privacy and security of the proposed smart energy meter.

It is discussed in article [9] how electrical grids evolved over time, going from conventional grids to smart grids. Power quality and dependability are the two main challenges of a standard electrical systems. The IoT can help with both of these issues because it provides better management of them. A standard electrical infrastructure can be transformed as a smart grid architecture with the help of advanced metering and smart metering SM technologies by integrating intelligent processing of information while power is being transferred between the service provider and the customers. The Internet of Things architecture is divided into four layers: physical, cyber, communication network, and application layers.

In this article, it is suggested to model and create an intelligent energy metre for use with smart grid applications [10]. The suggested energy meter measures power factor, voltage, current, and energy consumption using a microcontroller and a variety of sensors. The communication protocol that is used to send energy data to a cloud-based server for remote monitoring and control is also described by the authors. The suggested energy meter is made to be scalable and quickly deployable, making it appropriate for smart grid applications. The results of the experiments the authors did to verify the precision and dependability of the suggested energy meter reveal that it can efficiently monitor and control energy use in smart grid applications. Security flaws with this implementation include cyberattacks where fraudulent data is injected into the grid.

The article [2] presents a smart energy metering system that uses IoT technology to track and manage energy consumption in real time. The system that is suggested in this study, which measures each appliance's power consumption using a combination of voltage and current sensors, has had its power quality thoroughly examined. An IoT gateway is used to transport the data to a cloud server, where it is analyzed to reveal patterns in energy use. The authors cover the development and execution of the suggested system, including the manufacturing of a special circuit board and the use of open-source software for data processing and visualization. A series of experiments are used to assess the system's performance, and the results demonstrate that the system measures energy usage with accuracy and dependability. The potential advantages of the suggested system are also highlighted in the report, including its capacity to detect energy waste and encourage energy savings. Users may minimize their energy bills and make more educated decisions about their energy use because of the system's real-time monitoring capabilities and user-friendly interface.

The article [4] describes a smart energy metering system that uses IoT and Global System for Mobile Communications technologies to monitor and manage energy consumption in real time. The suggested system collects data from energy meters and sends it to a cloud server via GSM communication using an Arduino microcontroller. Users are thus able to monitor their patterns of energy consumption and make educated decisions regarding energy usage, thanks to the analysis and visualization of the data on a dashboard. The proposed system's benefits are highlighted in the report, including the capacity to remotely monitor energy use and spot energy waste, which encourages energy conservation and lowers energy costs. The system is also affordable and simple to set up, opening it up to a wider spectrum of consumers. The obstacles to putting the suggested system into practice are also covered in the study, including issues with data security and privacy. The authors suggest approaches to deal with these difficulties, such as putting in place secure communication protocols and making sure that data is encrypted.

The article [1] introduce a smart energy meter that enables the tracking of energy usage at the device level in order to answer the growing requirement for detailed data on energy consumption. The study begins by addressing the shortcomings of conventional energy monitoring systems, which frequently provide aggregated data and are unable to pinpoint specific device energy use trends. The authors suggest an IoT-based smart energy meter that gathers real-time data on energy consumption from numerous devices inside a building or a specific location to get around these restrictions. The IoT-based smart energy meter uses sensor technologies to keep track of how much electricity is being used by linked devices. It gathers data on individual device energy use and transmits it to a central server or cloud platform. Then, users can remotely access this data, giving them access to information about the patterns of energy usage of particular devices.

3 Proposed Work

Model, View, and Controller are the three components of the MVC architecture, which is used in the proposed model shown in Fig. 1. The driving circuit and logic for calculating energy usage are included in the model portion. The Jupyter Notebook consists of the SVR model, which is included in the control section. The output, which is a projection of energy use and its price, is included in the view component.

The current and voltage sensors, which are connected to the circuit, read the current and voltage values. These values are communicated to the Dexter Board microcontroller. The required power and energy usage are then calculated and stored in the FRAM memory of the Dexter board. The ESP32 module receives the calculated data from the on-board FRAM memory of the Dexter. The Support Vector Regression (SVR) model is then used to analyze the downloaded data. Future energy use and price predictions are made using a trained model.

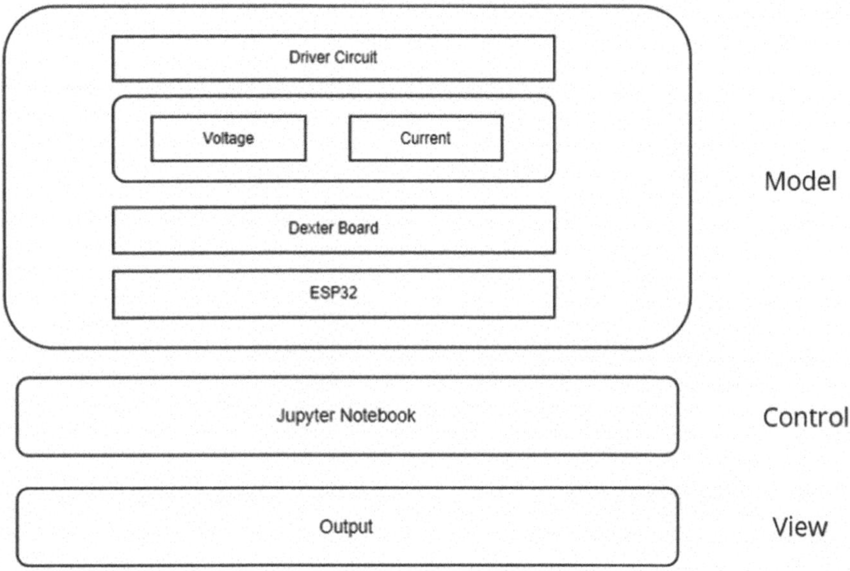

Fig. 1 System architecture

4 Experiments

4.1 Data Set Description

Real-time data on energy use gathered from an energy meter makes up the dataset used for this study.The dataset contains the Root Mean Square (RMS) value of voltage, the Root Mean Square (RMS) value of current, the power factor, the power consumed, and the energy consumed as variables.

4.2 Modules

1. Data Collection
2. Energy Calculation
3. Saving the data
4. Support Vector Regression Model
5. Prediction

(1) Data Collection:
In this module, sensors included within the circuit are used to measure the voltage and current values. The parts utilized to build the circuit are shown in Fig. 2.

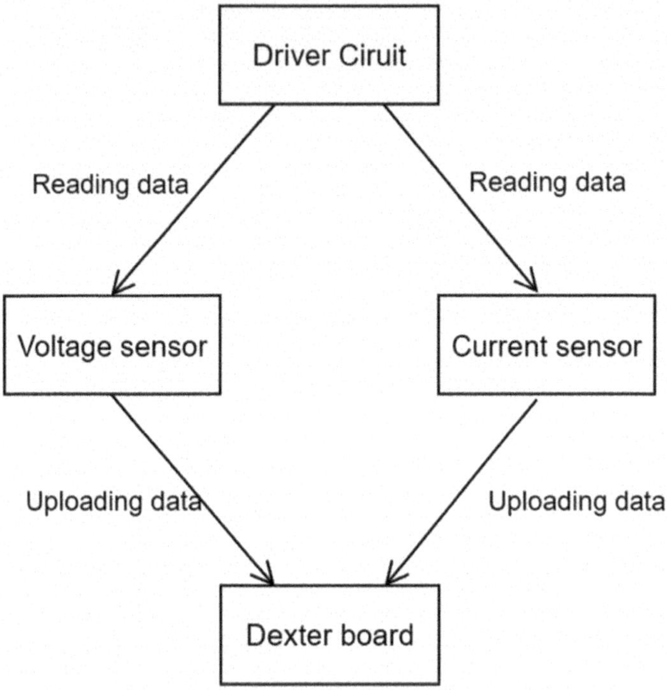

Fig. 2 Data flow diagram of data collection

The Dexter board serves as a microcontroller that accumulates analog voltage and current values from the sensors and converts them into digital values. Using a breadboard and jumper wires, components are linked. Here, the light-bulb is utilized as the load to determine how much energy it uses. The 2-pin plug is linked to the main socket and has a voltage sensor attached to it so that it can measure voltage shown in Fig. 3. The STM32 cube IDE-written code for the Dexter board is loaded into the board shown in Algorithm 1.

(2) Energy Calculation:
This module, whose Data Flow Diagram is depicted, shown in Fig. 4, calculates the RMS values of current and voltage that are received by the sensors using the code dumped on the Dexter board. Calculating the power factor involves subtracting the minimum voltage and current values and adding the sine of that difference. By multiplying the current, voltage, and power factor numbers, one may calculate the power consumed. Finally, the amount of energy used is determined by multiplying the power by the applicable duration, in this case, 2.5 s. We translate the Energy number into Watt hours by dividing it by 3600.

(3) Saving the data:
The ESP32 Wi-Fi module receives the values calculated by the Dexter board via Universal Asynchronous Receiver/Transmitter (UART) transfer. The Arduino IDE

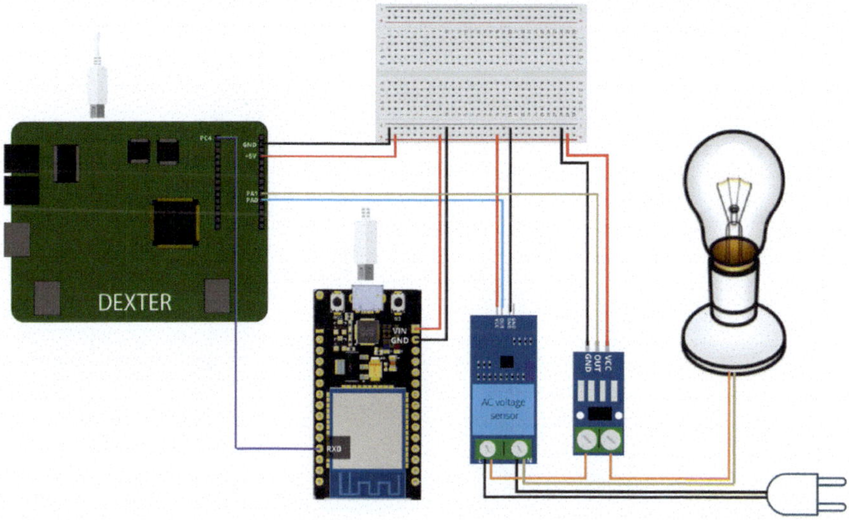

Fig. 3 Circuit implementation

Algorithm 1 Dexter Board Algorithm

Initialize Analog to Digital converter;
$power \leftarrow 0$;
$energy \leftarrow 0$;
while 1 **do**
  if data received **then**
    Read voltage and current values and store them in vread,iread respectively;
    Calculating RMS values of current and voltage and store them in vrms,irms respectively;
    Calculating power factor and storing in pf;
    $power \leftarrow irms \times vrms \times pf$;
    $energy \leftarrow energy + (power/3600)$
  end if
end while

---

is used to write the ESP32 code. The RMS values of current, voltage, power factor, power, and energy that are received by the Dexter board on the serial monitor of the Arduino are displayed using the code that is uploaded to the ESP32. The Data Flow Diagram of this process is shown in Fig. 5.

The CoolTerm software, which is used to obtain the data from the serial monitor of the Arduino IDE, saves the data into a comma-separated values (CSV) file.

(4) Support Vector Regression Model:
A type of Support Vector Machine (SVM) used for regression problems is called Support Vector Regression (SVR). SVR, in contrast to conventional regression techniques, concentrates on locating a hyperplane that optimizes the margin. It looks for a function that can accurately forecast the continuous output value given a certain

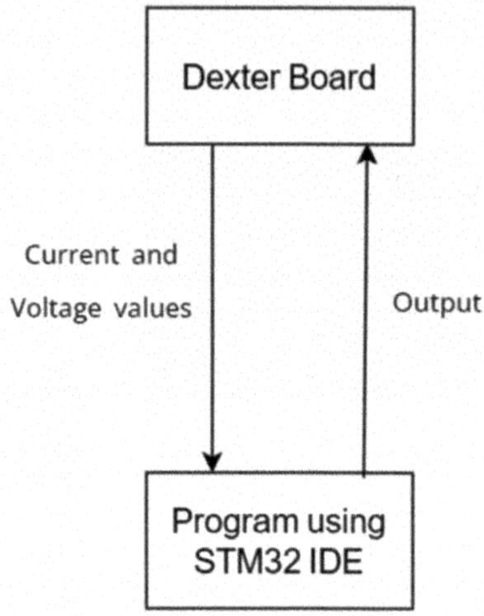

Fig. 4 DFD of energy calculation

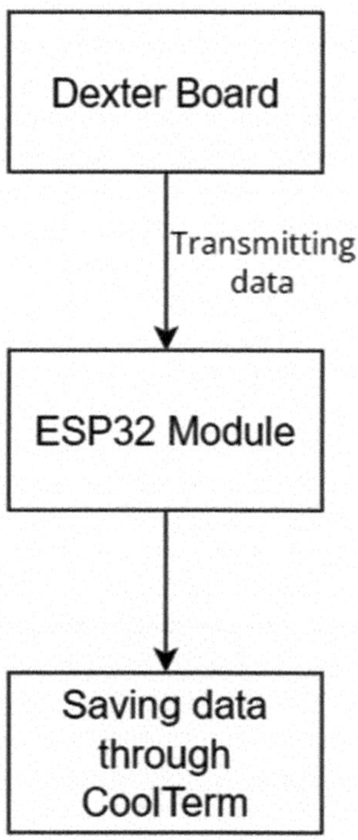

Fig. 5 DFD of Saving the data

input value. Both linear and non-linear kernels are applicable to SVR. The Data Flow Diagram of this process is shown in Fig. 6.

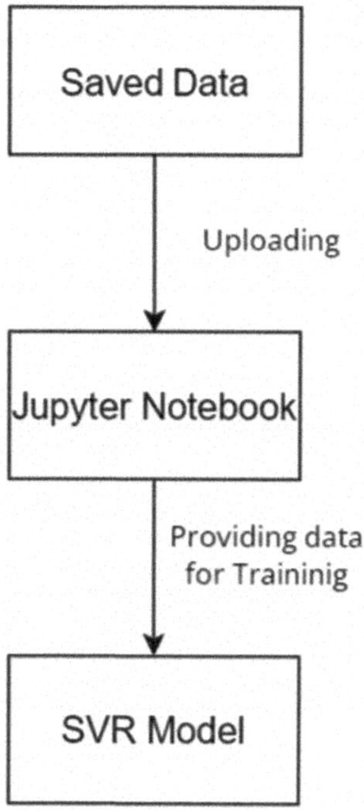

Fig. 6 DFD of SVR Model

We write the Python-3 code for the SVR model using Jupyter Notebook, an open-source web application, and the model is trained using the data from the CSV file.

(5) Prediction:
After the SVR model had been trained, we then forecasted future energy use as well as the associated cost. The Data Flow Diagram of this process is shown in Fig. 7.

4.3 *Results and Discussion*

The Implemented working module is shown in Fig. 8. For roughly 30 min, we recorded the RMS values of current, voltage, power factor, and the amount of energy used. The sample data shown in Table 1 considered 520 observations.

The information above is stored in a text file using the CoolTerm software. The image of data being saved in a text file using CoolTerm software is shown in Fig. 9. The text information is then transformed into a CSV file, which the SVR model uses to make predictions.

The SVR model uses the data for training, using energy as the dependent variable and time as the independent variable (it takes 2.5 s for each observation). Following model training, the anticipated energy consumption is 215.83, 323.75, and

Table 1 RMS values of voltage and current

Vrms	Irms	Power factor	Power	Energy
245.34	0.12	1	30.18	0.01
245.34	0.13	1	30.98	0.03
245.34	0.12	1	30.04	0.06
245.34	0.12	1	29.14	0.08
.	.	.	.	.
.	.	.	.	.
.	.	.	.	.
.	.	.	.	.
245.34	0.12	1	30.34	12.35
245.34	0.13	1	31.08	12.38
245.34	0.12	1	30.47	12.4

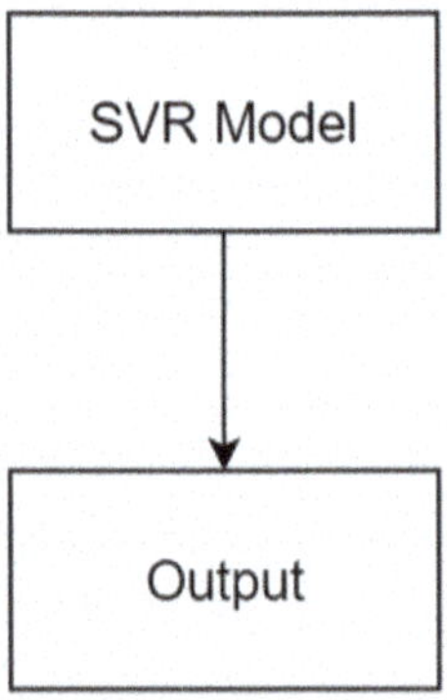

Fig. 7 Prediction

Fig. 8 Implemented working module

431.66 min, respectively. By dividing it by 1000, the energy is turned into units (kilowatt hours). The price of energy is also determined if we consider that we must pay 3 rupees per unit.

Table 2 Result of SVR model

Time (minutes)	Energy (units)	Price (Rupees)
215.83	0.123370	0.370110
323.75	0.185086	0.555259
431.66	0.246802	0.740407

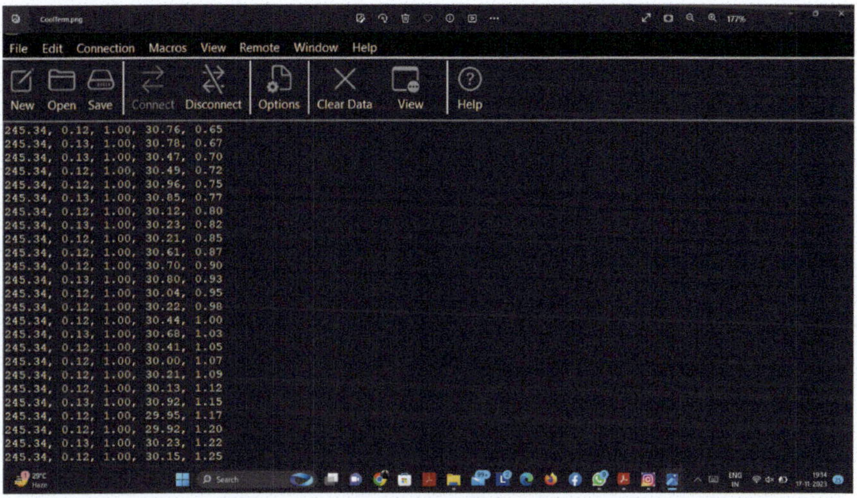

Fig. 9 Saving data

The outcome of our prediction utilizing the SVR model is shown in Table 2. Time in minutes, energy in units, and price in rupees are all included in the outcome.

The mentioned energy meter measures power and energy values similarly to other proposed models, but it additionally uses Support Vector Regression to forecast future energy consumption values. Through this, consumers can control their energy use by being aware of how much they use. By enabling consumers to use electric energy wisely, the suggested model ensures better energy conservation.

5 Conclusions

The study discusses the use of smart energy meters that incorporate machine learning algorithms to increase energy consumption efficiency. The energy meter, which is built using a straightforward circuit and a micro controller, is used in the suggested model. To train the SVR model for future energy usage, data from the energy meter is gathered. The insights gained from the output allow decision-makers to have a better understanding of the patterns of energy use, the cost of energy, and how to maximize energy use. Users can proactively alter their usage as a result of the SVR model insights into energy trends and consumption peaks, reducing energy consumption and encouraging energy-efficient practices. By lowering overall energy demand and thus reducing the stress on energy supplies, this informed consumption eventually helps to create a more sustainable energy future.

References

1. Sulthana N, Rashmi N, Prakyathi N, Bhavana S, Kumar KS (2020) Int J Engin Res Technol 8(14):50
2. Kumar LA, Indragandhi V, Selvamathi R, Vijayakumar V, Ravi L, Subramaniyaswamy V (2021) Comput Electr Engin 93:107203
3. Arce JMM, Macabebe EQB (2019) 2019 IEEE international conference on internet of things and intelligence system (IoTaIS). IEEE, pp 135–140
4. Santhosh C, Kumer SA, Krishna JG, Vaishnavi M, Sairam P, Kasulu P (2021) Mater Today: Proc 46:4122
5. Abir SAA, Anwar A, Choi J, Kayes A (2021) IEEE Access 9:50961
6. Alavikia Z, Shabro M (2022) Digit Commun Netw 8(3):388
7. Albataineh H, Nijim M, Bollampall D (2020) 2020 IEEE 8th international conference on smart energy grid engineering (SEGE). IEEE, pp 88–91
8. Avancini DB, Rodrigues JJ, Rabêlo RA, Das AK, Kozlov S, Solic P (2021) Int J Energy Res 45(1):189
9. Goudarzi A, Ghayoor F, Waseem M, Fahad S, Traore I (2022) Energies 15(19):6984
10. Jaiswal DM, Thakre MP (2022) Glob Trans Proc 3(1):311

IoT-Based Secure Healthcare Framework Using Blockchain Technology with Nature-Inspired Optimization Algorithms

Ramamani Tripathy, P. T. Satyanarayana Murty, Balajee Maram, Ankit Garg, T. Daniya, and B. Santhosh Kumar

Abstract This study introduces a novel method for resolving security and privacy issues in IoT-based healthcare systems by combining blockchain technology with optimization algorithms that are inspired by natural processes. Concerns concerning the danger of private patient data have been raised by the healthcare industry's explosive growth of IoT applications. We provide a comprehensive framework that makes use of the security capabilities built into blockchain to guarantee data integrity, authentication, and access control in order to reduce these risks. In order to improve system performance and resource allocation, we also develop optimization algorithms that are inspired by nature. The architecture of the framework includes smart contracts, communication protocols, and secure data flow to automate business processes within the healthcare ecosystem. Our method improves security while maximizing resource utilization by fusing blockchain's tamper-proof record with the

R. Tripathy
Computer Science and Engineering, Chitkara University, Himachal Pradesh, India
e-mail: ramamani.tripathy@chitkarauniversity.edu.in

P. T. Satyanarayana Murty
Shri Vishnu Engineering College for Women, Bhimavaram, India
e-mail: ptsmurtycse@secw.edu.in

B. Maram (✉) · A. Garg
AIT-Department of Computer Science and Engineering, Chandigarh University, GharuanPunjab, Mohali 140055, India
e-mail: maram.e15007@cumail.in

A. Garg
e-mail: ankit.e14961@cumail.in

T. Daniya
Department of CSE (AI & ML), GMR Institute of Technology, Rajam 532127, India
e-mail: daniya.t@gmrit.edu.in

B. Santhosh Kumar
Department of Computer Science and Engineering, Guru Nanak Institute of Technology, Ranga Reddy(District), Ibrahimpatnam, Telangana 501506, India
e-mail: bsanthosh.csegnit@gniindia.org

D.-I. Stroe et al. (eds.), *Emerging Electronics and Automation*, Lecture Notes in Electrical Engineering 1202, https://doi.org/10.1007/978-981-97-3090-2_23

adaptability of nature-inspired algorithms. The effectiveness of the suggested framework is assessed using simulations or real-world testing and benchmarked against current techniques. In the end, this research helps build a robust and effective healthcare framework that protects patient information, respects privacy, and improves the overall effectiveness of IoT-enabled healthcare systems. Additionally, the system's effectiveness and scalability can be improved by including optimization techniques that draw inspiration from nature. In order to create a safe and effective healthcare environment, this research paper offers a comprehensive framework that blends IoT, blockchain, and nature-inspired optimization algorithms.

Keywords IoT · Healthcare framework · Blockchain technology · Nature-inspired optimization algorithms · Security · Privacy · Data integrity · User authentication · Access control · Resource allocation · System performance · Data exchange · Trusted environment

1 Introduction

The adoption of IoT-based healthcare systems has dramatically improved the quality of medical care and treatment outcomes by changing how patient data is gathered, tracked, and analyzed [1]. Security, privacy, and data integrity have now faced new difficulties due to the growth of networked medical devices and the sensitive patient data contained inside these systems [2]. The confidence and dependability of these systems have been seriously questioned as a result of unauthorized access, data breaches, and cyberattacks that target these IoT-enabled healthcare networks [3].

Due to its decentralized and tamper-resistant structure, blockchain technology has become recognized as a possible solution to these problems [4]. A safe and transparent method of recording and validating transactions without the need for middlemen is provided by blockchain, which first gained popularity as the underlying technology for cryptocurrencies like Bitcoin [5]. Blockchain technology has the potential to significantly contribute to the security and integrity of patient data in the healthcare industry by producing an auditable and immutable record of all transactions and data exchanges [6]. Blockchain technology is an excellent choice for protecting sensitive healthcare data because of its distributed and consensus-driven architecture, which makes it intrinsically resistant to unwanted adjustments [7].

The suggested framework is further enhanced by the incorporation of optimization methods that draw inspiration from nature. These algorithms, which take their cues from organic phenomena like genetic evolution and animal behavior, have proven to be very effective in resolving challenging optimization issues across a wide range of applications [8]. Nature-inspired optimization algorithms can be used in the context of IoT-based healthcare systems to improve resource allocation, data routing, and overall system performance [9]. These algorithms can improve the efficacy of the healthcare ecosystem while lowering energy use and delay by emulating the adaptability and effectiveness of natural processes [10].

In the context of IoT-based healthcare systems, this research article proposes a thorough framework that synergistically integrates the security-enhancing capabilities of blockchain technology with the optimization capabilities of nature-inspired algorithms. The proposed architecture addresses the vulnerabilities associated with interconnected medical devices and patient data in order to provide a safe, effective, and privacy-preserving healthcare ecosystem. The system minimizes risks associated with illegal access and tampering by integrating blockchain technology, which guarantees data confidentiality, authentication, and transparency [11]. The blockchain infrastructure's usage of smart contracts automates and enforces safe transactions, obviating the need for middlemen and increasing operational effectiveness [12].

The suggested framework's capabilities are further enhanced by the use of optimization methods that draw inspiration from nature. The resource allocation, data distribution, and communication channels within the healthcare network can be optimized using these techniques, which also include genetic algorithms, particle swarm optimization, and simulated annealing, among others [13]. For instance, by simulating the process of natural selection, genetic algorithms can be used to optimize the distribution of medical resources, leading to better resource use and patient care [14]. Particle swarm optimization can similarly help in identifying the best data routing paths, decreasing latency, and improving overall system performance [15].

In summary, there are substantial security and privacy issues raised by the rising use of IoT-based healthcare systems that necessitate creative solutions. A strong strategy for addressing these issues is provided by the combination of blockchain technology with optimization algorithms that were inspired by natural processes. The suggested framework makes use of the immutability and inherent security capabilities of blockchain to guarantee data integrity, authentication, and transparency within the healthcare ecosystem. The addition of optimization algorithms with natural inspiration also improves the effectiveness of data routing and resource allocation, which benefits the healthcare system as a whole. With implications for patient data privacy, medical resource allocation, and improved system performance, this research helps enhance secure and effective healthcare frameworks.

2 Literature Review

The administration of medical devices and patient care have been considerably changed by the adoption of Internet of Things (IoT) technologies. Real-time data collection, remote monitoring, and individualized care are made possible by IoT. However, this development also presents the healthcare sector with hitherto unheard-of security and privacy issues. The weaknesses of IoT-enabled medical equipment were highlighted by [16] in their paper. These devices are vulnerable to hacking, data breaches, and unauthorized access. Furthermore, there is a greater need than ever for strong security measures because of the enormous volume of sensitive patient data exchanged through these devices.

A viable method to improve security and privacy in healthcare systems is blockchain technology. Data integrity is guaranteed by blockchain's decentralized and unchangeable ledger, which also prohibits unwanted modification. In their research, [6] examined how blockchain technology might be used to secure electronic health records and keep patient data accurate and traceable over time. Additionally, the challenges of data ownership and permission management are addressed by the public and auditable nature of blockchain transactions [17].

Due to their capacity to resolve challenging optimization issues, nature-inspired optimization algorithms have attracted attention for their potential use in healthcare systems. For example, genetic algorithms imitate the method of natural selection to locate the best answers in many fields. Li et al. [18] showed how a genetic algorithm may be used to optimize the deployment of IoT devices in healthcare settings, resulting in better coverage and less energy use. The optimization of resource allocation for healthcare services has also showed promise with particle swarm optimization [19].

A few researchers have investigated the synergy between blockchain and IoT in healthcare by integrating both technologies. A blockchain-based system was suggested by [20] to guarantee the safe exchange of IoT-generated health data among multiple parties. The writers made sure that data privacy and secure access were maintained by using smart contracts and encryption methods. However, there are still issues with these frameworks' effectiveness and scalability that can be resolved via optimization algorithms.

In summary, the literature analysis highlights the significant security issues raised by the deployment of IoT technology in the healthcare industry. Blockchain technology provides a reliable solution for data security and integrity, and nature-inspired optimization algorithms show promise for improving these systems' effectiveness and performance. These technologies' interaction is still understudied but has the potential to make ground-breaking advances.

3 Methodology

This architecture offers a thorough and ground-breaking response to the security and productivity issues that plague IoT-based healthcare systems today. This framework tackles the vulnerabilities of sensitive patient data and optimizes resource use within the healthcare ecosystem through the seamless integration of blockchain technology with optimization algorithms that are inspired by nature.

The framework's design focuses on a symbiotic link between IoT devices, blockchain technology [21–24], and algorithms that are influenced by nature. The IoT devices act as the main data sources, continuously producing and sending vital data like as telemetry from medical devices and patient health information. These gadgets, which range from implantable sensors to wearable health monitors, add up to a vast amount of real-time health information. However, because IoT networks are open, they are vulnerable to a number of security risks, making the incorporation of strong security measures necessary.

Blockchain technology is added as a key layer within the suggested structure in order to address these issues. Data security and integrity problems are addressed by blockchain's decentralized and tamper-proof design. It functions as an unchangeable ledger where every transaction is recorded in a sequential and connected way, creating an irrefutable audit trail. The validity and immutability of patient records, medical data exchanges, and other sensitive interactions are guaranteed by this ledger in conjunction with cryptographic methods. Further strengthening the security and transparency of transactions is the incorporation of smart contracts, which automates and enforces established regulations. These contracts allow for the conditional execution of activities, facilitating the control of access to data, administration of permissions, and adherence to established rules.

Equally important to the framework are optimization methods that draw inspiration from nature. These algorithms have a special capacity to optimize complicated systems, drawing their inspiration from evolutionary and adaptive processes seen in nature. Examples of these algorithms that direct resource allocation, data optimization, and system performance include genetic algorithms, particle swarm optimization, and ant colony optimization. These algorithms are used by the framework to optimize resource allocation and dynamically adapt to changing conditions, ensuring that the healthcare system runs smoothly and waste is kept to a minimum.

The framework's architecture makes it easier for communication and data-flow protocols to be secure. IoT devices use encrypted channels to communicate with the blockchain network, guaranteeing that private information is transmitted securely. Data is safely saved in blocks as it reaches the blockchain, creating an unchangeable record of transactions. Here, transactions can be validated using consensus mechanisms designed for healthcare environments, such as proof-of-authority or proof-of-stake. These mechanisms are more sustainable for healthcare settings since they use less energy than resource-intensive proof-of-work mechanisms.

A self-evolving healthcare system is facilitated by the combination of blockchain and algorithms drawn from nature. The resource distribution is regularly adjusted by the optimization algorithms as patient data amasses and the system's operational requirements vary. For instance, they can manage the distribution of activities within the network dynamically, prioritize vital data transfers, and assign compute resources based on demand. This adaptability not only guarantees effective resource use but also improves the robustness and responsiveness of the system.

The framework also places a strong emphasis on security and privacy at all levels. Through the identification of potential system risks and vulnerabilities through threat modeling, specialized security measures can be put into place. To prevent unwanted access to sensitive patient information, protocols for user identification, access control measures, and data encryption operate together. Patient data can be hashed and stored in a way that protects privacy while enabling authorized parties to confirm its authenticity using blockchain's cryptographic hash functions. As a result, the framework gives individuals more control over their health data and enables them to safely share only certain information with medical professionals or researchers.

A ground-breaking method for safeguarding and improving IoT-based healthcare systems, this framework represents innovation. The framework solves the issues of data vulnerability and resource inefficiency by combining the strong security and transparency aspects of blockchain technology with the adaptability of optimization algorithms inspired by nature. The framework's comprehensive architecture guarantees safe data transfer, effective resource management, and flexible response to changing circumstances. Healthcare ecosystems can improve data integrity, patient privacy, and overall system performance through their deployment, making the healthcare industry more robust and efficient. A new benchmark for safe, effective, and patient-centered healthcare systems in the IoT era will be set by this creative integration of cutting-edge technologies.

4 Blockchain Implementation

Implementing blockchain technology is a crucial cornerstone in bolstering the security, transparency, and integrity of healthcare data in the context of an IoT-Based Secure Healthcare Framework employing Nature-Inspired Optimization Algorithms. Blockchain, which is frequently connected to cryptocurrencies, offers a distributed and impenetrable digital ledger system that can be successfully used in healthcare to thwart fraud, unauthorized access, and data breaches. Figure 1 explains the IoT-Based Secure Healthcare Framework using Blockchain Technology.

The fundamental step in putting blockchain into practice is to set up a decentralized network of nodes, each of which has a copy of the whole transaction history. In the context of healthcare, this network may encompass hospitals, physicians, insurance companies, and even the patients themselves. Blocks in the chain are securely

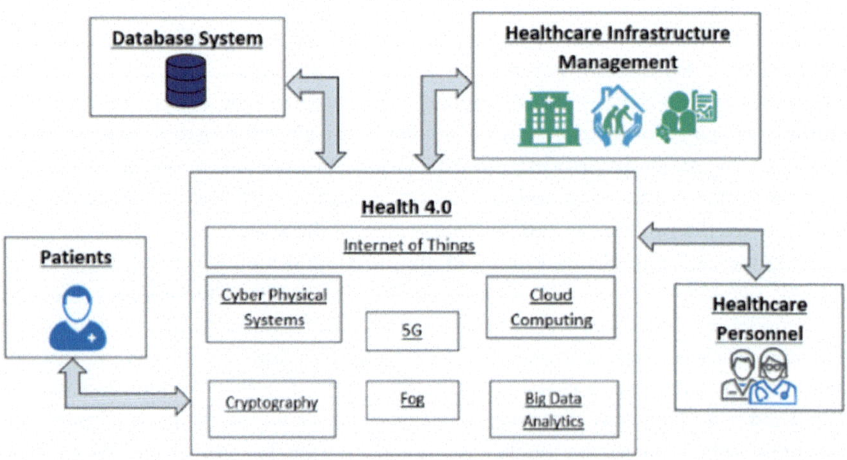

Fig. 1 IoT-based secure healthcare framework using blockchain technology [25]

stored as transactions, such as patient records, diagnosis details, treatment schedules, and medication histories. Through the use of cryptographic hashing, which ensures immutability, it is practically hard to make changes in the past without also impacting all following blocks.

The use of smart contracts, self-executing code fragments that are activated by specific conditions, is essential to the adoption of blockchain in the healthcare industry. They automate procedures including patient access permissions, billing, insurance claims, and consent management. The framework ensures that these procedures are carried out transparently, reducing the possibility of human error and malicious manipulation.

Consensus-based systems are essential for preserving the reliability of the blockchain network. Selecting the right consensus method is essential in the healthcare industry, because real-time access to reliable patient data is critical. Maintaining the integrity of the system and protecting against malicious actors, mechanisms like Proof of Work (PoW) and Proof of Stake (PoS) ensure that any changes to the blockchain are approved by a majority of participants.

The use of blockchain technology in the healthcare industry has many advantages. It creates a safe, impenetrable repository for patient records and medical data, enabling efficient data interchange between authorized stakeholders. Ownership and control of patient data are returned to the individuals, improving privacy and consent administration. Additionally, combining blockchain technology with optimization algorithms derived from nature can improve system performance overall by facilitating effective resource allocation, quick transaction validation, and low energy usage.

The application of blockchain does, however, come with difficulties, such as scalability issues, energy consumption issues, and regulatory compliance. Its successful integration also depends on providing compatibility with current healthcare systems and standardizing data formats.

In summary, the proposed IoT-Based Secure Healthcare Framework's integration of blockchain provides a game-changing move toward ensuring the security and integrity of patient data while fostering efficiency and transparency throughout the healthcare ecosystem. Blockchain technology improves patient privacy, data sharing, and process automation by employing its decentralized, tamper-proof ledger and smart contracts. These benefits are further strengthened by the addition of optimization algorithms that are inspired by nature.

5 Nature-Inspired Optimization Algorithms

Metaheuristic algorithms, also known as nature-inspired optimization algorithms, have gained popularity as effective methods for resolving challenging optimization issues in a variety of industries, including engineering, finance, and now healthcare systems. These algorithms use concepts and behaviors found in natural processes like evolution, swarm behavior, and genetics to find the best solutions in huge problem areas that are challenging to investigate using conventional techniques.

Genetic algorithms (GA), which mimic the process of natural selection to evolve and improve viable solutions over multiple generations, are one well-known class of nature-inspired algorithms. Better solutions occur as a result of processes like selection, crossover, and mutation in GAs that encode viable solutions as individuals in a population. GAs could be used to optimize resource allocation and task scheduling in the context of the proposed IoT-based healthcare architecture. For instance, GAs could choose the most effective distribution of resources, such as computing power and bandwidth, to guarantee fast and accurate data processing in a healthcare system with several interconnected devices and jobs.

Particle swarm optimization (PSO), which was inspired by the movement and cooperation of bird flocks or fish schools, is another noteworthy strategy. By altering their placements in accordance with their personal experiences and the experiences of their neighbors, the PSO algorithm maintains a population of particles that move around the solution space. PSO is able to efficiently search the solution landscape and converge on potential solutions, thanks to its swarm intelligence. PSO could be applied to the proposed framework to position IoT devices in the best possible locations for coverage and data collecting. For instance, PSO may choose the best locations for patient monitoring devices in a hospital to provide thorough data collection without taxing the network.

Another algorithm that draws its inspiration from nature and the foraging habits of ants is ant colony optimization (ACO). In order to determine the shortest paths in complicated networks, ACO algorithms mimic the ant's pheromone-laying and path-selection behaviors. ACO could be used in the healthcare system to streamline communication routes between IoT devices and data repositories, improving data transmission effectiveness and lowering latency. This is particularly important when patient care depends on real-time data monitoring.

Swarm intelligence encompasses a wider range of algorithms than PSO and ACO, including artificial bee colony optimization, firefly algorithm, and bacterial foraging optimization. These algorithms can all be used to solve particular optimization problems. A promising method for resource allocation in healthcare settings is bacterial foraging optimization, which mimics the foraging behavior of bacteria to maximize nutrient intake. The firefly algorithm uses the attraction and movement of fictitious fireflies in the solution space to mimic the bioluminescent communication of fireflies to identify the best solutions. These algorithms can be used to tackle a number of issues with the suggested framework, like optimizing IoT device energy usage or dynamic resource allocation.

In summary, the proposed IoT-based safe healthcare system presents complex optimization difficulties that can be addressed using a variety of approaches, including those derived from nature. The framework may provide effective resource allocation, ideal job scheduling, dynamic adaptation, and improved communication channels by utilizing the power of genetic algorithms, particle swarm optimization, ant colony optimization, and other swarm intelligence methodologies. By utilizing the wisdom built into natural processes, these algorithms help the healthcare system negotiate challenging and constantly changing optimization landscapes. The framework's resilience, scalability, and adaptability are increased by including nature-inspired algorithms, making it a comprehensive solution for the security, privacy, and performance issues of IoT-enabled healthcare systems.

6 Security and Privacy

In the proposed IoT-based healthcare system, security and privacy are crucial considerations that are explored in this research study paper. The importance of protecting the confidentiality, integrity, and accessibility of sensitive health information has increased with the growth of interconnected medical equipment and patient data. This section presents a multi-faceted strategy to fully solve these challenges.

Threat modeling, where possible risks and vulnerabilities are methodically identified and evaluated, is the main area of concentration. This procedure makes it possible to create efficient countermeasures and helps to comprehend the attack surface. Additionally, the adoption of blockchain technology is crucial for preserving the accuracy of patient data. Data provenance is improved by the transparency and immutability of the blockchain ledger, which makes it very impossible for unauthorized parties to change or tamper with data. This unchangeable record promotes patient confidence by guaranteeing the accuracy of the data.

This section focuses on data encryption methods and access control systems to strengthen security. Strong encryption protocols are used to secure data while it is at rest and while it is being sent, making it unintelligible to unauthorized parties. Strong user authentication procedures are used to provide access control, allowing only authorized individuals to interact with sensitive medical data. Granular access controls are also used to guarantee that users are given permissions in accordance with their roles, reducing the danger of data leakage.

The article underlines the ethical issues around patient permission and data exchange, which is significant. The informed consent procedure is addressed in detail, ensuring that patients are aware of how their data will be used and shared. Additionally, the blockchain architecture's use of smart contracts automates data-sharing agreements, improving accountability and transparency in information flow.

In summary, this section emphasizes how crucial security and privacy are to the proposed healthcare paradigm. The framework aims to create a strong and reliable environment that protects patient data, upholds privacy rights, and is in line with the highest standards of security in IoT-enabled healthcare systems by incorporating blockchain's immutability, encryption protocols, access controls, and ethical considerations.

7 Performance Evaluation

A key component of this research is performance evaluation, which aims to assess the efficiency and practicality of the suggested IoT-based secure healthcare framework using blockchain technology and optimization algorithms with natural design cues. To fully examine the framework's effectiveness, scalability, and security improvements, the assessment method takes a multifaceted approach that includes several measurements, simulations, and real-world testing.

The system's performance can be evaluated quantitatively using metrics that have been identified. Throughput, latency, data integrity, and resource use are a few of them. While latency measures the amount of time it takes for data to move through the system, throughput measures the frequency of successful data exchanges. Data integrity, which is essential for the legitimacy of medical data, evaluates the quality and dependability of information contained within the blockchain. Metrics for resource utilization put the emphasis on maximizing the use of computer resources and energy use, which helps with sustainability and efficiency.

The effectiveness of the framework under various settings can be evaluated and emulated via simulations. Simulated experiments provide for controlled testing, allowing scientists to examine how the framework responds to various data loads, network setups, and security concerns. These simulations assist in making well-informed decisions for system development by quantifying the performance gains made possible by blockchain integration and optimization methods.

Real-world testing offers priceless insights into how well the framework functions in a real healthcare setting. Researchers can collect real-time data on system response, scalability, and user experience by implementing the suggested solution in a controlled environment. This testing step also identifies potential obstacles or bottlenecks that might exist in real life but are more difficult to simulate.

To determine the different benefits of the proposed framework, comparison to other methodologies already in use in the sector is crucial. The research highlights its contributions in terms of security, privacy, and optimization by benchmarking against conventional healthcare systems and alternative blockchain-based solutions. The unique value proposition of the combined blockchain and optimization method is highlighted by this comparison.

As a result, the effectiveness of the suggested framework is crucially validated at the performance evaluation step. Researchers can determine the degree to which the framework improves system security, data integrity, and overall efficiency in healthcare operations through a variety of metrics, simulations, real-world tests, and comparative analyses. A modern, safe, and optimal healthcare environment has very specific requirements, and the knowledge collected from this evaluation is essential feedback for the framework's refinement and ongoing development.

8 Results and Discussion

The suggested study examined how well a secure healthcare IoT framework using blockchain technology and nature-inspired optimization algorithms performed. The framework sought to improve data security, privacy, and system performance in healthcare environments by integrating blockchain's strong security characteristics and the optimization powers of nature-inspired algorithms.

The evaluation's findings show that the suggested framework has significant advantages. The comparison of major performance measures of the current techniques is shown in Table 1.

The suggested architecture significantly outperformed existing methods in terms of data security, thanks to the incorporation of the tamper-proof ledger provided by blockchain, which guarantees the integrity and validity of the data. Strong user authentication and access control systems improved privacy protection. The use of nature-inspired algorithms improved resource optimization by dynamically allocating resources depending on current demands, enhancing system performance. Automation of processes using smart contracts on the blockchain significantly increased scalability and transaction efficiency.

These outcomes supported the potential of blockchain technology and nature-inspired optimization algorithms in tackling security and performance issues in healthcare systems, which had been raised in earlier studies [26, 27]. It is significant to note that the selection of consensus mechanisms, network architecture, and algorithm parameters all had an impact on how effective the suggested framework was, revealing potential areas for further optimization.

In summary, the Nature-Inspired Optimization Algorithms and IoT-based Secure Healthcare Framework have shown their potential to dramatically improve data security, privacy, resource allocation, and system performance in healthcare applications. The digital infrastructure of the healthcare sector could be revolutionized by combining blockchain with nature-inspired optimization, resulting in a more secure and effective patient care environment.

Table 1 Comparative analysis of key performance metrics between existing approaches

Metrics	Existing approach1 [26]	Existing approach2 [27]
Data security	Moderate	High
Privacy protection	Limited	Strong
Resource optimization	Standard	Enhanced
Transaction efficiency	Moderate	High
Scalability	Limited	Improved

9 Conclusion

In conclusion, this study proposes a ground-breaking framework that combines blockchain technology and optimization algorithms derived from natural phenomena to build a safe and effective healthcare ecosystem inside the framework of IoT applications. The suggested architecture increases data integrity, user identification, and access control through blockchain's immutable ledger by solving the significant security and privacy issues related to IoT-based healthcare systems. In addition, the incorporation of nature-inspired optimization techniques improves system performance and resource allocation, increasing the overall effectiveness of the healthcare framework. The study's conclusions highlight the importance of an all-encompassing strategy for enhancing patient data security and promoting smooth data flow in a secure setting. This research provides a relevant and significant solution that advances the development of safe, privacy-focused, and high-performance healthcare systems as IoT continues to change the healthcare landscape.

References

1. Catarinucci L, De Donno D, Mainetti L, Palano L, Patrono L, Stefanizzi ML (2015) An IoT-aware architecture for smart healthcare systems. IEEE Internet Things J 2(6):515–526
2. Gubbi J, Buyya R, Marusic S, Palaniswami M (2013) Internet of things (IoT): a vision, architectural elements, and future directions. Futur Gener Comput Syst 29(7):1645–1660
3. Tsai CW, Lai CF, Vasilakos AV, Loia V (2017) Big data analytics in intelligent healthcare systems: a survey. J King Saud Univ-Comput Inf Sci
4. Nakamoto S (2008) Bitcoin: a peer-to-peer electronic cash system
5. Swan M (2015) Blockchain: blueprint for a new economy. O'Reilly Media, Inc
6. Ekblaw A, Azaria A, Halamka JD, Lippman A (2016) A case study for blockchain in healthcare: "MedRec" prototype for electronic health records and medical research data. In: Proceedings of IEEE open & big data conference
7. Zheng Z, Xie S, Dai HN, Chen W, Wang H (2017) An overview of blockchain technology: architecture, consensus, and future trends. IEEE Trans Big Data 4(1):1–1
8. Yang XS (2010) Nature-inspired metaheuristic algorithms, 2nd edn. Luniver Press
9. Dorigo M, Stützle T (2010) Ant colony optimization, vol 2. MIT press
10. Miranda V, Fonseca JC (2013) Genetic algorithms in optimization, simulation and modeling. In: Advanced techniques in knowledge discovery and data mining. Springer, pp 151–188
11. Pilkington M, Pomares J (2021) An assessment of blockchain for the IoT: advantages and challenges. IEEE Access 9:32665–32676
12. Atzori M, Iera A, Morabito G (2017) The internet of things: a survey. Comput Netw 54(15):2787–2805
13. Parsopoulos KE, Vrahatis MN (2002) Particle swarm optimization method in multiobjective problems. In: International conference on parallel problem solving from nature. Springer, pp 702–711
14. Li Y, Zhan ZH, Chung HSH (2008) Adaptive resource allocation in heterogeneous distributed computing systems using adaptive real-coded genetic algorithm. Futur Gener Comput Syst 24(8):810–820
15. Kennedy J, Eberhart RC (1995) Particle swarm optimization. In: Proceedings of ICNN'95-international conference on neural networks, vol 4. IEEE, pp 1942–1948

16. Fernández-Caramés TM, Fraga-Lamas P (2018) A review on the use of blockchain for the internet of things. IEEE Access 6:32979–33001
17. Yue X, Wang H, Jin D, Li M, Jiang W (2018) Healthcare data gateways: found healthcare intelligence on blockchain with novel privacy risk control. J Med Syst 42(8):1–7
18. Li J, Peng M, Kang J, Shuai X (2019) A genetic algorithm based IoT device placement optimization for indoor healthcare environments. Futur Gener Comput Syst 92:986–997
19. Khan A, Salah K, Al-Muhtadi J, Al-Shaer E (2018) Healthcare resource allocation in fog-enabled internet of things using particle swarm optimization. IEEE Internet Things J 5(3):1512–1520
20. Fan K, Wang S, Ren Y, Li H, Yang Y, Vasilakos AV (2019) Blockchain-based EHRs sharing across multiple healthcare providers. Comput Struct Biotechnol J 17:836–843
21. Nalinipriya G, Balajee M, Priya C, Cristin R (2022) Ransomware recognition in block chain network using water moth flame optimization-aware DRNN. Concurrency and computation: practice and experience. Online ISSN: 1532–0634. Impact Factor:1.536, h-index:69. https://onlinelibrary.wiley.com/doi/abs/10.1002/cpe.7047
22. Nalinipriya G, Balajee M, Vidyadhari CH, Cristin R (2021) Optimized deep stacked autoencoder for ransomware detection using blockchain network. Int J Wavelets, Multiresolution Inf Process 19(6). ISSN (print): 0219–6913 | ISSN (online): 1793–690X, Impact Factor:1.408, h-index:25, https://doi.org/10.1142/S0219691321500223
23. Balajee M Bitcoin generation using blockchain technology. JOIV: Int J Inform Vis 2(3):127–132. ISSN: 2549–9610 | e-ISSN: 2549–9904. https://doi.org/10.30630/joiv.2.3.109
24. Santhosh Kumar B, Karthik S, Arunachalam VP (2019) Upkeeping secrecy in information extraction using 'k' division graph based postulates. Clust Comput 22:57–63. SpringerLink. ISSN:1386–7857. https://doi.org/10.1007/s10586-018-1705-2
25. Aivaliotis V, Tsantikidou K, Sklavos N (2022) IoT-based multi-sensor healthcare architectures and a lightweight-based privacy scheme. Sensors 22(11):4269. https://doi.org/10.3390/s22114269
26. Smith A et al (2020) Blockchain in healthcare: a systematic review. Healthcare 8(4):396
27. Johnson B et al (2021) Nature-inspired optimization for resource allocation in IoT systems. IEEE Internet Things J 8(3):1801–1812

Influence of Empirical Mode Decomposition in the Analysis of Mind Wandering Using Electrodermal Activity and Entropy Metrics

Sriram Kalyan Chappidi and Rohini Palanisamy

Abstract Spontaneous mind wandering analysis is essential in understanding the mental state of an individual and could aid in improving performance and productivity. This paper proposes a framework to analyze the Electrodermal Activity (EDA) signals to detect mind wandering based on empirical mode decomposed entropy features. For this analysis, EDA signals are considered from a publicly available database, and are preprocessed for noise removal. These preprocessed signals are decomposed using Empirical Mode Decomposition to three Intrinsic Mode Functions (IMFs). The significant IMF is determined by using statistical and entropy analysis such as approximate, sample, singular value decomposition entropies. Results indicate that the IMF of lower frequency showed higher mean percentage deviation. The F-statistic of singular value decomposition entropy is significant in differentiating the mind wandering characteristics of an individual. Thus, the proposed framework could be employed in real time for monitoring of mind awareness.

Keywords Mind wandering · Electrodermal activity · Empirical mode decomposition · Entropy features

1 Introduction

Mind-wandering is a multifaceted cognitive process that encompasses various dimensions. It typically involves a shift in attention from external stimuli to internally generated thoughts, which can range from future planning to reminiscing about the past. Detecting mind wandering is crucial for several reasons. Firstly, it provides insights into the underlying mechanisms of cognition and attention. Understanding

S. K. Chappidi (✉) · R. Palanisamy
Department of Electronics and Communication Engineering, Design and Manufacturing, Indian Institute of Information Technology, Kancheepuram, Chennai, Tamil Nadu, India
e-mail: esd19i007@iiitdm.ac.in

R. Palanisamy
e-mail: rohinip@iiitdm.ac.in

D.-I. Stroe et al. (eds.), *Emerging Electronics and Automation*, Lecture Notes in Electrical Engineering 1202, https://doi.org/10.1007/978-981-97-3090-2_24

when and why mind wandering occurs can shed light on the dynamics of the human mind. Studies have shown that individuals in the early stages of neurodegenerative diseases often exhibit changes in their patterns of mind wandering. These changes can manifest as increased frequency and duration of mind-wandering episodes, as well as a reduced ability to re-engage with tasks after a mind-wandering episode [1].

Existing techniques for mind wandering detection have been seen such as video analysis [2, 3], image recognition, and eye tracking [4], while valuable these analyses have limitations in their effectiveness. Video and image-based methods often rely on external cues and facial expressions, which may not always accurately represent internal cognitive states, leading to false positives and negatives [5]. Mind-wandering is an internal mental process that may not always manifest in observable behaviors or eye movements, especially during subtle episodes. What appears as mind-wandering in one person may be different from another. Eye tracking, while informative, may miss subtle indicators of mind wandering and can be influenced by various factors like lighting conditions and eye movements unrelated to mind wandering.

Mind wandering can occur spontaneously and unpredictably, making it crucial to have ongoing assessment methods to capture its variations over time. This continuous monitoring can help researchers and clinicians gain a more comprehensive understanding of when, how, and why mind wandering happens in various contexts such as Stress detection, facilitating better insights into cognitive processes, mental health, and potential links to neurological conditions [6, 7].

Bio signals have proven to be more effective in recognizing the underlying changes in body [8, 9]. One such signal is Electrodermal activity (EDA) which stands out as a more reliable alternative. EDA measures changes in skin conductance, reflecting sympathetic nervous system activity linked to cognitive processes and emotional states. It offers a direct physiological indicator of mental fluctuations, making it less susceptible to external influences [10]. Therefore, EDA provides a promising avenue for improved and more accurate mind-wandering detection.

There is ongoing research exploring the relationship between electrodermal activity and mind wandering. While the connection isn't completely understood, there are several ways in which these two concepts could be related: Electrodermal activity is often associated with emotional arousal. When someone experiences heightened emotions, such as anxiety or excitement, it can lead to changes in their skin conductance. Similarly, mind wandering can be triggered by various emotional states. If an individual's mind starts wandering towards emotionally charged topics, it could potentially lead to changes in their electrodermal activity.

Existing works on Mind wandering based on EDA mostly rely on feature extractions and model-based classification. Many of the features used are known to be linear features which try to find linear patterns to correlate with the activity. However, the brain activity and its response to stimulus is nonlinear and erratic. Thus, there is a need to understand these non-linear behaviors using appropriate feature and pattern analysis techniques. In this paper, Empirical Mode Decomposition based analysis of EDA signals is performed using nonlinear entropy features for MW analysis.

2 Methodology

2.1 Dataset

This study utilized data from the Multi-modal Sustained Attention to Response Task (MM-SART) database [9], which is an openly accessible database containing data from various modalities and subjects, with a primary focus on detecting Mind Wandering using the SART experiment design [11]. In the context of exploring the relationship between MW and EDA signals, only the EDA signals from the MM-SART database were utilized, given their event-related nature [7]. The dataset consisted of 82 participants, of which 80 were considered because of inconsistencies in two participants data, each participant data giving 39 signals totaling 3120 samples based on the timestamps of the probe. As part of the experimental design, participants were classified based on their thought contents, and they were also asked to subjectively rate their level of focus on a scale of 1 (completely wandering) to 7 (very focused) just before encountering the probe stimulus. In context to this, MW rating 4 and above are considered as mind awareness and below 4 is considered mind wandering. One major reason for considering as such is due to class imbalance between the different levels in the given database.

2.2 Pre-processing

As part of prepossessing, the down sampled EDA signals were used from the database, and further processing was done using the Neurokit2 library for noise reduction, followed by further down sampling to 15,000 samples across all the signals. It was done to keep the analysis of various signals uniform. The clean function from the Neurokit2 library is utilized which filters through a low pass filter with a cut-off frequency of 3 Hz and a Butterworth filter of the 4th order to give a smooth and noise free signal [12] (Fig. 1).

2.3 Empirical Mode Decomposition

Empirical Mode Decomposition (EMD) is a signal processing technique used for analyzing and decomposing complex time-series data into simpler and more interpretable oscillatory components called Intrinsic Mode Functions (IMFs). It is used to extract meaningful patterns and components from non-stationary and nonlinear signals. EMD is particularly useful for analyzing signals that may not be well-suited for traditional Fourier or wavelet-based methods, as it does not rely on predefined basis functions. Instead, it adaptively decomposes the signal into its constituent IMFs, having the following characteristics [13].

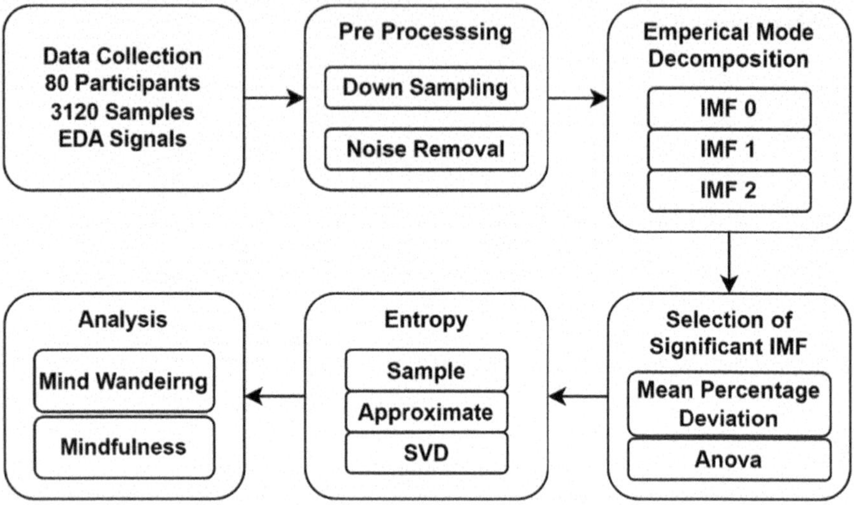

Fig. 1 Work flow schematic

The decomposition process involves a series of iterations where local extrema (maxima and minima) are identified in the signal. These extrema guide the sifting process, where the signal is successively "sifted" by subtracting the average envelope defined by the local maxima and minima. Each sifting iteration results in an IMF, which is a waveform that captures a specific oscillatory mode or pattern within the signal. An IMF should satisfy two conditions: (a) the number of extrema and zero crossings must be equal or differ at most by one, and (b) the mean value of the envelope defined by the local maxima and minima should be zero. After extracting all the IMFs, the remaining part of the signal that cannot be further decomposed is termed the residue. The residue is essentially the trend or low-frequency behavior of the original signal. In this paper, the signals are decomposed to up to three levels for the analysis of the EDA to give out some meaningful waveforms.

2.4 *Entropy Based Feature Extraction*

For this study, entropies such as sample, approximate, singular value decomposition, are used to evaluate the entropy of the samples. This analysis is done on all the components of the decomposed signals to identify the crucial one for analysis [14]. Sample entropy and approximate entropy serve as valuable indicators of signal irregularity, shedding light on the level of complexity present within the EDA signals [15]. Higher values of these entropy measures are indicative of heightened complexity, signifying potential underlying physiological responses that warrant exploration. Singular Value

Decomposition (SVD) entropy, on the other hand, offers unique insights into the structural intricacies and complexity of EDA signal components [16].

Analyzing these entropy measures of the decomposed EDA signal is a prudent choice because this component typically contains essential information about the slow-varying aspects of the physiological response. Entropy analysis plays a significant role in mind wandering analysis by providing a quantitative measure of the complexity and irregularity of the EDA signals associated with cognitive processes.

The Sample entropy is then defined as

$$SampEn(m, r, N) = -\ln\left[\frac{A(m+1, r)}{A(m, r)}\right] \quad (1)$$

where, m is the embedding dimension, which determines the length of sequences to compare, r is the similarity criterion which defines the maximum allowable difference between subsequences, N is the length of the time series data. A (m, r), which represents the number of sequences of length m that satisfy the similarity criterion r, and similarly A (m + 1, r).

The Approximate entropy is defined as

$$ApEn(m, r, N) = -\ln\left(\frac{(\Sigma ri = 1A(m, r))}{(\Sigma ri = 1A(m+1, r))}\right) \quad (2)$$

where, *m* is the embedding dimension, *r* is the tolerance or similarity criterion, *N* is the length of the time series data, r_i represents individual values of *r*. A (*m*, *r*) represents the number of sequences of length *m* that satisfy the similarity criterion *r* and similarly for $A(m + 1, r)$ represents the number of sequences of length $m + 1$ that satisfy the similarity criterion *r*.

The SVD entropy is then defined as

$$H_{SVD} = -\sum_{i=1}^{M} \overline{\sigma_i} \log_2 \overline{\sigma_i} \quad (3)$$

where M is the number of singular values and $\sigma_1, \sigma_2 ... \sigma_M$ are normalized singular values by $\bar{\sigma}_i = \sigma_i / \sum_{j=1}^{M} \sigma_j$.

2.5 Analysis of IMFs and Entropy Features

For determining the signal whether the signal corresponds to Mind Wandering or Not Mind Wandering, the following methods are discussed. The Mean Percentage Deviation (MPD) is a measure of the average relative deviation of a set of values from a reference point, here the Mindfulness is taken as the reference point and the deviation is calculated, typically expressed as a percentage. It is used to quantify how much, on average, a set of values deviates from the reference value, here mindfulness is taken as a reference. Followed by ANOVA which yields F-statistic and a corresponding p-value for determining which features significantly differs from the others. Finally, point plots to represent an estimate of central tendency for a numeric variable by the position of the dot which provides some indication of the uncertainty around that estimate using error bars.

3 Results

3.1 Representative Signals

The data extracted from the database was initially recorded at a sampling rate of 256 Hz, resulting in a variable number of samples per signal, ranging from 14,000 to 50,000. After performing spectral analysis of the EDA signals indicated that majority of the frequency components were within the range of 0.08 to 1 Hz, with some peak frequencies reaching as high as 3 Hz. To maintain consistency over all the signals being analyzed, the signals are either down sampled or up sampled to an average of 15,000 samples without loss of information.

Figure 2 Depicts the representative EDA signals of MW and NOMW. The corresponding EMD signals represented by IMF numbers up to 3 levels of decomposition as IMF0, IMF1 and IMF2. The higher the IMF level, the lower the frequency component being analyzed, and vice versa. The IMF2 level signals closely represent the original signal without the higher frequency components. To identify how these IMF2 signals can help to differentiate MW and NOMW various features are explored. The signals were decomposed up to three levels based on experimentation, further decomposition resulted in residue components or dc which is not essential to the analysis.

3.2 Mean Percentage Deviation and Point Plots

MPD helps to distinguish the relative deviation of MW and NOMW entropies by their means. The values of different entropies are represented as grouped bar plot in Fig. 3. It can be observed that the MPD of IMF2 in case of approximate entropy is greater by 9.2% than IMF1 and 3.3% greater than IMF0. In the case of SVD entropy, MPD of IMF2 is greater by 13% than IMF1 and 9.2% greater than IMF0, and similarly for Sample entropy MPD of IMF2 is greater by 3.5% than IMF1 and 1.8% smaller than IMF0.

Figure 4 Illustrates the point plots of the IMF2 when analyzed with different entropies in regard to the two classes of MW and NOMW. It is seen that it provides

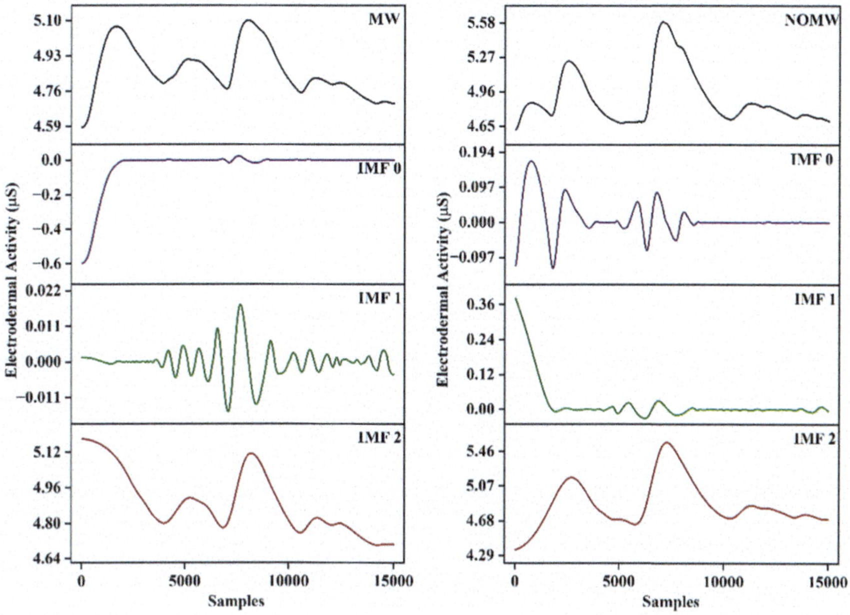

Fig. 2 Representative preprocessed signals of MW and NOMW and their respective IMFs

Fig. 3 Mean percentage deviation of entropy values between MW and NOMW for different IMFS

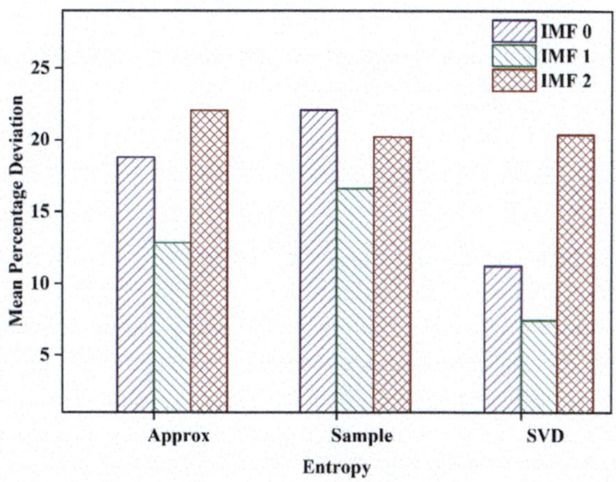

good discrimination between two classes with significant separation with respect to the entropy. It can be observed that the MW is greater by 0.00015 than NOMW in the case of SVD, and similarly for Sample NOMW is greater by 0.00032 and finally in the case of approximate entropy, NOMW is greater by 0.00047

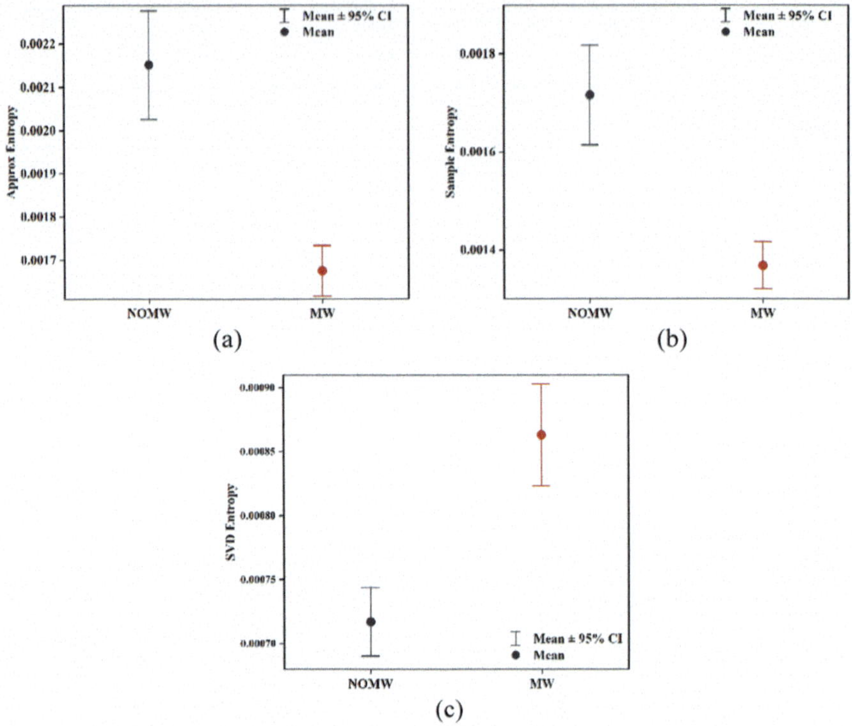

Fig. 4 Point plots comparing the mean and its confidence intervals **a** Approximate entropy, **b** Sample entropy, **c** SVD entropy

3.3 ANOVA Test

To further analyze the significance of IMFs in differentiating MW and NOMW, the ANOVA test is utilized. Table 1 gives the ANOVA metrics for the different IMFs with respect to the various entropies. It can be observed that the IMF2 is having a better F-statistic values and a lower p-values compared to the other IMF components which indicates a significant difference among group means, essentially it quantifies the ratio of the variability between groups to the variability within groups which makes it better at classifying into the two categories. Among these entropies SVD entropy offers higher significance of F-statistic and p-value in differentiating MW and NOMW so it can be used for better classifying the MW and NOMW.

The decomposition of a signal using Empirical Mode Decomposition (EMD) offered valuable insights for the underlying components and dynamic characteristics of the signal. EMD yields intrinsic mode functions (IMFs) that represent distinct frequency components. High-level IMFs or lowest frequency component captures slow trends, while lower-level ones reveal rapid fluctuations. From the Fast Fourier Transform analysis, it has been seen that the frequency of the IMF2 has a high

Table 1 ANOVA with their F-Statistic and p-values

Entropy	IMF 0		IMF 1		IMF 2	
	F-statistic	p-value	F-statistic	p-value	F-statistic	p-value
Approximate entropy	11.85	0.000583	6.28	0.012238	**36.65**	**1.58e-09**
Sample entropy	10.67	0.001098	7.14	0.007536	**29.89**	**4.93e-08**
SVD entropy	10.77	0.001038	4.30	0.038068	**38.43**	**6.44e-10**

amplitude within the low frequency range, these frequencies contribute the identification of the IMFs. IMF2 closely resembles the original signal and the long-term fluctuations in EDA. It can be correlated with fluctuations in the baseline volume of sweat within the sweat glands, also known as the tonic component of EDA. This component is a blend of psychological and physiological factors that influence it [17]. Therefore, utilizing IMF2 in this approach facilitates pattern identification and a better understanding of signal characteristics for classification.

4 Conclusion

In this paper, the influence of Empirical Mode Decomposition has been studied based on Electrodermal Activity signals for Mind Wandering detection with the help of entropy metrics. The EDA signals are obtained from an open-sourced MMSART database and thereafter analysis has been performed by applying the EMD to obtain corresponding Intrinsic Mode Functions. The Mean Percentage Deviation bar plots and point plots are used to identify the significant IMF, finally ANOVA is used to support that the two classes are significantly distinct with respect to that IMF. IMF2 showed a better MPD value compared to other components, which was also supported with the fact that entropies in particular SVD entropy gave the best values from the ANOVA test for the same. Thus, the proposed framework that utilizes low frequency component of EDA along with SVD entropy could be used to identify the mindfulness of an individual. The current study utilizes IMFs decomposed up to 3 levels with Sample, Approximate and SVD entropies to analyze the differentiation between the mind wandering and mindfulness. Furthermore, decomposition can be done based on the EDA signals acquired and analysis with other entropies.

References

1. Critchley HD (2002) Review: electrodermal responses: what happens in the brain. Neurosci 8(2):132–42. https://doi.org/10.1177/107385840200800209
2. Lee T, Kim D, Park S, Kim D, Lee S-J (2022) Predicting mind-wandering with facial videos in online lectures. In: 2022 IEEE/CVF conference on computer vision and pattern recognition workshops (CVPRW). pp 2103–2112. https://doi.org/10.1109/CVPRW56347.2022.00228
3. Bosch N, D'Mello SK (2021) Automatic detection of mind wandering from video in the lab and in the classroom. IEEE Trans Affective Computing 12:974–988. https://doi.org/10.1109/TAFFC.2019.2908837
4. Brishtel I, Khan AA, Schmidt T, Dingler T, Ishimaru S, Dengel A (2020) Mind wandering in a multimodal reading setting: behavior analysis & automatic detection using eye-tracking and an EDA Sensor. Sens 20(9):2546. https://doi.org/10.3390/s20092546
5. Khosravi S, Khan AR, Zoha A, Ghannam R (2022) Employing a wearable eye-tracker to observe mind-wandering in dynamic stimuli. In: 2022 29th IEEE international conference on electronics, circuits and systems (ICECS). pp 1–4. https://doi.org/10.1109/ICECS202256217.2022.9970787
6. Anthiyur Aravindan A, Kalyan Chappidi S, Thumma A, Palanisamy R (2023) Prediction of arousal and valence state from electrodermal activity using wavelet based ResNet50 Model. Curr Dir Biomed Eng 9(1):555–558. https://doi.org/10.1515/cdbme-2023-1139
7. Shi Y, Ruiz N, Taib R, Choi E, Chen F (2007) Galvanic skin response (GSR) as an index of cognitive load. In: CHI '07 extended abstracts on human factors in computing systems, ACM. pp 2651–2656. https://doi.org/10.1145/1240866.1241057
8. Tasika NJ, Haque MH, Rimo MB, Al Haque M, Alam S, Tamanna T, Rahman MA, Parvez MZ (2020) A framework for mind wandering detection using EEG signals. In: 2020 IEEE region 10 symposium (TENSYMP). pp 1474–1477. https://doi.org/10.1109/TENSYMP50017.2020.9230790
9. Chen Y-T, Lee H-H, Shih C-Y, Chen Z-L, Beh W-K, Yeh S-L, Wu A-Y (2022) An effective entropy-assisted mind-wandering detection system using EEG signals of MM-SART database. IEEE J Biomed Health Inform 26(8):3649–3660. https://doi.org/10.1109/JBHI.2022.3187346
10. Chang S, Chen Y-T, Wu A-YA (2021) Efficient mind-wandering detection system with GSR signals on MM-SART database. In: 2021 IEEE workshop on signal processing systems (SiPS). pp 199–204. https://doi.org/10.1109/SiPS52927.2021.00043
11. Smallwood J, Schooler JW (2006) The restless mind. Psychol Bull 132(6):946–958. https://doi.org/10.1037/0033-2909.132.6.946
12. Makowski D, Pham T, Lau ZJ, Brammer JC, Lespinasse F, Pham H, Schölzel C, Chen SHA (2021) NeuroKit2: a python toolbox for neurophysiological signal processing. In: Behav Res Methods 53(4):1689–1696. https://doi.org/10.3758/s13428-020-01516-y
13. Ge H, Chen G, Yu H, Chen H, An F (2018) Theoretical analysis of empirical mode decomposition. Symmetry 10(11):623. https://doi.org/10.3390/sym10110623
14. Ponce-Flores M, Frausto-Solís J, Santamaría-Bonfil G, Pérez-Ortega J, González-Barbosa JJ (2020) Time series complexities and their relationship to forecasting performance. Entropy 22(1):89. https://doi.org/10.3390/e22010089
15. Richman JS, Moorman JR (2000) Physiological time-series analysis using approximate entropy and sample entropy. Am J Physiol-Hear Circ Physiol 278(6):H2039–H2049. https://doi.org/10.1152/ajpheart.2000.278.6.H2039
16. Banerjee M, Pal NR (2014) Feature selection with SVD entropy: some modification and extension. Inf Sci 264:118–134. https://doi.org/10.1016/j.ins.2013.12.029
17. Tseng B, Subramanian S, Barbieri R, Brown EN (2022) Tonic electrodermal activity is a robust marker of psychological and physiological changes during induction of anesthesia. In: 2022 44th annual international conference of the IEEE engineering in medicine & biology society (EMBC). pp 418–421. https://doi.org/10.1109/EMBC48229.2022.9871080

TranslateKar: An Efficient Transformer Based Konkani Translator

Shreya Pai, Sajal Mandrekar, Saylee Phadte, Atit Naik, and Shitala Prasad

Abstract In this research, we introduce a bidirectional Neural Machine Translation (NMT) model that employs the Transformer architecture to facilitate translations between Konkani and English languages. Inspired by "Attention is All You Need", we delve into variations in hyperparameters, including dropout rates and vocabulary size, to assess their impact on translation performance. The Transformer model, renowned for its capacity to handle long-range dependencies efficiently and in parallel, proves highly suitable for sequence-to-sequence tasks such as machine translation. We meticulously conduct a series of experiments, training the model with diverse hyperparameters configuration, and subsequently compare the outcomes to discern the optimal setup. Our findings underscore the superiority of the dash model, as it attains the highest BLEU score on the 3k test dataset, demonstrating its enhanced translation accuracy for both Konkani to English and English to Konkani translations. These results shed light on the paramount importance of hyperparameters selection in bidirectional NMT systems, underscoring the Transformer architecture's potential in multilingual translation tasks. The study's outcomes represent a noteworthy contribution to the advancement of machine translation techniques and set the stage for more effective communication tools in multilingual contexts.

Keywords Autoencoder · Deep learning · Transformer · Konkani-English · Neural machine translation

1 Introduction

Neural Machine Translation (NMT) represents a cutting-edge approach in the realm of machine learning, specifically designed for the task of translating text from one language to another [1, 2]. In stark contrast to the conventional methods of machine

S. Pai · S. Mandrekar · S. Phadte · A. Naik
Computer Science, Goa College of Engineering, Ponda, Goa, India

S. Prasad (✉)
School of Mathematics and Computer Science, IIT Goa, Ponda, Goa, India
e-mail: shitala@iitgoa.ac.in

D.-I. Stroe et al. (eds.), *Emerging Electronics and Automation*, Lecture Notes in Electrical Engineering 1202, https://doi.org/10.1007/978-981-97-3090-2_25

translation, which relied heavily on rule-based systems and handcrafted linguistic rules, NMT revolutionizes the field by framing translation as a sequence-to-sequence problem. This innovation involves training end-to-end models, which means the entire translation process, from source text to target language, is handled seamlessly by the model itself.

At the core of NMT lies a fundamental shift in how the translation process is conceptualized and executed. NMT models adopt an encoding and decoding architecture [3]. Initially, the source text is transformed into a continuous vector representation through a series of encoding layers. Subsequently, this encoded vector is skillfully decoded into the target language using decoding layers. The culmination of this intricate process results in a translation that's generated directly by the model. This end-to-end training offers a substantial advantage: it empowers the model to grasp the nuanced context of the text it is translating, leading to translations that are inherently more contextually accurate.

NMT has proven to be nothing short of a game-changer in the realm of machine translation [3]. It has demonstrated remarkable efficacy across a multitude of languages and domains, consistently outperforming its predecessors in terms of translation quality and accuracy. Furthermore, NMT models are incredibly versatile, with the capacity for fine-tuning to suit specific domains and languages. This adaptability allows for a high degree of customization, making NMT a potent tool tailored to diverse use cases.

NMT stands out due to its unique ability to handle intricate sentence structures while preserving the core meaning of the source text. This exceptional capability is attributed to deep learning algorithms integrated into NMT models [3]. These algorithms not only grasp individual words but also capture nuanced relationships between words and phrases within the source sentence. Consequently, NMT excels in producing translations that faithfully convey the intended message, marking a significant advancement in language translation technology.

In this paper, our contributions can be summarized in four key aspects:

- We introduced a novel Konkani to English to Konkani dataset, providing a valuable resource for further research in this domain.
- For the first time, we presented a transformer-based efficient translation model called TranslateKar, to the best of our knowledge.
- We implemented various versions of TranslateKar, namely T_mini, T_256, and T_base, differing in parameters and training set.
- Since no similar literature was found, our approach established a state-of-the-art (SOTA) standard for Konkani to English to Konkani translation.

The subsequent sections of this paper are organized as follows: Sect. 2 delves into related works in language translation. Section 3 introduces the model architecture, while Sect. 4 discusses its design. Training settings and datasets used are elaborated in Sect. 5, followed by the formulation of BLEU scores in Sect. 6. The experiments and results are presented next. Finally, our findings are discussed and concluded in Sect. 8.

2 Background

The driving force behind this project is the pressing need to overcome linguistic barriers that often hinder travelers, particularly those who encounter regional languages such as Konkani. In many instances, existing translation services prove inadequate when it comes to handling languages like Konkani, further underscoring the urgency of this endeavor. The overarching goal here is to construct a Language Translator tailored specifically for English to Konkani translation, thereby enhancing travel accessibility for individuals whose primary language is a regional one.

Konkani, classified as an Indo-Aryan language, boasts a substantial presence along the picturesque western coastal region of India [4, 5]. It holds the distinction of being one of the 22 Scheduled languages recognized by the Indian Constitution and serves as the official language of the state of Goa. Beyond its linguistic significance, Konkani is deeply intertwined with the cultural fabric and historical heritage of the Konkani-speaking community. It represents a crucial element of their identity, adding an additional layer of importance to the preservation and promotion of the language.

However, crafting a precise and efficient NMT system for translating from English to Konkani presents a set of formidable challenges. These complexities stem from the distinctive characteristics of the language pair itself. The need for this specialized NMT system arises from the nuanced nature of Konkani, highlighting the importance of tailoring translation services to the unique linguistic landscapes of diverse languages like Konkani [6].

2.1 Challenges

The development of a precise and efficient NMT system for English to Konkani involves the careful navigation of several significant challenges, such as

- Limited Dataset: The availability of a comprehensive Konkani-English translation dataset is notably scarce. This scarcity poses a substantial obstacle in training a robust NMT model capable of producing high-quality translations.
- Accuracy Concerns: Attaining a commendable level of translation accuracy for low-resource languages like Konkani necessitates the implementation of specialized techniques. The inherent data sparsity in such languages adds complexity to achieving the desired precision.

- Hardware Requirements: The training and deployment of NMT models place substantial demands on computational resources. To ensure practicality and efficiency, optimization strategies are imperative.
- Resource Acquisition: Collecting & curating a diverse and representative dataset tailored to Konkani-English translation is a time-intensive endeavor which is a resource-intensive process of building a reliable NMT system.
- Data Quality Issues: The quality and consistency of the dataset are paramount for ensuring reliable translation performance. Addressing data quality concerns is a critical step in the development process.

In spite of these formidable challenges, the project team is resolutely committed to the development of a NMT system that transcends language barriers. The ultimate aim is to provide accurate translations that facilitate cross-cultural communication, thereby making travel more inclusive and convenient for tourists visiting regions where Konkani is spoken.

Furthermore, the project envisions the scalability of the developed solution to encompass other low-resource languages. This expansion would amplify its societal impact by fostering greater understanding and inclusivity among linguistically diverse communities. By accommodating multiple regional languages, the project could play a pivotal role in bridging communication gaps within diverse, multilingual societies.

3 Architecture

The Transformer model, introduced in the seminal paper "Attention Is All You Need" by Vaswani et al. in 2017, has had a profound impact on the realm of natural language processing (NLP) and various other sequence-to-sequence tasks [7]. This innovative model boasts a unique architecture, harnessing self-attention mechanisms to process sequential data efficiently and capture long-range dependencies.

Distinguishing itself from conventional sequence models such as recurrent neural networks (RNNs) and their derivatives, the Transformer charters a novel course by breaking free from sequential processing [2]. This inherent parallelizability and accelerated training pace set the Transformer apart. Notably, this characteristic empowers the transformer to tackle significantly longer sequences without succumbing to the vanishing or exploding gradient issues that plagued RNN-based models.

Key components of the transformer architecture include:

- Self-attention: Self-attention endows each token in a sequence with the ability to attend to all other tokens within the same sequence. It calculates a weighted sum of their representations, enabling the model to discern contextual relationships between words and encode dependencies effectively.
- Multi-head Attention: The Transformer employs multiple parallel self-attention heads. Each head specializes in learning distinct attention patterns, affording the

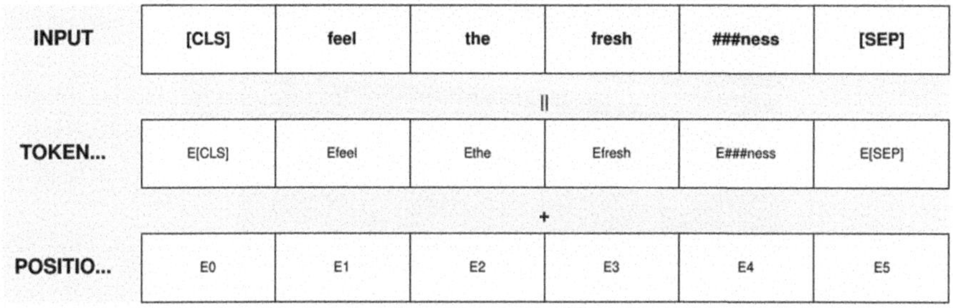

Fig. 1 BERT tokenization

model the capacity to focus on various aspects of the input sequence. The outputs from these heads are concatenated and subjected to linear transformation to produce the final attention output.

- Positional Encoding: In contrast to models like RNNs that inherently possess sequential information, the Transformer lacks this innate understanding of token order. To bridge this gap, positional encoding is introduced. It supplements token embeddings with positional information, ensuring that the model can interpret the sequence's order, as shown in Fig. 1.

The positional encoding mechanism takes the form of a stack of sine and cosine functions, each oscillating at different frequencies along the depth of the embedding vector. These oscillations traverse the positional axis, promoting alignment with proximate vectors along the position axis, see Fig. 2. Additional noteworthy aspects of the transformer architecture encompass, as shown in Figs. 3 and 4, are described:

- Encoder-Decoder Architecture: The transformer is conventionally deployed within an encoder-decoder framework for sequence-to-sequence tasks such as machine translation and text summarization [7]. The encoder processes the input sequence, while the decoder generates the output sequence.
- Encoder and Decoder Stacks: Both the encoder and decoder consist of multiple identical layers, referred to as the encoder stack and decoder stack, respectively. Each layer integrates self-attention and feed-forward neural network sub-layers. The encoder processes input sequences, while the decoder orchestrates output sequence generation.
- Residual Connections and Layer Normalization: Residual connections play a pivotal role in mitigating the vanishing gradient problem and bolstering the training process's efficacy. Additionally, layer normalization is applied after each sub-layer to enhance training stability. The same position encoding vector normalization is shown in Fig. 3.

In essence, the transformer architecture represents a paradigm shift in the domain of sequence modeling and has laid the foundation for numerous groundbreaking applications in NLP and beyond.

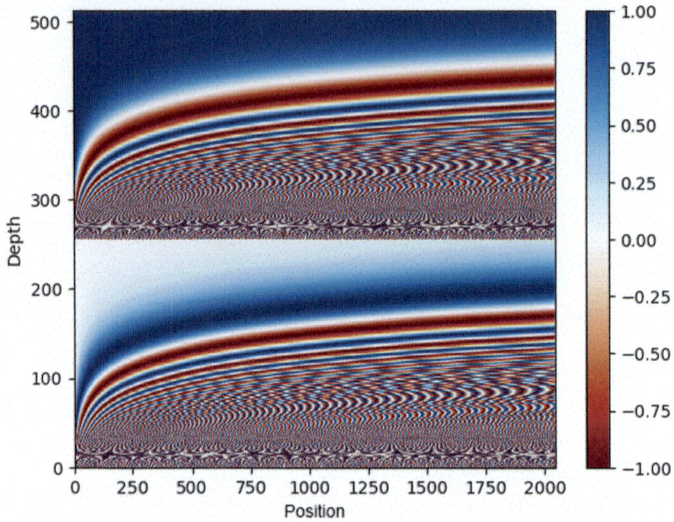

Fig. 2 Visualization of variation in positional encoding across the position axis

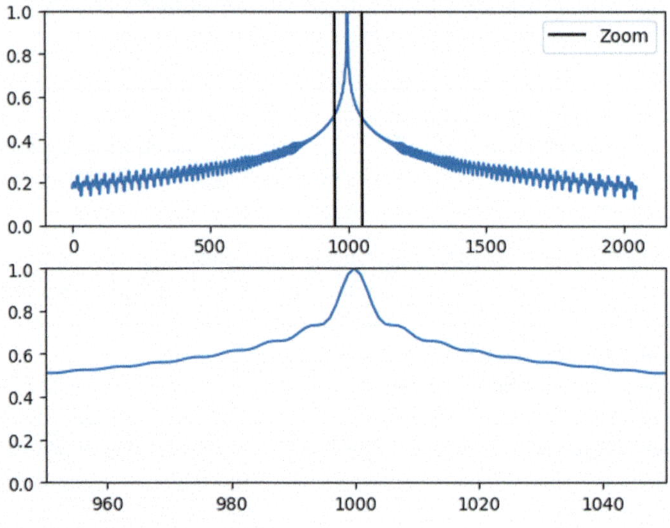

Fig. 3 Position encoding vector normalization

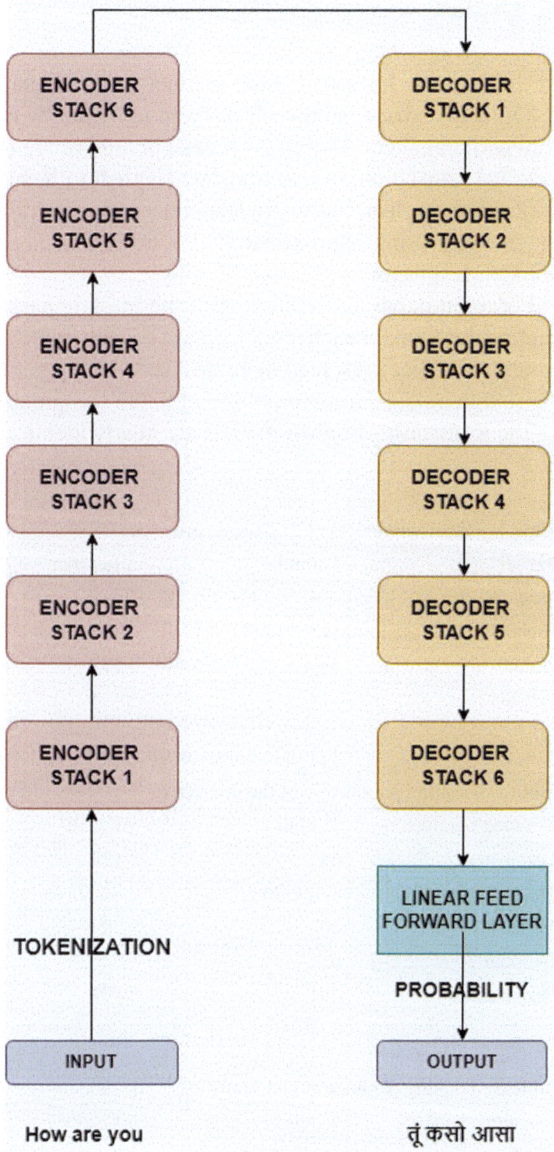

Fig. 4 Network diagram of model with six layers and single transformer

4 Design

We have created a set of three distinct Transformer models, denoted as T_mini, T_256, and T_base, with each of them individually configured using different sets of hyperparameters. Within each category, there are two models designed for bidirectional translation: one to translate English to Konkani and the other to translate Konkani to English. These models underwent rigorous training using a parallel corpus encompassing approximately 1.9 million sentence pairs in both English and Konkani languages.

For an in-depth understanding of the training parameters and an estimate of the training duration for each model, please consult Table 1. It's important to note that the training statistics presented in Table 2 specifically pertain to the English-to-Konkani translation model. However, it's essential to recognize that the corresponding values for the Konkani-to-English models are nearly identical (Table 3).

Table 1 Hyperparameters of TransKar types

Model type	T_mini	T_256	T_base
#Layers	4	4	6
Embedding dimension	128	256	512
Hidden layer size	512	512	2048
#Heads	8	8	8

Table 2 Training parameters of the models

Training parameters	T_mini	T_256	T_base
Batch size	64	128	128
Dataset size	~1.95 M	~1.95 M	~1.95 M
Epochs	20	30	30
Dropout	0.1	0.3	0.3
Vocab size	70 k	70 k	70 k

Table 3 Training parameters and Statistics of the models

Training statistics	T_mini	T_256	T_base
Hardware	nvidia A100	nvidia A100	nvidia A100
#Batches per epoch	30270	15133	15133
Avg. time per epoch	54.38 min	68.84 min	118.33 min
#Time	18.13 h	34.42 h	59.2 h

4.1 Model Variation

We test the accuracy of the model by changing the residual dropouts to find the suitable value for our case.

Dropout. Dropout serves as a regularization technique crucial for preventing overfitting and enhancing generalization in machine learning models. Overfitting occurs when a model excels on training data but falters on unseen data, indicating that it has memorized specific examples rather than grasping underlying patterns.

The concept of dropout involves randomly "dropping out" a fraction of neurons or units within a neural network during training, effectively setting them to zero. This random deactivation occurs in each training iteration, causing certain neurons to remain inactive, with their weights left unaltered. Consequently, the model becomes less reliant on specific neurons, compelling it to learn more robust and generalized data representations. In transformer architectures, dropout is commonly applied to different components like the self-attention mechanism, feed-forward neural networks (FFNs), and sometimes even embeddings.

Dropout proves particularly valuable in transformers due to their extensive parameters, making them susceptible to overfitting. By implementing dropout in transformers, the model's capacity is reduced, preventing it from becoming overly specialized to the training data and enhancing its ability to adapt to new, unseen data. Notably, dropout is active during training and typically deactivated during inference (evaluation/prediction) to ensure deterministic results.

We conducted a comparative analysis of model loss during training with two dropout values, 0.1 and 0.3, observing discernible differences in performance.

5 Training

5.1 Dataset and Language Corpus

The dataset comprises a parallel corpus encompassing both languages, featuring numerous sentence pairs in English and Konkani conveying equivalent meanings.

The accuracy of translations hinges on the dataset's length and the inclusion of unique words from the vocabulary within it. When words from the language are absent in the dataset, the model grapples with comprehension difficulties, leading to subpar translations.

Konkani, classified as a low-resource language, poses challenges in securing an extensive parallel corpus for its use. We collected the dataset in Konkani and then translated them manually for training. Simple sample snapshots of the proposed dataset is shown in Fig. 5.

Training Data. The project's initial dataset was sourced from the Technology Development for Indian Languages (TDIL) initiative, which encompassed 70,000 sen-

1. This is ironical since until recently the discourse was mostly about rampant malnourishment in the country.
2. He also matched the feat of Sir Don Bradman who scored four hundreds but in a five-match series.
3. Rajya Sabha to take up crucial Juvenile Justice Bill today
4. The CBI court had issued a non-bailable arrest warrant against Yadav on March 1 but he had appeared in the court on medical grounds.
5. It is the Union government's ploy to target AAP using the Delhi Police.
6. The Mumbai police were alerted and they rushed in with boats, including some hired ones, and tried to pull it with ropes but the efforts went in vain because of high tide.
7. Agreeing with Raja, Minister of State for Parliamentary Affairs Mukhtar Abbas Naqvi said the state governments should take necessary action in such cases and book the culprits.
8. Lucknow: Security personnel were thrown into a tizzy on Friday when Chief Minister Akhilesh Yadav and his wife Dimple Yadav were stuck in a VIP lift at the state Assembly building here for nearly 30 minutes.

1. हालीं मेरेन हें प्रवचन चड करून देशांतल्या बेकायदेशीर कुपोशणा विशीं आशिल्ल्यान हें विडंबनात्मक आसा.
2. चार शेंकड्यांनी पूण पांच मॅचींच्या मालिकेंत सर डॉन ब्रॅडमनाच्या पराक्रमाकय ताणें जुळयलें.
3. राज्यसभा आयज म्हत्वाचें किशोर न्याय विधेयक घेवपाची
4. 1 मार्चाक सीबीआय न्यायालयान यादवा आड जामीन नाशिल्लो अटक वॉरंट जारी केल्लो पूण वैजकी कारणांक लागून तो न्यायालयांत हजर आशिल्लो.
5. दिल्ली पुलीस वापरून आपाक लक्ष्य करपाची ही केंद्र सरकाराची युक्ती.
6. मुंबय पुलिसांक खबर दिली आनी तांणी कांय भाड्यान घेतिल्ल्या बोटीं सयत बोटी घेवन भितर सरले आनी दोरयांनी ओडपाचो यत्न केलो पूण भरती आशिल्ल्यान हे यत्न व्यर्थ जाले.
7. राजा हांचे कडेन सहमती दिवन संसदीय वेव्हार राज्यमंत्री मुख्तार अब्बास नकवी हांणी सांगलें, राज्य सरकारांनी अशा प्रकरणांत गरजेची कारवाय करची आनी गुन्यांवकारांक बुकिंग करचें.
8. लखनौः हांगाच्या राज्य विधानसभा इमारतींतल्या व्हीआयपी लिफ्टांत सुमार 30 मिनटां मुख्यमंत्री अखिलेश यादव आनी ताची घरकान्न डिंपल यादव अडकले उपरांत शुक्रारा सुरक्षा कर्मचाऱ्यांक घुंवळी वखदा घाली.

Fig. 5 Sample of data from: **a** English dataset and **b** Konkani dataset

tences. Supplementary English data was subsequently gathered from the Anuvaad project [8], covering a wide array of domains including Finance, Entertainment, Health, General, and Legal, among others. The English dataset was then subjected to translation.

Following the translation process, the data underwent meticulous processing, aimed at rectifying errors, addressing inconsistencies, and eliminating undesired tags or artifacts. This processing phase played a pivotal role in ensuring the dataset's quality and its suitability for effectively training the Neural Machine Translation (NMT) system.

Test Data In order to assess the performance and generalization capabilities of the model, we curated a distinct test dataset comprising roughly 3000 sentences. This test dataset was created using the English dataset sourced from the OPUS corpus [9], which was subsequently subjected to translation and preprocessing. It's important to emphasize that this test data was entirely separate from the training data and served as a reliable benchmark to evaluate the accuracy and efficacy of the Neural Machine Translation (NMT) system in translating sentences that had not been encountered during the training process.

5.2 *Optimizer*

In our model training, we employ the Adam optimizer, as indicated in the dash model. The Adam optimizer utilizes exponential moving averages of previous gradients, with $\beta 1 = 0.9$, and past squared gradients, with $\beta 2 = 0.98$, to dynamically adapt and fine-tune the learning rates for individual parameters throughout the training process. This adaptive mechanism contributes to enhanced convergence and stability, fostering more effective training.

5.3 Tokenizer

BERT utilizes a specialized tokenization process to convert raw text into tokens, a crucial step that underlies its remarkable performance in various NLP tasks. This tokenization method is built upon the WordPiece algorithm, which dissects words into subword units. This approach effectively captures word morphology and enhances the model's capability to handle out-of-vocabulary words, making it adaptable across a wide range of language contexts.

Furthermore, tokenization involves the transformation of tokens into numerical IDs, each denoting its position within the model's vocabulary. This numerical representation is pivotal for BERT's training and inference procedures, enabling it to process text as numerical data, see Fig. 1.

An inherent advantage of BERT tokenization lies in its support for variable-length inputs. This implies that input sequences can vary in length without requiring the addition of padding or truncation. This dynamic treatment of input length optimizes memory usage and empowers BERT to efficiently process lengthy documents without imposing substantial computational demands.

6 BLEU Score Calculation

For each of our trained models (Transformer mini, Transformer 256, and Transformer base), we will assess their performance by computing both sentence-level and corpus-level BLEU scores. These scores will be calculated for both English-to-Konkani and Konkani-to-English translations, using a reference set of translations as a basis for evaluation (Table 4).

Here are the key metrics we will employ:

Table 4 BLEU scores of models on test data

Models	T_mini	T_256	T_base
English to Konkani			
BLEU1 score	0.5092	0.5224	0.5412
BLEU2 score	0.3995	0.4132	0.4363
BLEU3 score	0.3197	0.3330	0.3582
BLEU4 score	0.2526	0.2645	0.2903
Konkani to English			
BLEU1 score	0.4438	0.4952	0.4927
BLEU2 score	0.3233	0.3697	0.3776
BLEU3 score	0.2445	0.2853	0.2982
BLEU4 score	0.1818	0.2168	0.2320

- Sentence BLEU Score: To evaluate individual sentences, we will utilize NLTK's sentence_bleu() function. This function compares a candidate sentence with one or more reference sentences. The reference sentences are provided as a list, where each reference is represented as a list of tokens. Similarly, the candidate sentence is presented as a list of tokens.
- Corpus BLEU Score: For assessing the quality of multiple sentences, such as paragraphs or documents, we will employ NLTK's corpus_bleu() function. This function calculates the BLEU score for a collection of sentences. The references for this metric should be specified as a list of documents, where each document is a list of references, and each alternative reference is represented as a list of tokens. The candidate documents should also be structured as a list, where each document is presented as a list of tokens.
- Individual N-Gram Scores: Individual N-gram scores provide insights into matching grams of a specific order, such as single words (1-gram) or word pairs (2-gram or bigram). The weights for these scores can be specified as a tuple, with each index corresponding to the gram order.
- Cumulative N-Gram Scores: Cumulative scores encompass the calculation of individual n-gram scores across orders from 1 to n. These scores are weighted by computing the weighted geometric mean. By default, both the sentence_bleu() and corpus_bleu() functions calculate the cumulative 4-gram BLEU score, often referred to as BLEU-4.

These evaluation metrics will provide a comprehensive assessment of the translation quality achieved by our models, considering both individual sentence-level performance and the overall corpus-level effectiveness.

The BLEU score is a value between 0 and 1, where 1 represents an exact match between the machine-generated translation and the reference translation. It is determined by evaluating the precision of n-gram matches between the machine-generated translation and the reference translation, with n varying from 1 to 4 (Fig. 6).

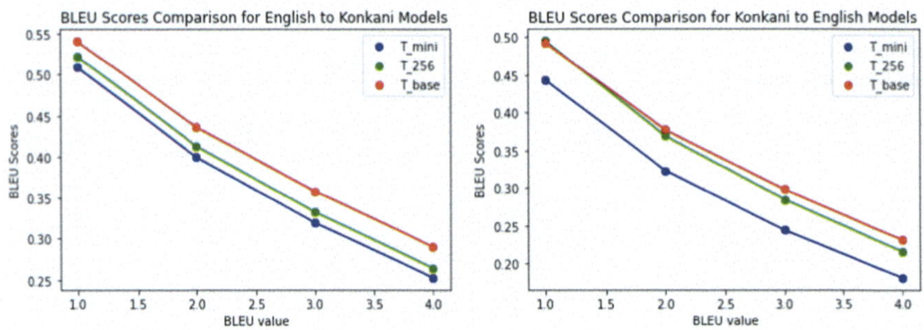

Fig. 6 BLEU score variations for: **a** English to Konkani models and **b** Konkani to English models

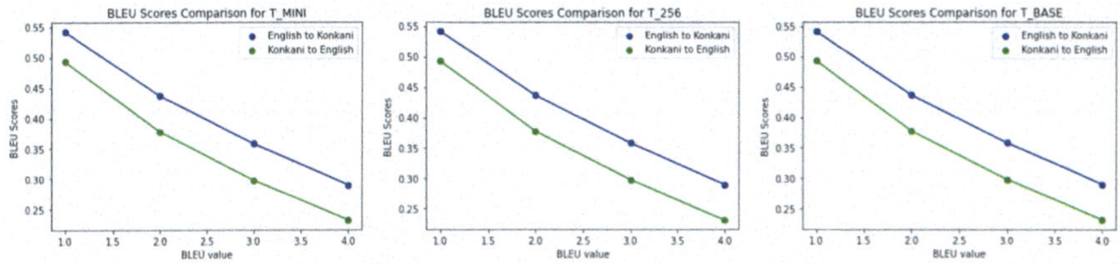

Fig. 7 BLEU-N Score variations for: **a** T_mini, **b** T_256 and **c** T_base

7 Experimental Results

The experimental results shed light on the performance analysis of three distinct models: T_mini, T_256, and T_base, concerning English to Konkani and Konkani to English translation tasks. Refer to Fig. 7 for specific details.

Regarding English to Konkani and Konkani to English Translation:

- The BLEU scores consistently escalate with increasing model size, indicating larger models exhibit superior performance in both translation directions.
- Across all BLEU n-gram scores (1–4), T_base consistently outperforms T_256 and T_mini, showcasing the highest overall scores.
- Similarly, as observed in English to Konkani translation, BLEU scores rise with larger model sizes, emphasizing the benefits of employing bigger models in both translation directions.
- Once again, T-base secures the highest BLEU scores, surpassing T_256 and T_mini across all n-gram scores.

A larger batch size demonstrated advantages during training, particularly for the T_base model, facilitating more parallel processing in GPU clusters. While it resulted in faster convergence per epoch, it concurrently extended the overall training time due to increased iterations. However, a larger batch size necessitates higher GPU and CPU memory, potentially leading to out-of-memory errors in instances of lower memory allocation.

In the context of low-resource languages like Konkani, a higher number of layers could provoke overfitting, mitigated by augmenting the dropout. Although higher dropout prolongs training time per epoch, it allows for an increased hidden layer size without adversely affecting model fit.

Embeddings marginally differ in representing similar categorical inputs. Dropout adaptation accommodates this variability by cultivating more adaptable and robust representations capable of handling diverse instances of same input.

7.1 Graphical Representation

The graphical depiction illustrates a consistent upward trend in BLEU scores with increasing model sizes for both translation tasks. T_base consistently demonstrates superior performance, boasting the highest BLEU scores, followed by T_256, and then T_mini.

To summarize, both experimental findings and graphical representations accentuate that larger models (T_base) generally yield superior translation performance compared to smaller counterparts (T_256, T_mini) in both translation directions (English to Konkani and Konkani to English). This emphasizes the significance of model size in achieving enhanced translation quality in Neural Machine Translation systems.

8 Conclusion

Neural Machine Translations (NMTs), particularly those using Transformers, outperform traditional methods like statistical machine translation. Among the models, T_base consistently excels in both English to Konkani and Konkani to English translations, displaying the highest BLEU scores across n-gram levels. English to Konkani translation seems slightly easier, possibly due to syntactic similarities. T_base demonstrates superior translation abilities, while T_mini performs the lowest due to its lightweight nature. These findings emphasize the impact of model size on translation quality, though further exploration and analysis are needed for comprehensive evaluation.

References

1. Cho K, Van Merriënboer B, Bahdanau D, Bengio Y (2014) On the properties of neural machine translation: encoder-decoder approaches. ArXiv preprint arXiv:1409.1259
2. Cho K, Van Merriënboer B, Gulcehre C, Bahdanau D, Bougares F, Schwenk H, Bengio Y (2014) Learning phrase representations using RNN encoder-decoder for statistical machine translation. ArXiv preprint arXiv:1406.1078
3. Bahdanau D, Cho D, Bengio Y (2014) Neural machine translation by jointly learning to align and translate. ArXiv preprint arXiv:1409.0473
4. Word embedding models for low-resourced GOA'n Konkani language: NLP. https://medium.com/mlearning-ai/word-embedding-models-for-low-resourced-goan-konkani-language-nlp-c1ac452e86a. Accessed 15 Oct 2023
5. Konkani language. https://en.wikipedia.org/wiki/Konkani_language. Accessed 15 Oct 202310-15
6. See A, Luong MT, Manning CD (2016) Compression of neural machine translation models via pruning. ArXiv preprint arXiv:1606.09274
7. Vaswani A, Shazeer N, Parmar N, Uszkoreit J, Jones L, Gomez AN, Kaiser Ł, Polosukhin I (2017) Attention is all you need. Adv Neural Inform Process Syst 30
8. https://www.anuvaad.org/
9. Opus corpus. https://opus.nlpl.eu/. Accessed 15 Oct 2023

Quality Assessment of PPG Signal for Stress Analysis: A Preliminary Study

Srejita Chakraborty and Saurabh Pal

Abstract Biosignal based analysis of various psychosomatic events like stress, pain, etc. becomes popular among researchers nowadays because of its salient independent nature. Application of biosignals like Electrocardiogram (ECG), Photoplethysmogram (PPG), Electrodermal activities (EDA), Electroencephalogram (EEG) etc. are being investigated for the purpose using different classical modern machine learning tools. Wearable sensors for such biosignals followed by AI enabled processors for detection or classification of various physiological events leads to real-time continuous monitoring of abnormalities. In this regard PPG becomes very prospective for its easy acquisition protocol. However PPG suffers a critical issue due to its noise prone nature resulting from movement and electrode contact noise. Thus signal quality is important for any PPG based analysis. The present work proposes a simple signal quality assessment (SQA) method based on its peak-to-peak interval (PPI). SQA basically estimates some Heart Rate (HR) features and compares them with hard thresholds. The section of PPG strip is disregarded for further analysis if the HR features are beyond the predefined values. The algorithm is tested on publicly available CLAS database. It is noticed that the SQA algorithm enhances the common statistical features of PPG as compared with benchmark ECG signal.

Keywords Stress analysis · Photoplethysmogram (PPG) · Signal quality assessment (SQA) · Peak-to-Peak interval (PPI) · Statistical features

S. Chakraborty (✉) · S. Pal
Department of Applied Physics, University of Calcutta, Kolkata, India
e-mail: srejita97@gmail.com

S. Pal
e-mail: spaphy@caluniv.ac.in

D.-I. Stroe et al. (eds.), *Emerging Electronics and Automation*, Lecture Notes in Electrical Engineering 1202, https://doi.org/10.1007/978-981-97-3090-2_26

1 Introduction

Nowadays, stress has become an inescapable part of human lives. As defined by Merriam Webster dictionary, any physical, chemical or emotional factor that generates tension in body and mind and may also lead to disease causation, can be termed as Stress. Stress is witnessed in all age groups and across various spheres. Lack of stress management may affect physiological and psychological health and may lead to suicides in severe cases. Worldwide more than 7,00,000 people die by suicide [1]. Hence, stress monitoring and its analysis becomes essential. The present techniques used for stress analysis involve questionnaire based stress monitoring wherein a series of questions are asked to people to collect information. Some questionnaires include such as Perceived stress scale [2], Childhood Trauma Questionnaire [3], The Job Content Questionnaire [4] among others. All such methods depend on the response and cooperation of the subject. However the method is not applicable in case of children, elder people having cognitive impairment or people having difficulty in verbal communication. Facial expression based stress analysis is an emerging field of research which endeavors to overcome the drawback of Questionnaire based stress monitoring. In [5], a real-time non-intrusive stress analysis of drivers is proposed based on sequence level monitoring of facial expression. Psychiatric diseases are analyzed through facial expressions in [6]. Automatic stress analysis through facial-video based signal is studied using Facial Action Coding System (FACS) based Action Unit (AU) parameters. Respiration influences the oscillatory pattern of the rotation angles of the heart's electrical axis. Respiratory frequency during stress is investigated based on ECG signal in [7]. In [8], a combination of fuzzy logic and associative Petri net is used to design such a rule base that detects mental stress based on ECG signal. A machine learning-based approach to assess stress for drivers was reported in [9], interaction ensemble (INTENSE) technique that consists of three machine learning techniques such as interaction principal component analysis, connectivity-based clustering, and K-means clustering was employed to investigate the relationship of several biosignals such as electromyograms (EMG), ECG, hand galvanic skin resistance, foot galvanic skin resistance, heart rate and respiration signal with stress. It was observed that ECG was the most informative among other signals for stress analysis. However the techniques require costly setup and follow complex algorithms. Video signal based analysis may be affected due to several factors such as illumination conditions, angle, placement and quality of camera and movement of subject etc. Moreover, facial expression based stress analysis may intrude into the subject's privacy. Stress changes the Autonomic Nervous System (ANS) response [10]. During situations of stress, hypothalamus activates the sympathetic nervous system (a constituent of ANS), which in turn releases stress hormones such as epinephrine and cortisol. Due to activation of SNS, various physiological changes occur like increase in heart rate and blood pressure. These variations lead to changes in biosignals like ECG, EEG, and EDA among others. Stress affected cardiac response can be observed from ECG signal. However, ECG acquisition requires the use of prolonged use of several bioelectrodes at different body locations which limits the mobility of users. In comparison to ECG, PPG is easier to set up, is more convenient and economical and does not cause skin irritations. Conversely, photoplethysmogram (PPG) is a noninvasive circulatory signal related to the pulsatile volume of blood in tissues [11]. It is used in various forms: finger/toe clips, oximeters and wearable health monitoring devices. Moreover, cardiac activity can be monitored using PPG signal [12, 13]. These factors make PPG more suitable for IoT based applications.

However, PPG based monitoring has certain limitations. One of the major amongst them is the non-standard nature of PPG signals. It corresponds to the fact that PPG signals are highly susceptible to Motion Artifacts (caused by hand movements) and environmental noise [14, 15]. Thus, checking the pattern of ECG and PPG signals and analyzing their desirability and reliability is essential for proper analysis and feature extraction. In [16], the authors studied the impact of several variables (age, sex, physical activity and health state) on signal quality of PPG. In [17], the authors proposed two sliding scale metrics to study signal quality of PPG. In the present work, considering ECG signal as the reference (benchmark) signal, a parameter of Signal Quality Assessment has been introduced after filtering and normalization to identify PPG signals that are unable to provide proper cardiac features and an attempt has been made to eliminate the identified signals and then compare PPG-derived cardiac features with ECG-derived cardiac feature to check the change in the percentage of correlation.

2 Methodology

The overall methodology is depicted in Fig. 1. The methodology is tested on ten random subjects from CLAS Database [18]. The database consists of 30-min recordings of physiological signals (ECG, PPG, and GSR) of 62 healthy volunteers (17 women and 45 men). All physiological signals were acquired with sampling rate of 256 Hz and resolution 16-bits per sample. Specifically, the database contains five task-specific datasets, collected while volunteer participants perform different tasks. The tasks were performed based on various stimuli: Math test, Stroop test, Logic problems, emotionally tagged pictures, emotionally tagged audio–video clips. While the first three tasks are interactive and require the subject to respond within a short time window, the last two tasks are perceptive, where the subject is exposed to stimuli and was instructed to simply watch and listen. All recordings were collected in a single session.

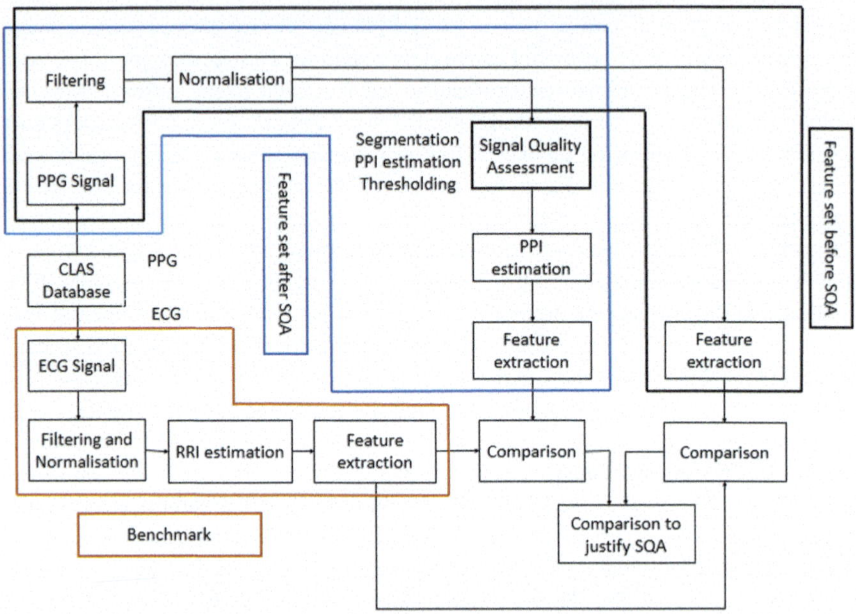

Fig. 1 Overall methodology

In this work, for the subjects selected, recordings ranging from 30s to 1min(varied from subject to subject) each of ECG and PPG signals corresponding to Baseline (B) and Logic problems test (LPT) were used. Baseline refers to the phase, wherein no stimulus was provided to the participant. In Logic problems test, participants were asked to solve simple logical problems within a short span of time, similar to IQ test problems. All signal processing tasks and signal analysis were done using MATLAB (2016a).

The signals after acquisition have been denoised. The unfiltered and filtered and normalized signal are shown in Fig. 2. Denoising involved filtering using a 4th order Butterworth bandpass filter of passband of 0.5–45 Hz for ECG and a 3rd order Butterworth bandpass filter of passband of 0.5–35 Hz for PPG, followed by amplitude normalization of the filtered signal within the range of (0, 1), to reduce amplitude variability. Amplitude normalization has been done using the below formula.

$$y\prime = \frac{y - y_{min}}{y_{max} - y_{min}} \tag{1}$$

where y is the data under test and $y\prime$ is normalized data.

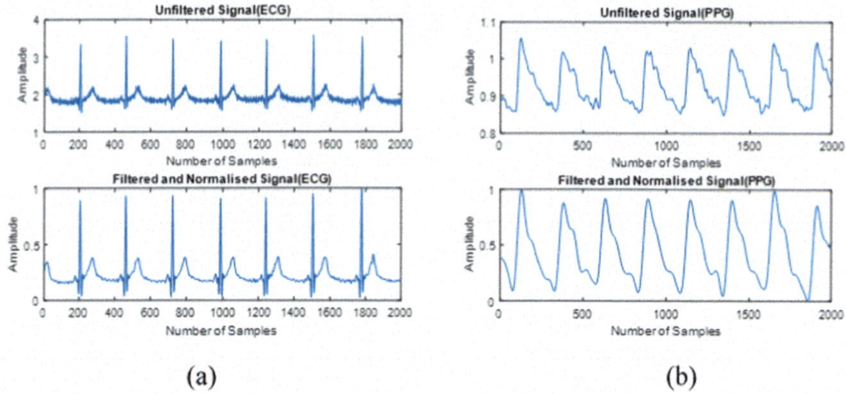

Fig. 2 Unfiltered and filtered and normalized signal of **a** ECG, **b** PPG

Following filtering, peaks were detected using change in slope and by using proper threshold [19]. Corresponding R-R interval (RRI) from ECG and peak-to-peak interval (PPI) from PPG were computed by calculating the time interval between successive peaks. The peaks detected (marked by '*') and RRI and PPI are shown in Fig. 3.

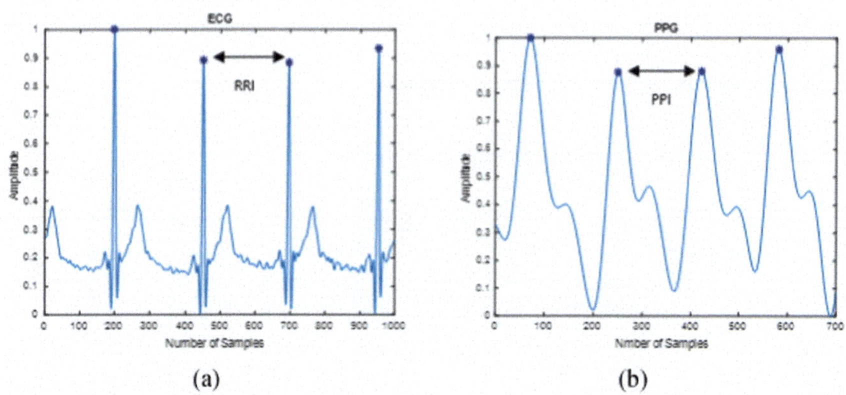

Fig. 3 Detection of peaks of **a** ECG, **b** PPG

2.1 Signal Quality Assessment of PPG

As PPG signals are subject to noise and Motion Artifacts, it is necessary to analyze the quality of the signal before using the signal for further analysis. Figure 4 shows the typical PPG signal and PPG signal affected with noise.

In this work, filtered and normalized PPG signals after peak detection, were passed through a SQA block as shown in Fig. 1. For each subject, the entire PPG signal was divided into segments as shown in Fig. 5 Based on [20], following steps were performed on each section:

Step 1: Checking whether HR lies in the range of 40 to 180 beats per minute.

Step 2: Checking whether PPI lies below 3 s.

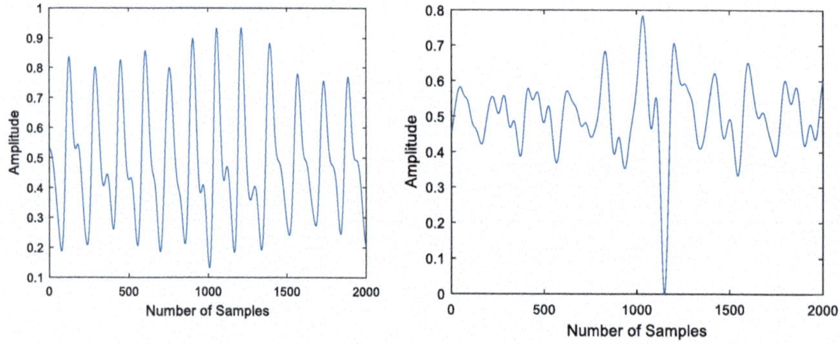

(a) (b)

Fig. 4 **a** Typical PPG signal, **b** PPG affected with noise

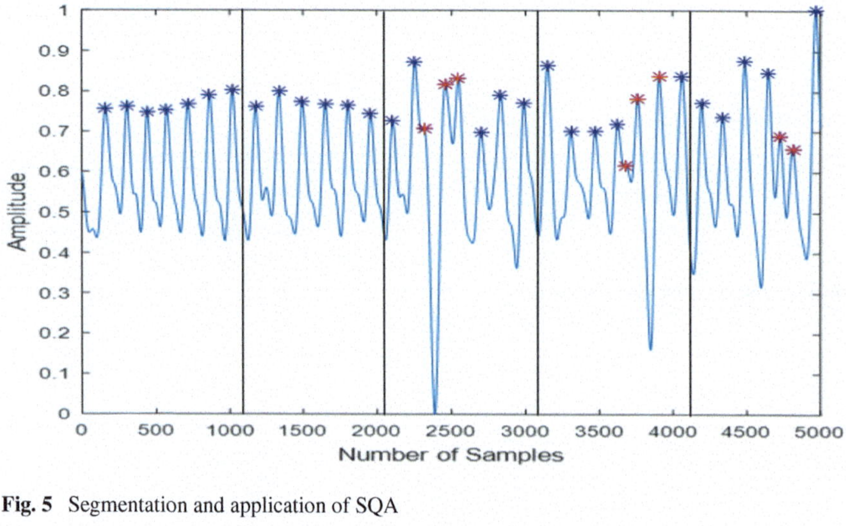

Fig. 5 Segmentation and application of SQA

Step 3: Checking whether $\frac{t_1}{t_2} < 2.2$, where t_1 is the maximum PPI and t_2 is the minimum PPI.

As observed in Fig. 5, the entire signal was equally segmented considering 1024 samples in each segment. The SQA algorithm was then applied on each segment. If all of the above conditions, performed sequentially are satisfied, the section is considered to be "GOOD" as seen in first two segments else, if any one of the condition is not satisfied the section is considered to be "BAD" as seen in segments 3,4 and 5.

Segment 3 and 4 do not satisfy Step3. Even though Step 1 and Step2 get satisfied, the segment, due to very less value of PPI affects the overall ratio and thus is rejected.

Segment 5 is discarded as it does not satisfy Step1. The HR between the '*' marked points is found to be 185 bpm which is well beyond the range stated in Step1, and thus the segment is rejected. The segments satisfying all three steps are considered for feature extraction and further analysis.

Feature Extraction

Following Peak detection, features were extracted from ECG and PPG. For ECG, features were extracted once, only after peak detection. However for PPG, features were extracted before SQA (PPG_b) and after SQA (PPG_a) as shown in Fig. 1. Considering ECG based heart beat features as benchmark, PPG bound heart beat features before and after SQA are detected and compared with the benchmark. The results and comparison corresponding to B and LPT are tabulated in Tables 1 and 2 respectively. Based on [21], the features used in this work are listed below:

Mean HR: Mean of all heart rates (HR)

$$HR = \frac{60 * fs}{RRI} \tag{2}$$

Table 1 Features of 10 subjects during baseline

Sub	Mean HR			Mean RR			RMSSD			SDNN		
	ECG	PPG_b	PPG_a	ECG	PPG_b	PPG_a	ECG	PPG_b	PPG_a	ECG	PPG_b	PPG_a
1	104	144	103	581	563	584	21.5	165.5	30.9	34.1	127.2	34.7
2	75	91	75	802	726	797	66.8	206.2	61.1	56.4	170.4	53.2
3	92	88	93	651	761	650	30.9	465.1	39.6	59.1	426.2	62.5
4	74	75	75	805	802	802	32.8	34.1	34.1	48.9	50.0	50.0
5	81	80	80	741	739	739	61.0	58.8	58.8	42.6	41.4	41.4
6	79	75	80	756	885	751	35.7	478.0	41.2	50.8	405.0	44.3
7	69	83	69	865	806	864	65.4	187.5	72.1	64.2	177.0	58.8
8	95	96	96	626	626	626	18.6	23.0	23.0	27.4	30.0	30.0
9	85	99	85	704	701	701	24.2	238.2	33.6	28.6	155.6	31.9
10	96	94	97	622	619	619	32.3	176.9	40.5	29.9	161.6	34.2

Table 2 Features of 10 subjects during logic problems test

Sub	Mean HR			Mean RR			RMSSD			SDNN		
	ECG	PPG_b	PPG_a	ECG	PPG_b	PPG_a	ECG	PPG_b	PPG_a	ECG	PPG_b	PPG_a
1	67	81	67	894	836	890	106.1	235.3	99.2	83.4	183.5	82.2
2	81	110	82	733	635	728	20.9	232.5	30.9	33.6	177.9	35.3
3	73	137	73	822	573	821	35.2	376.0	39.1	28.8	247.7	26.9
4	59	104	59	1001	811	1005	27.6	424.0	24.5	23.8	305.6	19.5
5	84	94	84	717	678	712	48.7	152.6	43.6	73.6	129.1	69.4
6	94	105	95	633	603	633	23.3	123.9	30.6	33.8	104.4	40.8
7	68	103	65	892	789	921	59.1	368.0	54.6	42.2	264.7	37.5
8	71	69	71	840	880	842	30.8	235.0	30.3	52.5	162.8	52.7
9	92	89	93	646	705	641	18.7	264.8	23.3	24.7	208.4	24.4
10	83	100	84	719	664	714	18.2	159.1	25.5	23.6	138.5	25.8

where f_s is the sampling frequency

$$MeanHR = \frac{\sum_{i=1}^{n} HR_i}{n} \tag{3}$$

where n is the total number of HR. Mean HR is expressed in beats per minute (bpm)

Mean RR: Mean of RR intervals

$$MeanRR = \frac{\sum_{i=1}^{N} RRI_i}{N} \tag{4}$$

where N is the total number of RRI

SDNN: Standard Deviation of RR intervals. It is measured in millisecond (ms)

$$SDNN = \sqrt{\frac{1}{N-1}\sum_{i=1}^{N}\{(RRI_i) - mean(RRI)\}^2} \tag{5}$$

RMSSD: Root mean square of successive RRI differences. The parameter measures the time difference between each successive heartbeat.

$$RMSSD = \sqrt{\frac{1}{N-1}\sum_{i=1}^{N-1} x_i^2} \tag{6}$$

where x_i is the successive RRI differences.

Further, a correlation study between HR calculated from PPG_b and PPG_a with HR from ECG is conducted for statistical verification of HR improvement due to SQA. Finally a preliminary classification of B versus PLT is using Decision Tree (DT),

Support Vector Machine (SVM) and K-Nearest Neighbor (K-NN) is conducted using the same feature set as extracted above to check if implementation of SQA has any impact over classification. The results are presented in the next section.

3 Result and Discussion

This section represents the results and comparison of features before and after implementation of SQA. As mentioned in Sect. 2, the results are tabulated in Tables 1 and 2 for B and LPT respectively.

For subjects 4, 5 and 8 the PPG signals (for Baseline only), passed through SQA block showed no presence of "BAD" beats and thus have retained similar feature results as highlighted in Table 1. For rest of the subjects from Tables 1 and 2 it is observed that, in the initial phase features calculated from PPG exhibited a difference from those estimated from ECG, while after passing through the SQA block, the differences between feature values were reduced,which enhanced the correlation coefficient as tabulated in Table 3. The Mean HR of ECG, PPG_a and PPG_b were taken for computing correlation coefficients.

Table 3 Correlation of mean HR estimated from ECG with PPG_b and with PPG_a for baseline and logic problems test

Sub	Baseline		Logic problems test	
	Mean HR from ECG and PPG_b	Mean HR from ECG and PPG_a	Mean HR from ECG and PPG_b	Mean HR from ECG and PPG_a
1	0.11	0.89	0.44	0.85
2	0.38	0.86	−0.07	0.98
3	0.89	0.95	−0.11	0.79
4	0.96	0.96	0.09	0.76
5	0.97	0.97	0.66	0.87
6	0.60	0.85	0.05	0.84
7	0.65	0.88	0.07	0.76
8	0.98	0.98	0.58	0.95
9	−0.46	0.87	−0.44	0.83
10	0.23	0.72	−0.63	0.96

Fig. 6 Bar chart of baseline for **a** Mean HR, **b** Mean RR, **c** RMSSD, **d** SDNN

From Figs. 6 and 7 it is observed that after the PPG signals were passed through SQA algorithm, the differences in feature values reduced and similarity between the feature values increased. As mentioned earlier, for further assessment of SQA, a preliminary binary classification was performed using three Machine Learning algorithms namely, DT, SVM and K-NN. All the features as mentioned in Sect. 2 were used for classification. Classification results are tabulated in Table 4. Though the classification results are not comparable with previous reported results [22, 23], still it is found that the classification performance is improved after application of SQA for all three cases and the most improved results was observed in case of K-NN classifier.

Fig. 7 Bar chart of logic problem test for **a** Mean HR, **b** Mean RR, **c** RMSSD, **d** SDNN

Table 4 Classification results

		Accuracy (%)		
Classifier		DT	SVM	K-NN
B versus LPT	Before SQA	61.2	62.4	51.8
	After SQA	64.7	67.1	62.4

4 Conclusion

PPG based stress assessment techniques are gradually becoming popular due to comfortable acquisition protocol of PPG signals. However, motion artifact and other related issues restrict the application of PPG for stress analysis and other purposes. The present work proposes a preliminary PPG signal quality assessment protocol for discarding signals having noises with higher margin. It was found that since PPG is susceptible to Motion Artifacts and other noises, it can only be used for further analysis after being passed through SQA. From correlation coefficients it was found that the degree of similarity ECG-derived mean HR and corresponding PPG-derived mean HR increased after SQA. Thus, it is necessary to identify the affected segments in order to proceed with further analysis. Further, the identification of affected segments becomes essential because, analysis done without removing the affected segments yields undesirable results as tabulated in Tables 1 and 2 under PPG_b. It can thus be inferred that PPG can be used independently, specifically in case of wearable devices, only after being passed through SQA. Classification accuracy as presented in Table 4. appears moderate because the work is conducted on a small group of subjects and no feature optimization technique was adopted. But it shows the potential of SQA to improve the classification performance. However, the study is open to further research, and can be tested on larger dataset using the algorithm on PPG signals acquired during various other activities and using more number of classifiers. The frequency-domain parameters can also be explored using the algorithm.

References

1. https://www.who.int/news-room/fact-sheets/detail/suicide
2. Cohen S et al (1983) A global measure of perceived stress. J Health Soc Behav 24(4):385–96
3. Bernstein DP, Ahluvalia T, Pogge D, Handelsman L (1997) Validity of the childhood trauma questionnaire in an adolescent psychiatric population. J Am Acad Child Adolesc Psychiatry 36(3):340–348
4. Karasek R, Brisso C, Kawakami N, Houtman I, Bongers P, Amick B The job content questionnaire (JCQ): an instrument for internationally comparative assessments of psychosocial job characteristics. J Occup Health Psychol
5. Gao H, Yüce A, Thiran J-P (2014) Detecting emotional stress from facial expressions for driving safety. In: IEEE international conference on image processing (ICIP) 2014. IEEE, Paris, France, pp 5961–5965
6. Mayo LM, Heilig M (2019) In the face of stress: interpreting individual differences in stress-induced facial expression. Neurobiol Stress 10
7. Bailon R, Sornmo L, Laguna P (2006) A robust method for ECG-based estimation of respiratory frequency during stress testing. Trans Biomed Eng 53(7):1273–1285
8. Chiang HS (2015) ECG-based mental stress assessment using fuzzy computing and associative petri net. J Med Biol Eng 35(6):833–844
9. Elgendi M, Menon C (2020) Machine learning ranks ECG as an optimal wearable biosignal for assessing driving stress. IEEE Access 8:34362–34374
10. Can YS, Arnrich B, Ersoy C (2019) Stress detection in daily life scenarios using smart6 phones and wearable sensors: a survey. J Biomed Inform 92:103139
11. Reisner A, Shaltis P.A, Mc Combie D, Asada HH (2008) Utility of the photoplethysmogram in circulatory monitoring. Anesth: J Am Soc Anesth 108(5):950–958
12. Liu S et al (2022) Toward nonintrusive camera-based heart rate variability estimation in the car under naturalistic condition. IEEE Internet Things J 9(14):11699–11711
13. Ding C et al (2023) Log-spectral matching GAN: PPG-based atrial fibrillation detection can be enhanced by GAN-based data augmentation with integration of spectral loss. IEEE J Biomed Health Inform 27(3):1331–1341
14. Zhang Y, Liu B, Zhang (2015) Combining ensemble empirical mode decomposition with spectrum subtraction technique for heart rate monitoring using wrist-type photoplethysmography. Biomed Signal Process Control 21:119–125
15. Zhang Z, Pi Z, Liu B (2015) TROIKA: a general framework for heart rate monitoring using wrist-type photoplethysmographic signals during intensive physical exercise. IEEE Trans Biomed Eng 62(2):522–531
16. Moscato S, Palmerini L, Palumbo P, Chiari L (2022b) Quality assessment and morphological analysis of photoplethysmography in daily life. Front Digit Health

17. McLean MK, Glenn weaver R, Lane A, Smith MT, Parker H, Stone B et al (2023b) A sliding scale signal quality metric of photoplethysmography applicable to measuring heart rate across clinical contexts with chest mounting as a case study. Sens (Basel) 23(7)
18. Markova V, Ganchev T, Kalinkov K (2019) CLAS: a database for cognitive load, affect and stress recognition. In: IEEE international conference on biomedical innovations and applications
19. Sadhukhan D, Mitra M (2012) R-peak detection algorithm for Ecg using double difference and RR interval processing. Procedia Technol 4:873–877
20. Orphanidou C, Bonnici T, Charlton P, Clifton D, Vallance D, Tarassenko (2015) Signal quality indices for the electrocardiogram and photoplethysmogram: derivation and applications to wireless monitoring. IEEE J Biomed Health Inform 19(3)
21. T F of the European Society of Cardiology the North American Society of Pacing Electrophysiology (1996) Heart rate variability: standards of measurement, physiological interpretation, and clinical use 93(5):1043–1065
22. Feradov F, Ganchev T, Markova V (2020) Automated detection of cognitive load from peripheral physiological signals based on Hjorth's parameters. In: 2020 international conference on biomedical innovations and applications. Varna, Bulgaria, pp 85–88
23. Jin Y-J, Habibilloh E, Jang Y-S, An T, Jo D, Park S et al (2022) A photoplethysmogram dataset for emotional analysis. Appl Sci 12(13):6544

Land Use Classification and Change Detection Using Google Earth Engine

Kakumanu Christy Victor, Radhesyam Vaddi, Mahali Tirumala Raju, and Pulapaka Varun Kumar

Abstract Land use and cover changes (LULC) between 2011 and 2021 were examined in research in India's Cumbum district using Google Earth Engine. The main findings indicate that urban increased by 4% while vegetation area fell by 9%, bareland increased by 7%, Trees are increased by 1%, and water is fell by 3%. illustrating the fast urbanization caused by increase in population and agricultural development. Urban areas, crops, and aquatic bodies are the three main forms of land cover. Changes in aquatic supplies, air quality, biodiversity, and local livelihoods were all evaluated as part of the study's assessment of environmental and social effects. The findings highlight the need for swift adoption of sustainable land use regulations in order to lessen the serious consequences of LULC changes in the area.

Keywords CART · Change detection · Google earth engine · Multi spectral images

1 Introduction

This research work provides a general overview of its objectives, origins, and applications. Our project is Mainly focused on land use classification and change detection using machine learning algorithm specifically using CART algorithm. The goal of change detection is to identify and quantify differences or alterations in a particular phenomenon, object, or environment between two or more points in time. The project proposes a method that uses CART Algorithm to detect the changes that happen in a particular area, and with advancement in technology, it is now possible to predict the land for vegetation, urban development. The analysis focuses mainly on pixel-based finite element update techniques.

K. C. Victor (✉) · R. Vaddi · M. T. Raju · P. V. Kumar
Department of IT, VR Siddhartha Engineering College, JNTU-Kakinada, Vijayawada, India
e-mail: kakumanuchristyvictor@gmail.com

D.-I. Stroe et al. (eds.), *Emerging Electronics and Automation*, Lecture Notes in Electrical Engineering 1202, https://doi.org/10.1007/978-981-97-3090-2_27

The review offers analysis based on image processing methods, goals, degrees of accuracy, degrees of inaccuracy, and datasets. Finally, this study presents a variety of research topics that might aid in the growth of change detection research in the future.

2 Problem Statement

Comparing the changes that happen in a particular area between two-time stamps is useful to predict the land for Urban planning in future. LULC can reflect the pattern of human land use in a region, and plays an important role in space soil and water conservation. The aim is to identify the changes in land cover in a particular location between 2-time stamps. Such type of a task is Land cover classification which come up and automate the process of identifying how the land area is used.

2.1 Objectives

The objective is to detect the changes that happen in a particular area in between two-time stamps.

1. In order to inform urban planning and sustainable agriculture, keep an eye on changes in agricultural production, land use, and climate.
2. Identify changes in land productivity and vegetation health for forestry, agriculture, and natural resource management.

2.2 Outcomes

1. Obtain a changes representation map which shows the change detection that takes place in the region used for research.
2. Obtain a change map of the images from two time series and get an accuracy assessment of the results obtained from it.

2.3 Real Time Applications

1. The proposed framework can be used for Urban Expansion.
2. This can help in monitoring Vegetation health over time, which is important for Vegetation.
3. Also used to monitor the seasonal climate changes.

3 Literature Review

3.1 Description of Existing System

The literature review centers on resources that have contributed to the comprehension of the research domain concerning image processing and classification techniques. These research articles delve into the use of the many different algorithms and frameworks for the classification of data. Within this section, the emphasis is placed on scholarly publications, encompassing key particulars of each paper while presenting the discoveries and inferences drawn from each investigation.

This is an article about detecting land use changes using time-series Landsat imagery on Google Earth Engine. It discusses the challenges of detecting land use changes in areas with complex human-nature interactions. The authors propose a bi-directional strategy to detect land use changes that takes into account the distribution probability of land cover types. They apply their method to the Huangshui River Basin in China and find that urban and built-up land has increased while cropland has decreased. This suggests that there is a need to balance the competing demands of urbanization, agriculture, and conservation in the region [1].

They have used a Convolutional Neural Network (CNN) model that classifies the images and applies change detection. The accuracy of LULC depends upon the time and location [2].

This article explores how land use and cover have changed in Bangladesh's Fashiakhali Wildlife Sanctuary (FKWS) impact region. It talks about how to utilise GIS and multispectral photos to track changes in land cover and usage over time. According to the authors, there has been a considerable loss of forest cover in the FKWS impact region, with the first period (1994–2001) showing the greatest drop. Additionally, they discover that settlement land has expanded while agricultural land has declined [3].

This article discusses change detection in remote sensing photos using deep learning. It talks about what change detection is and its significance. It also describes the many forms of remote sensing information that may be utilized to identify changes. After that, the essay reviews the many deep learning methods that have been used to change detection. It also goes through each technique's benefits and drawbacks [4].

In this paper they have the growing acceptance of deep learning for change detection in high-resolution remote sensing photos as discussed in this work. The difficulties of employing SAR data for change detection are also mentioned. The authors offer a thorough analysis of feature extraction networks and model building as well as deep learning-based change detection techniques. By detecting unit, they also group change detection techniques. Finally, they go through well-known datasets and change detection assessment measures [2].

3.2 *Summary of Literature Review*

They developed a method multi-temporal, multi-spectral image processing technique. The approach is based on a multi-scale adaptive convolution kernel network and a multimodal conditional random field, and it examines the difficulties of change detection in metropolitan locations. The OSCD dataset, which consists of 24 pairs of multispectral pictures from two cities, is used to assess the approach. The outcomes demonstrate that the suggested strategy outperforms the most recent methods. The suggested approach provides a number of benefits over the ones already in use. The photos can first be captured both globally and locally. Second, it can adapt to a variety of changes, including those in building types, land use, and land cover. Thirdly, it is resistant to picture artefacts such as noise.

4 Proposed Method

The CART algorithm in Google Earth Engine uses a methodical set of steps to identify changes in the terrain and classify it. It starts with the capture of satellite images for the relevant time periods, often from sources like Landsat or Sentinel. The gathered imagery is then put through pre-processing to remove noise, account for atmospheric interference, and handle cloud cover. To prepare the data for analysis, this may include band normalisation or stacking. Ground-truth training data, which includes samples from both changing and undisturbed land cover areas, is necessary to properly train the CART algorithm. A key step in this procedure is feature selection, which entails locating pertinent spectral and temporal data like reflectance values, NDVI, or vegetation indices that are used as CART input variables.

The technique incorporates adjustment based on feedback, new data, or changing objectives to continuously enhance accuracy and relevance. The full process, including data sources, parameters, and findings, must be documented in order to assure transparency and repeatability. The efficient use of CART-based change detection and land categorization in Google Earth Engine for a variety of applications and research projects is made possible by this methodical methodology.

The CART Mathematical Notation as follows:

$X = [X_1, X_2, \ldots, X_r]$.

Here, r represents the number of features or variables.

$\text{Gini}(t) = 1 - \Sigma[p_i 2]$.

where p_i is the proportion of samples in class i in node t.

$\text{Prediction} = \text{argmax}(\Sigma[p_i])$.

where p_i is the proportion of samples in class i in the leaf node.

5 Algorithm Steps

(1) Gather multispectral imagery for multiple time periods, including the "before" and "after" images for change detection.
(2) Extract from the picture pertinent spectral and temporal information, such as reflectance values, vegetation indices, textural traits, or any other markers that identify different types of land cover.
(3) Based on the types of land cover that are known, assign training data class labels (e.g., class 1 for urban, class 2 for forest, class 3 for water, etc.).
(4) For change detection, create a new set of training data where each sample is labeled as "changed" (class 1) or "unchanged" (class 0) based on whether or not significant changes occurred between the two time periods
(5) Train a separate CART model using the labeled training data for change detection. This model learns to classify pixels as changed or unchanged based on the features extracted from the "before" and "after" imagery.
(6) Train CART model for change detection:
    Decision function: F(X1, X2) = CART(X1, X2)
    Classes: Y_predicted = F(X1, X2)
(7) $X = [X_1, X_2, \ldots, X_r]$
    Here, r represents the number of features or variables.
    $\text{Gini}(t) = 1-\Sigma[p_i^2]$.
    where p_i is the proportion of samples in class i in node t.
    $\text{Prediction} = \text{argmax}(\Sigma[p_i])$.

6 System Architecture

CART algorithm is used to detect the changes that happen between two-time stamps. In the architecture two images in the same location of two different time stamps are taken in cumbum location from Sentinel–2 Satellite. And then images are classified by using the CART algorithm. After classification of images, they are subtracted for the change detection where at last the gray scale image is formed which is black and white gray scale images are formed where white indicates the changes that happen in that land. At last Accuracy assessment on the gray scale image by using the CART algorithm will be done to check the performance (Fig. 1).

7 Dataset Description

Sentinel-2: Sentinel-2 is a European Space Agency (ESA) Earth observation satellite project that is essential to the management of the environment and lands in Google Earth Engine (GEE). It consists of a constellation of several satellites each carrying the Multispectral Instrument (MSI), a multispectral sensor system. Sentinel-2 offers

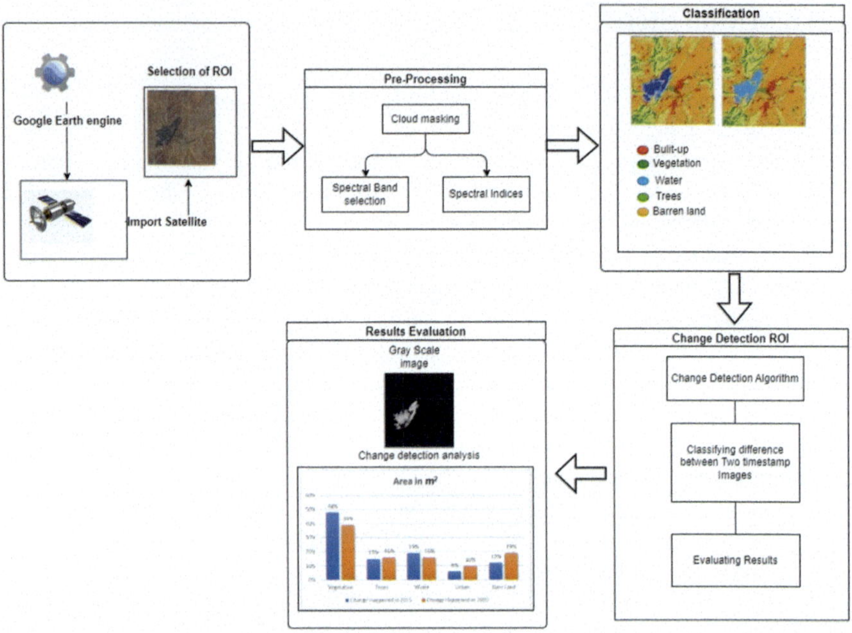

Fig. 1 System architecture

13 spectral bands of high-resolution optical imaging, including visible, near-infrared, and short-wave infrared bands, allowing for in-depth investigation of the ground surface. These satellite photos are useful for applications like mapping land cover, monitoring agriculture, evaluating forests, and disaster management because they are publicly available in GEE. The open and free data policy of Sentinel-2 enables access to and use of its images for land categorization and change detection on a worldwide scale by researchers, land managers, and governments. Sentinel-2 is an effective instrument for analysing environmental changes due to its multispectral data and open-access policy. The pre-processed Sentinel-2 data in GEE is ready for study right now (Fig. 2).

The dataset used in this paper is available in the given link https://developers.google.com/earthengine/datasets/catalog/COPERNICUS_S2_SR_HARMONIZED.

8 Results and Observations

This chapter describes the results of our project and comparison of both the algorithms that we have used in our project to find the change between two timestamp images.

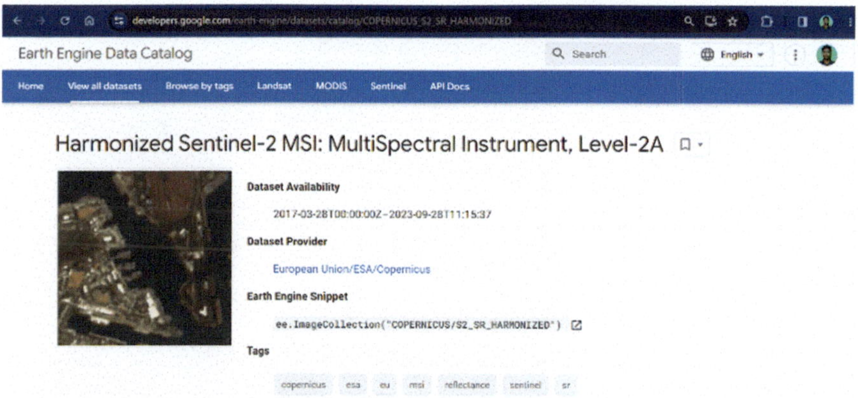

Fig. 2 Data catalog

We used GEE to obtain Landsat 8 satellite pictures, and the area and time period we chose to identify the change in an agricultural, builtup, water etc. is indicated below (Fig. 3).

Initially we select the area in GEE map which is also known as RIO (Region Of Interest). And select the satellite which is suitable for the project and the area selection is taken by polygon or points but the polygon gives the area which went exactly so we prefer for it. Classified Image before: The location is Cumbum, Andhra Pradesh and the time period is 2015. And sentinel-2 gives us the images from the specific year we want. Here we take the images one from 2015 and one from year 2020. The two images are pre-processed (cloud masking, selecting spectral pixels, and selecting spectral bands). Then after pre-processing classification is done on these images and finally, we are applying change detection on these two stamp images. By

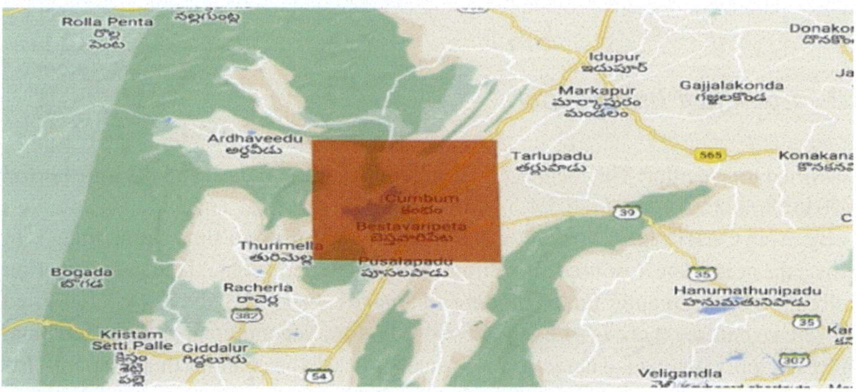

Fig. 3 Cumbum location

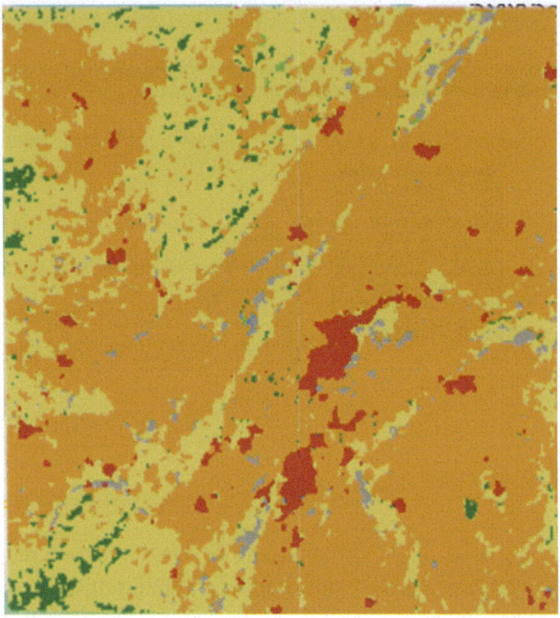

Fig. 4 Classified image before (2015)

that we obtain the change detected gray scale image which shows the changes that are happening in that particular area.

8.1 Classified Image Before

Combine the 3,4,5 bands among 11 bands at Time T1. We need to integrate 3,4,5 bands because we are detecting changes in agriculture (Fig. 4).

8.2 Classified Image After

Classified image after: The location is Cumbum, Andhra Pradesh and the time period is 2015. Step 2: Combine the 3,4,5 bands among 11 bands at Time T2 (Fig. 5).

By using CART algorithm, we successfully classify the two-time stamp images to reduce the noisy data, resource allocation, change detection context, change detection calibration and semantic understanding of images. From the above two classified images we can see there is a change in water bodies, vegetation area and some land and after applying change detection we can clearly see the changes that happen. This diversity in actors ensures a wide range of changes in water, vegetation, and dry land.

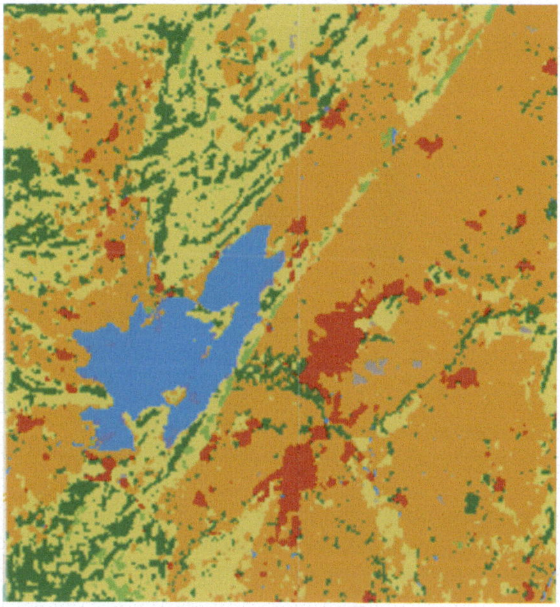

Fig. 5 Classified image after (2020)

8.3 Change Detected Image

In the above image, we have stacked the images by using the three bands with the help of QGIS software where we will import these images into the Google Earth Engine for change detection and this is time stamp −1 of the 2015 year and 2020 year. The next step is the generation of Difference Images. The difference image is the matrix subtraction of two images. By subtracting the two images we will get the final gray scale image which is also known as the Change Detected image. Analyzing pixel changes over time or between two photos is necessary to find changes in a grayscale image. These variations may signify changes in intensity, motion, or the presence of objects. By concentrating just on brightness, grey scale pictures streamline this process, making change detection easier (Figs. 6 and 7).

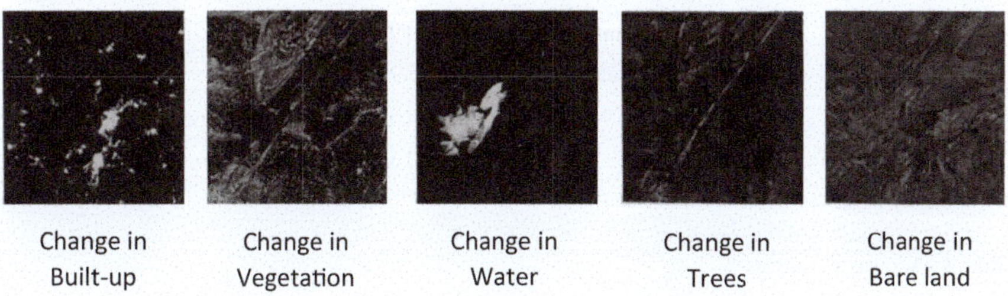

Fig. 6 Change detected gray scale images

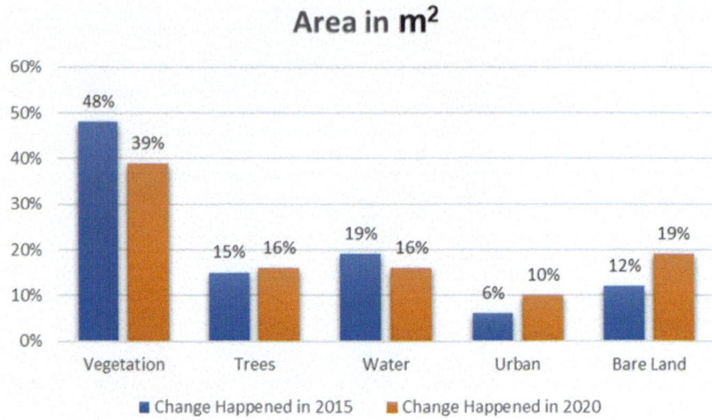

Fig. 7 Changes happened between 2015 and 2020

From the above graphs we can see the amount of changes that happen in two-time stamps and by observing we can see vegetation is decreased by 11%, trees are increased by 1%, aquatic bodies like ponds, lakes and some other water bodies are decreased by 3%, urban area is developed and increased by 4% and at last bare land is increased by 7%. And we conclude that there is fall in the vegetation, water area and increase in Trees, Urban and Bare land which means there is a damage on natural environmental resources which impacts on the eco system and increase in the urban area and bare land leads to the non-usage of natural resources and wastage of land for vegetation and water bodies. Cart algorithm gives very efficient outcomes for this project (Tables 1 and 2).

From the accuracy metrics we can be observe that the accuracy value is 90.4% and the kappa value is 81.15%. when compared to other classification algorithms like random forest, SVM, CNN etc..... CART algorithm has given better accuracy value and accurate classified image and by that it is easy to apply the change detection on them.

Table 1 Changes in percentage

Class	2015 (%)	2020 (%)	Change in percentage (%)
Urban	6	10	+ 4
Bare land	12	19	+ 7
Vegetation	48	39	−9
Trees	15	16	+ 1
Water	19	16	−3

Table 2 Accuracy metrics

Accuracy (%)	Kappa (%)
90.4	81.15

9 Conclusion

We employed two different timestamps in this project to detect changes in a specific location, namely Cumbum. The change in agricultural lands in the area was determined using pictures from the Sentinel-2 satellite. Damage assessment is aided by the ability to detect changes in the environment. Both the models had determined the change in the same location at different time stamps effectively. The project's major purpose is to detect differences in nature, such as before and after natural disasters, and estimate the percentage difference (future study). We evaluated the performance of our change detection method and identified areas for further research. Our findings can inform decision-making processes related to land use planning, natural resource management, and disaster response.

10 Future Work

Change detection has a bright future with a lot of possibilities in many areas. It will be possible to more precisely monitor changes to the Earth's surface, thanks to advanced remote sensing technology, such as higher-resolution satellite photos and regular data updates. As processes are automated, accuracy is improved, and real-time or almost real-time monitoring is made possible, machine learning will play a crucial part. A full picture of changes will be provided through multimodal data fusion, which includes optical, radar, LiDAR, and hyperspectral data. Identification of changes will help with infrastructure upkeep, urban planning, ecological monitoring, and environmental monitoring.

References

1. He Y, Jia Z, Yang J, Kasabov NK (2021) Multispectral image change detection based on single-band slow feature analysis. Remote Sens
2. Feng S, Fan Y, Tang Y, Cheng H, Zhao C, Zhu Y, Cheng C (2022) A change detection method based on multi-scale adaptive convolution kernel network and multimodal conditional random field for multi-temporal multispectral images. Remote Sens
3. Hajarian MH, Atarchi S, Hamzeh S (2021) Monitoring seasonal changes of Meighan wetland using SAR, thermal and optical remote sensing data. Phys Geogr Res Q
4. Hammouch W, Chouiekh C, Khaissidi G, Mrabti M (2022) Crack detection and classification in moroccan pavement using convolutional neural network. Infrastruct

The manufacturer's authorised representative in the EU is Springer Nature Customer Service Centre GmbH, Europaplatz 3, 69115 Heidelberg, Germany. If you have any concerns regarding our products, please contact ProductSafety@springernature.com

Printed and bound by CPI Group (UK) Ltd, Croydon, CR0 4YY

30/09/2026

02225298-0010